Energy Efficient Buildings

Energy Efficient Buildings

Fundamentals of Building Science and Thermal Systems

Zhiqiang (John) Zhai
Department of Civil, Environmental and Architectural Engineering
University of Colorado at Boulder
Boulder, CO, USA

Registered Office
John Wiley & Sons, Inc., 111 River Street, Hoboken, NJ 07030, USA

Editorial Office
111 River Street, Hoboken, NJ 07030, USA

For details of our global editorial offices, customer services, and more information about Wiley products visit us at www.wiley.com.

Wiley also publishes its books in a variety of electronic formats and by print-on-demand. Some content that appears in standard print versions of this book may not be available in other formats.

Library of Congress Cataloging-in-Publication Data

Names: Zhai, Zhiqiang (John), author.
Title: Energy efficient buildings : fundamentals of building science and
  thermal systems / John Zhai.
Description: Hoboken, NJ : Wiley, 2023.
Identifiers: LCCN 2022017503 (print) | LCCN 2022017504 (ebook) | ISBN
  9781119881933 (cloth) | ISBN 9781119881957 (adobe pdf) | ISBN
  9781119881940 (epub)
Subjects: LCSH: Architecture and energy conservation. | Heat engineering.
Classification: LCC NA2542.3 .Z42 2023 (print) | LCC NA2542.3 (ebook) |
  DDC 720/.472–dc23/eng/20220715
LC record available at https://lccn.loc.gov/2022017503
LC ebook record available at https://lccn.loc.gov/2022017504

Cover Image: © imamember/Getty Images
Cover Design by Wiley

Set in 9.5/12.5pt STIXTwoText by Straive, Pondicherry, India

Contents

1

Sustainable Building

1.1 Building Functions

Buildings are created for various purposes and functions, such as for living, production, retail, and storage. Without functions, buildings are purely pieces of art. Following the Maslow's hierarchy of needs theory (Figure 1.1), building functions can generally be grouped as follows:

- **Shelter**: provides basic living or working conditions that mitigate severe environmental influences such as rain, storm, and snow. Such examples include caves and original vernacular shelters.
- **Safety**: presents additional security to indoor environments with measures such as structure supports (e.g. columns and beams), doors, and windows.
- **Productivity**: ensures efficient and effective organization of spatial-temporal layout, circulation, and space operation.
- **Comfort**: delivers controllable indoor environments with desired temperature, humidity, air speed, and visual and acoustic qualities using various passive and active systems.
- **Health**: ensures proper air quality with minimum pollutions and adequate oxygen and other necessary supplies (e.g. CO_2 for plants).
- **Privacy**: fulfills phycological requirements for appropriate distance, boundary, independence, and connectivity.
- **Aesthetics**: meet aesthetic and emotional needs for spatial and spiritual joy.

It is important to note that many technical approaches can be used to meet these function requirements, in which most energy efficient and environmentally friendly solutions should take the priority. Achieving the desired building functions, rather than energy efficiency and environmental friendliness, is the ultimate goals of building design and construction. Sacrificing one of these functions, such as thermal comfort, to meet the energy efficiency target fully opposes the original motivations of sustainable building development. Buildings in poor areas without access to electricity and natural gas have a natural "net-zero energy" feature of buildings; however, these buildings with the same indoor–outdoor temperature (in both summer and winter) are surely not the target for sustainable development.

Energy Efficient Buildings: Fundamentals of Building Science and Thermal Systems, First Edition. Zhiqiang (John) Zhai.
© 2023 John Wiley & Sons, Inc. Published 2023 by John Wiley & Sons, Inc.

Figure 1.1 Maslow's hierarchy of needs theory (https://www.simplypsychology.org/maslow.html). *Source:* McLeod (2020)/Simply psychology.

1.2 Building Elements

Buildings require three primary elements during both construction and operation:

- Materials
- Water
- Energy

As one kind of ecosystem, buildings take inputs of energy, water, and materials while generating waste, pollution, and possible poor health (Figure 1.2). Both inputs and outputs have limitations on their capacities; therefore "sustainable" approaches are keenly sought after to minimize the impacts on both natural resources and sinks.

1.2.1 Input: Energy

According to the US Energy Information Administration (EIA) (https://www.eia.gov/tools/faqs/faq.php?id=86&t=1), buildings use 39% of the total primary energy and 74% of the electricity in the United States (Figures 1.3 and 1.4). The International Energy Agency (IEA) predicted that in the New Policies Scenario, global energy needs rise more slowly than in the past but still expand by 30% between today and 2040. This is the equivalent of adding another China and India to today's global demand. It was also indicated that most of the required energy was produced (and will continue to be produced) by fossil fuels (Figure 1.5), such as petroleum, coal, and natural gas, although renewables are rapidly rising with reduced costs and policy incentives.

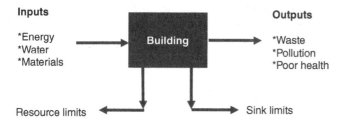

Figure 1.2 Architectural ecosystem.

US total energy consumption by end-use sector (1950-2020)
quadrillion British thermal units

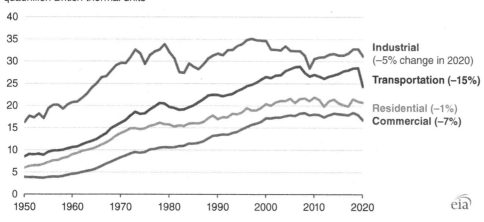

Industrial
(−5% change in 2020)

Transportation (−15%)

Residential (−1%)
Commercial (−7%)

eia

Figure 1.3 US total energy consumption by end-use sector (https://www.turbomachinerymag.com/view/us-energy-consumption-fell-by-record-7-in-2020). *Source:* U.S. Energy Information Administration/Public Domain.

1.2.2 Input: Water

The US Green Building Council (USGBC; https://www.usgbc.org/articles/benefits-green-building) stated that buildings use about 14% of all potable water (15 trillion gallons per year), and water efficiency efforts in green buildings might reduce water use by 15% and save more than 10% in operating costs. Retrofitting one out of every 100 American homes with water-efficient fixtures could avoid about 80 000 tons of greenhouse gas (GHG) emissions, which is the equivalent of removing 15 000 cars from the road for one year. In addition, buildings require a significant portion of the total water consumption for direct and indirect use by electricity generating plants. The US Geological Survey (USGS) estimated that total water withdrawals for thermoelectric power for 2015 were 133 000 Mgal/d, nearly 100% of which was withdrawn from surface-water sources, predominantly freshwater. Total withdrawals for thermoelectric power accounted for 41% of total water withdrawals, 34% of total freshwater withdrawals, and 48% of fresh surface-water withdrawals for all uses (https://water.usgs.gov/watuse/wupt.html).

Transporting water and cleaning wastewater also consume significant energy. The required energy for water supply systems ranges from below $0.30 \, \text{kWh/m}^3$ for developing countries (e.g. $0.29 \, \text{kWh/m}^3$ for

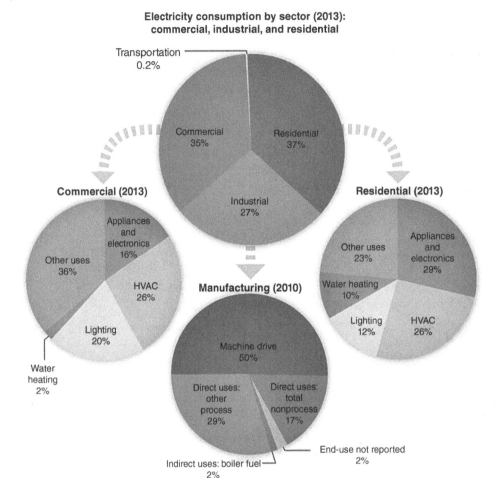

Figure 1.4 US electricity consumption by sector (https://www.epa.gov/energy/electricity-customers).

China (Smith et al. 2016) and 0.3 kWh/m³ for India (Miller et al. 2013), to 0.4–0.79 kWh/m³ for most developed countries, and even to 1.71 kWh/m³ for some countries such as Germany (Wakeel et al. 2016). The energy consumption by water supply systems will be rising along with the rapid progress of global urbanization.

In the United States, the energy used by water and wastewater utilities accounts for 35% of typical US municipal energy budgets (https://www.epa.gov/sites/default/files/2015-08/documents/wastewater-guide.pdf). Electricity use accounts for 25–40% of the operating budgets for wastewater utilities and approximately 80% of drinking water processing and distribution costs. Drinking water and wastewater systems account for approximately 3–4% of energy use in the United States, resulting in the emissions of more than 45 million tons of GHGs annually.

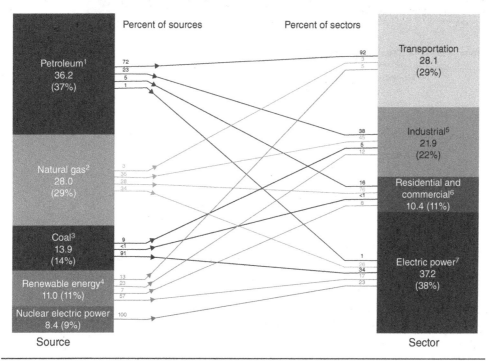

US primary energy consumption by source and sector, 2017
Total = 97.7 quadrillion British thermal units (Btu)

Figure 1.5 US energy consumption by source and sector (https://www.eia.gov/energyexplained/?page=us_energy_home).

1.2.3 Input: Materials

In their 1995 report to the Worldwatch Institute, Roodman and Lenssen (1995) claim that buildings use 40% of the global raw materials. Between 1974 and 1994, while the world's population increased by 40%, the world's consumption of energy-intensive cement increased 77%; the consumption of plastics (a man-made nonbiodegradable material) increased by a whopping 200% (UNEP 2004). Industrialized society continues to lean heavily on finite and nonrenewable resources. By weight, 95% of the raw materials used in the United States are nonrenewable, compared to just 59% 100 years ago (University of Michigan 2018). The building industry consumes 40% of the raw materials flow of the global economy every year (USGBC 2003). This amounts to about three billion tons annually, in large part composed of nonrenewable

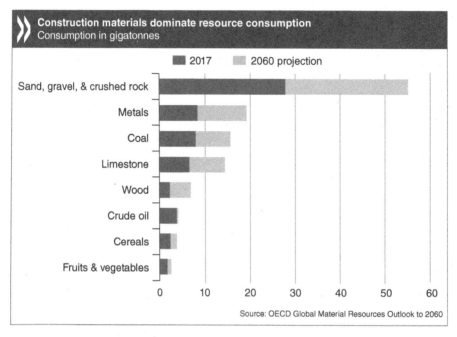

Figure 1.6 Construction material consumptions (http://www.oecd.org/newsroom/raw-materials-use-to-double-by-2060-with-severe-environmental-consequences.htm).

resources. Construction materials (including sand, gravel, and rock) make up the largest piece of overall material consumption (Figure 1.6).

Similar to water, the entire life cycle of materials also demands significant use of energy, from harvesting raw materials, to transporting materials, to manufacturing building elements, to demolishing waste materials.

1.2.4 Output: Waste

Waste is generated throughout the process of material procurement and throughout the life of a building. Brick, metals, wood, cardboard, and other waste are generated and landfilled during building construction, demolition, and renovation. In the United States, the Environmental Protection Agency (1998) estimated 136 million tons of construction and demolition waste were created per year, as compared to the Environmental Protection Agency (2005) estimate of 245.7 million tons of municipal solid waste. Construction and demolition in buildings account for half as much waste as produced by the municipal waste stream. During building operations, residents generate waste (although not directly caused by buildings), such as food, paper, packaging, and other disposable items, a further burden on landfills. The waste generated by buildings impacts land use by taking up room in landfills, as well as impacting water and air quality.

1.2.5 Output: Pollution

Building development, construction, operation, and demolition have both direct and indirect impacts on natural environments including air, water, and soil that are fundamental and crucial for supporting human living quality.

One direct example is airborne particles of dust from construction, such as hardwood or silica dust, that can cause severe health impacts including silicosis, asthma, and heart disease (https://www.safetyand-healthmagazine.com/articles/12507-silicosis-what-it-is-and-how-to-avoid-it). Silica dust is often produced in the creation of concrete, so exposure to this toxic substance can cause health risks across the built environment worldwide. Sediment pollution from construction sites contaminates drinking water, impacts recreational waters, diminishes commercial fisheries, and increases the risk of flood damage. Dirty runoff from construction sites can also carry other pollutants into waterways. The US construction industry accounts for 160 million tons, or 25%, of nonindustrial waste generation a year, according to the US EPA. According to new research by construction blog Bimhow, the construction sector contributes to 23% of air pollution, 50% of climatic change, 40% of drinking water pollution, and 50% of landfill wastes (https://gocontractor.com/blog/how-does-construction-impact-the-environment).

1.2.6 Output: Poor Health

Confined indoor environments, even with the use of sophisticated mechanical systems, do not always guarantee a comfortable and healthful indoor environment. Studies by the Harvard T.H. Chan School of Public Health have shown that workplaces with a specific focus on volatile organic compound (VOC) minimization and enhanced ventilation lead to superior cognitive functioning from the occupants than equivalent environments with higher indoor pollutants and lower fresh air intake. Cognitive scores were demonstrated in controlled trials to be 101% higher in these experiments, which reveal the potential impact on concentration, productivity, and work quality the polluted air could be having in a work environment (https://ehp.niehs.nih.gov/doi/10.1289/ehp.1510037).

1.3 Definition of Sustainable Building

Sustainable building is defined as "the creation and responsible management of a comfortable and healthy built environment based on resource efficient and ecological principles." This covers the entire life of building design, construction, operation, and maintenance. "Comfortable and healthy built environment" indicates the building development goal (i.e. function). "Resource efficient and ecological principles" emphasize the approach, the impact, and the limitation to the resource and environment.

Sustainable building may also be called "green building," "ecological building," "high performance building," or "environmentally responsible/friendly building," which all represent similar principles as described in the definition.

A few other terminologies are used to define sustainable buildings but with somewhat different emphases, such as:

- **Passive house**: which mainly focuses on super insulated and airtight envelopes combined with high efficiency heat recovery of ventilation air.

- **Solar house**: which mostly focuses on the utilization of renewable energy technologies, such as passive and active solar heating and solar photovoltaic (PV) cells.
- **Smart house**: which mainly focuses on advanced solutions for demand control and efficient use of fossil fuel technologies.
- **Adaptive building**: whose building elements are designed to actively respond to changing climate conditions and indoor environment conditions as required by users.

1.4   Origin and Significance of Sustainable Building

The beginning of the modern sustainability movement in the United States began with the environmental movement of the 1960s. Books, such as Rachael Carson's *Silent Spring*, made the public more aware of the dangers they faced from industrial and agricultural toxins and pollutants. Several years later, in 1972, the United Nations held the Conference on the Human Environment in Stockholm, Sweden to discuss environmental concerns. Following this conference, several national and international organizations, such as the United Nations Environmental Programme, were formed. The work of those in the 1960s and 1970s environmental movement laid the groundwork for the concept of sustainability. In 1987, the United Nations World Commission on Environment and Development (WCED) published the report *Our Common Future*, creating the definition of sustainable development that is most often used by the building industry. The WCED definition of sustainable development is, "development that meets the needs of the present without compromising the ability of future generations to meet their own needs" (World Commission on Environment and Development, 1987). While this definition is somewhat vague and has many interpretations, it is notable for providing a concept that individuals can use to construct more descriptive definitions.

Interest in sustainable building practices has grown significantly among the global building industry. A few countries and organizations have initiated a variety of green building development goals and guidelines, such as BREEAM by UK and CASBEE by Japan. In 2000, the USGBC launched its first green building rating system – Leadership in Energy and Environmental Design (LEED) – the most widely used green building rating system in the world. In two decades, USGBC's membership has increased by a factor of 100, and billion square feet of commercial and residential buildings are involved in the LEED certification program. The American Society of Heating, Refrigerating, and Air-Conditioning Engineers (ASHRAE) also released its *GreenGuide*, in 2006, which provides complete guidance from planning to operation for sustainable buildings and evolved into its 5th edition in 2018.

The reason why sustainable building has been receiving growing attention globally is threefold. First of all, construction and operation of buildings consume tremendous amounts of natural resources while producing wastes and pollutants that contribute to environmental damage and potentially compromise the health and productivity of building occupants. This situation becomes much more serious when urbanization is developing rapidly at both regional and global levels. In the twentieth century, global population increased 4 times, the economy grew 20 times, while energy consumption rose 40 times. This reality brings significant competition to the world resources and the challenges to the global environments. Sustainable building development can greatly reduce these impacts while keeping the momentum of urban, social, and economic development.

More specially, buildings in the United States collectively consume:

- 37% of all energy
- 68% of all electricity
- 12% of freshwater supplies
- 88% of potable water supplies
- 40% of raw materials

And these buildings generate:

- more than one-third of municipal solid waste streams
- 36% of total emissions of anthropogenic carbon dioxide (CO_2) emissions – the primary GHG associated with global climate change
- 46% of sulfur dioxide emissions (SO_2) – a precursor to acidic deposition – through the consumption of fossil-fuel-fired electricity
- 19% of nitrogen oxide emissions (NO_x)
- 10% of fine particulate emissions

Second, sustainable buildings deliver a better living and working environment, along with an improved quality of life, leading to higher comfort, health, and productivity. The UCGBC estimated considerable productivity-related benefits for all building applications (Figure 1.7). One case study conducted by West Bend Mutual Insurance Company also showed that moving into a sustainable building from an old building increased the productivity of employees by 16% (Figure 1.8).

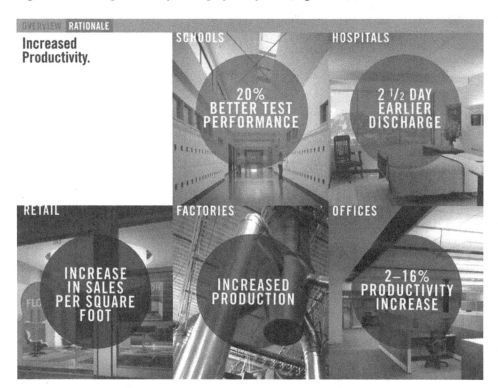

Figure 1.7 Estimated productivity-related benefits by green buildings. *Source:* USGBC.

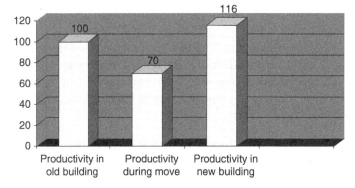

Figure 1.8 Productivity increase by moving into a green building (West Bend, WI).

Third, sustainability results in significant economic benefits at both the building and district level. At the individual building level, the perceived advantages of green buildings, estimated by the USGBC, include:

- 8–9% decrease in operating costs
- 7.5% increase in building values
- 6.6% improvement in return on investment (ROI)
- 3.5% increase in occupancy
- 3% increase in rent

Productivity and health impacts are extremely important for labor-cost-intensive industries, in which most of the business expenditures are salaries (Figure 1.9). Poor indoor environment, comfort, health, and productivity can result in tremendous economic costs. Most studies indicate an average productivity loss of 10% due to a poor indoor environment, although a conservative value of 6% is widely accepted (Dorgan et al. 1998). Fisk (2000) estimated that the overall economic losses due to poor indoor environments in US commercial buildings to be about $40 to $160 billion per year in lost wages and productivity, administrative expenses, and health care costs.

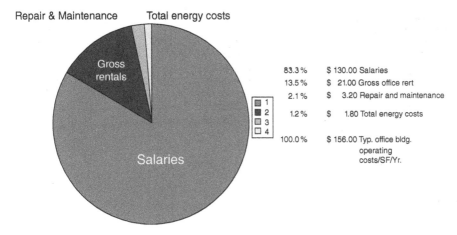

Figure 1.9 Typical expenditure categories for a US office business.

1.5 Sustainable Principles

Sustainable building development is the application of sustainability principles to the built environment. Sustainable building development considers the economic, social, and ecological impact of buildings on their surroundings (Figure 1.10). While this seems an easy concept to understand, it requires a far different approach to building design than has traditionally been used by architects and engineers.

The design process for sustainable buildings focuses on practices that minimize the use of nonrenewable resources, encourages the use of materials that are reused or recycled, and reduces energy consumption during the operation of the building. In this design process, the entire life cycle of the building is considered. Materials are tracked from resource extraction to disposal or reuse. Energy consumption considers not only the operational energy use but also the energy necessary to produce, dispose of, and transport building materials. While traditionally only first costs (costs directly related to construction) have been used to budget the building, the sustainable design process also considers the costs associated with the full life cycle of the building. The building life cycle costs include the energy, maintenance, replacement, and decommissioning costs. A life cycle cost analysis (LCCA) offers a far more realistic picture of the cost to implement sustainable building technologies. Chapter 2 of this book will introduce more fundamentals and practices about LCCA.

The following sustainable design principles are commonly promoted in both architecture and engineering, architects starting from good designs and engineers starting from good indoor environments. An integrative design team and process can therefore facilitate a collaborative design with transparent needs and supplies throughout the design procedure and using the same terminologies. A variety of solutions may be utilized to build a comfortable and healthful interior environment while those with higher energy efficiency using more ecologically benign materials will stand out under the sustainability principles. Certainly, the good-looking designs among these standout ones will ultimately be selected and remembered.

- Comfortable and healthful interior environment
- Energy efficiency

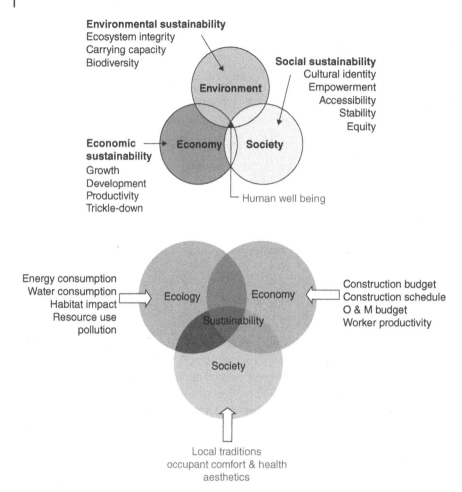

Figure 1.10 The principles of sustainability.

- Ecologically benign materials
- Environmental form
- Good design

Before any advanced technologies or sophisticated systems are considered and applied, there are four best design principles that should always be practiced first, which can be named as 4R principles.

1.5.1 Reduce

The very first consideration of a design is whether the building size, energy, and water system capacity to be designed can be reduced. For instance, the required building functional areas are obtained by creating alternative outdoor living spaces. The sizes of buildings and systems directly impact the energy, material, and water needs, and thus should be properly determined at the beginning. Most of these decisions cannot be changed after the planning phase and have direct consequences to the following design decisions (e.g. heating, ventilation and air-conditioning [HVAC] system sizing for a larger indoor space and water piping design for toilets).

Although the average number of people living in each household has decreased in the United States, houses are twice as large as they were in the 1960s, using even more material in their construction. New homes include, on average, ¾ of an acre of forest in trees per house. With nearly half of the planet's original forests gone, global wood consumption is projected to increase by 50% by 2050.

1.5.2 Reuse

When parts of existing buildings and facilities are still intact, e.g. the foundation and the main structure (beams and columns), saving the building and its elements (e.g. doors) can be a wise approach. This can significantly reduce the construction cost of the new building, ranging from 20 to 70% depending on the reused part percentage. Although reuse is a valuable solution, not all building elements can be reused. For instance, old-standard-based windows and glasses, insulations, toilets, HVAC ducts, etc. may not be good items for reuse. Meeting current design and construction standards should be kept as the priority when reusing existing building components.

Flexible space is another great aspect of reuse, which can alter building layouts and functions with minimum impacts to building main structures and systems. One example is the multifunction rooms in large hotels, which can quickly convert a big banquet ballroom into smaller conference spaces.

1.5.3 Recycle

Many buildings and their elements may not be reusable due to their age, condition, or old standards. However, many of these elements may be recyclable. For instance, old door frames may be recycled to produce new chairs to be used in a new building. Recycling avoids the harvest of new raw materials (e.g. trees) and the production of basic materials (e.g. wood panel).

For new construction, it is therefore important to select materials that can be easily recycled. For instance, in comparing concrete and wood, it is obvious that wood can be more easily recycled. However, the functionality of the materials should always be judged first, e.g. wood may not be suitable for mid-rise or high-rise buildings in terms of structural performance. In examining the cost of materials and their recyclability, the same consideration should be taken.

1.5.4 Regenerate

The end-life of buildings and materials implies a continuation of the destructive effects of material consumption. Reconsidering the nature of "waste," however, offers an opportunity to avoid the problem of disposal and minimizes impacts throughout a material's life cycle.

Landfill is the last process to handle nonrecyclable materials, with the hope to restore the waste to the Earth. However, some materials, such as concrete, will take an infinitely long period of time before they can be deteriorated or absorbed by the Earth, while some such as bamboo can be quickly regenerated. Therefore, selecting proper design modes, structures, and building materials can lead to an environmentally friendly end-life of the building.

Using the 4R principles can create different forms of sustainable buildings. There are some common misunderstandings about a sustainable or green building:

- **Green building = expensive buildings**: this is not true if the building is inherently designed and optimized, which can lead to a cost-less building in construction or operation or both. The lower cost is especially true for the life cycle of the building.
- **Green building = high-tech buildings**: high-tech may not provide the best ROI, although they may shed light on the future technologies. Most green buildings use off-the-shelf technologies and products to provide the best economic and technical performance.
- **Green building = luxury buildings**: green buildings target to deliver better living conditions in terms of safety, convenience, comfort, and health, which are not equal to the definition of luxury.
- **Green building = buildings with green roof/landscaping**: a green roof and landscaping could be a great feature of a green building, depending on the building's location and local climate. But the term "green building" has a much deeper meaning beyond simply a green roof and landscaping. Most of the green building features have to deal with building thermal and environmental systems.
- **Green building = buildings with green products**: green products often refer to energy efficient or environment-friendly products, some of which could be a great fit for a green building after a systematic evaluation. However, green building is not a simple stack-up of these green products, which usually do not guarantee a harmonious performance without a comprehensive and inherent design.
- **Green building = buildings with renewables**: renewables can be the last step to help reduce the energy demand of a building to the power supply and help achieve a net-zero energy goal. However, renewables will never work without substantially reducing the building energy need due to both the cost and size limitation of renewable technologies. Indeed, green building has more objectives beyond energy.

1.6 Three-Layer Design Approach

Building is a system that comprises many compartments and components. Generally, a building system can be divided into three layers as shown in Figure 1.11. Indoor environment (Layer-1) is where human (and other) activities take place, representing the functions of the space and therefore specific requirements on temperature, humidity, air movement, species concentration, lighting, noise, etc. ASHRAE Standard 55 "Thermal Environmental Conditions for Human Occupancy" specifies the conditions for acceptable thermal environments and is intended for use in design, operation, and commissioning of buildings and other occupied spaces. ANSI/ASHRAE Standards 62.1 and 62.2 "Ventilation for Acceptable Indoor Air Quality" (".1" is for commercial and institutional buildings and ".2" is for residential buildings) are the widely used standards for ventilation system design and acceptable indoor air quality (IAQ), which specify minimum ventilation rates and other measures in order to minimize adverse health effects for occupants. Indoor thermal comfort and IAQ will be introduced, respectively, in Chapters 6 and 7. Lighting and acoustic standards (e.g. IES RP-1-12 "American National Standard Practice for Office Lighting" and ASTM E1374 "Standard Guide for Open Office Acoustics and Applicable ASTM Standards") should also be applied to specify proper indoor conditions and are especially crucial for spaces of special

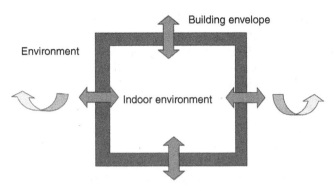

Figure 1.11 Three layers in a building system.

need such as a theater, museum, stadium. Lighting and acoustics design are not included in this book and deserve a separate book for each topic.

The indoor environment is formed by the envelope (Layer-2), separated from the outdoor environment by the envelope, and is connected with the outdoors through the envelope. Therefore, the building envelope plays a critical role in creating a safe, comfortable, healthy, and productive indoor environment. It directly determines the heating, cooling, and ventilation energy needs of indoor spaces, provides daylight to the indoors, and protects indoor environments from outdoor pollutants and noise. It also creates the indoor–outdoor connection physically, visually, and spiritually. Different materials can be adopted to create the functions of building envelopes, which have different implications in energy efficiency and environmental impacts. Proper design of the building envelope is thus the top task of both architects and engineers, starting from shaping and massing, to material selection and insulation sizing, to window-wall-ratio and shading design. Chapters 8 and 9 will focus on the concept, calculation, and discussion of these important topics.

The building's outdoor environment (especially the surrounding microenvironment) (Layer-3) has immediate bidirectional interactions with the building's indoor environment through the building envelope. The outdoor environment will determine the indoor environment's conditions in thermal comfort, air quality, light, and acoustic quality, and thus the energy requirement to maintain a comfortable and healthy indoor environment. The building's outdoor environment will reversely be affected by the building's construction and operation, such as from the exhausted heat from air-conditioning systems. The change of microenvironment will further impact the local and regional climates (e.g. urban heat island) and eventually contribute to the global warming. Climate-adaptive design is thus fundamental to developing a sustainable building that has a minimal environmental impact while achieving a pleasant indoor environment. Chapter 5 will start from the climate and site analysis to discuss and determine the best location-specific design strategies.

An integrative building design should consider all of these three layers, typically starting with climate and site analysis (Chapter 5) and indoor environment demand analysis (Chapters 6 and 7). This will help identify appropriate design strategies, for instance, to use natural ventilation and/or night cooling, or provide additional thermal insulation and/or mass, or use passive and active solar heating (and PV solar panels). With an understanding of indoor and outdoor design conditions, a proper building envelope (including the ceiling/roof, ground, wall, window, door) can be designed with suitable materials and optimal values. Upon that, active systems (for heating, cooling, ventilation, lighting, etc.) can be designed and optimized for the best performance with the least costs. Energy producing systems (e.g. PV solar panels and micro-wind turbines) can then be integrated and specified to provide the required power and reach the net-zero energy goal – the ultimate goal of sustainable building design.

1.7 Three-Tier Design Approach

A successful indoor environment can be achieved by both passive and active systems. A simple definition of "passive system" implies no direct use of energy (e.g. electricity and natural gas) to deliver heating and cooling via transfer media (e.g. hot air and water, cold air) to conditioned spaces, while "active system" is the opposite. It is often a debate whether a passive system that requires electricity can still be called "passive." Such examples include the use of a fan or a pump (both requires electricity) to accelerate the movement of air (e.g. in mechanical/hybrid ventilation and evaporative cooling) and water (e.g. in earth tube cooling), and the use of electricity-powered sensors and controllers to manage the operation of the systems (e.g. window open and close in natural ventilation). Most studies and designs still count these systems as "passive" as no direct heating or cooling media are generated by using electricity or gas.

The three-tier design approach starts from the basic building design (Tier 1 in Figure 1.12), which covers the shape, orientation, openings (windows and doors), exterior color, materials, shading, etc. This part of the design is usually conducted by architects and is therefore mostly experience or rule-of-thumb based. A more scientific based design and optimization by engineers can maximize the efficiency and comfort, creating more favorable rooms for design decisions at the following stages. In general, an optimized basic building design can save 5–20% energy compared to a conventional design without optimization.

Advanced passive design (Tier 2) can then facilitate further energy savings with superior indoor environmental conditions. Most design strategies as shown in Table 1.1 will require advanced engineering knowledge and computational simulation to compare and optimize solutions. A successful passive design with multiple passive features can significantly improve the quality of the indoor environment while reducing energy demand. It may reduce energy by another 20–50% compared to those without considering passive systems. Passive systems not only provide a human-preferred living environment but also

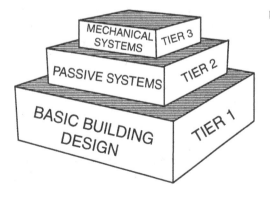

Figure 1.12 Three-tier design approach.

Table 1.1 Design strategies in three-tier design approach.

	Heating	Cooling	Lighting
Tier 1 Basic building design	*Conservation* 1) Surface-to-volume ratio 2) Insulation 3) Infiltration	*Heat avoidance* 1) Shading 2) Exterior colors 3) Insulation	*Daylight* 1) Windows 2) Glazing type 3) Interior finishes
Tier 2 Natural energies and passive techniques	*Passive solar* 1) Direct gain 2) Trombe wall 3) Sunspace	*Passive cooling* 1) Evaporative cooling 2) Convective cooling 3) Radiant cooling	*Daylighting* 1) Skylights 2) Clerestories 3) Light shelves
Tier 3 Mechanical and electrical equipment	*Heating equipment* 1) Furnace 2) Ducts 3) Fuels	*Cooling equipment* 1) Refrigeration machine 2) Ducts 3) Diffusers	*Electric light* 1) Lamps 2) Fixtures 3) Location of fixtures

are necessary to reach a net-zero energy building. Without these systems, renewable technologies (e.g. PV) will not be able to supply adequate energy for the entire building for the entire year due to the constraints in available field size and installation cost, etc. Chapter 10 will introduce some passive designs and systems.

Mechanical systems (Tier 3) consume energy but are inevitable for most buildings due to extreme weather conditions and preferred indoor conditions. After Tie-1 and Tie-2, the largely reduced demand for heating and cooling allows for the use of a much smaller mechanical system, implying a much lower first and operation cost. Advanced heating, cooling, and ventilation systems, such as a radiative heating system, active chilled beam system, or displacement ventilation system, can provide a more comfortable and healthier indoor environment while consuming less energy. Wise sizing and selecting proper mechanical systems may further reduce energy use by 5–20%, depending on location, building function, size, etc. Chapter 12 will introduce the basics of the mechanical systems, while in-depth knowledge of mechanical systems can be found in a few references such as Reddy et al. (2016) and McQuiston et al. (2005).

With a comprehensive and holistic three-tier design, a building is anticipated to have a tremendous energy cut (e.g. 50–80%) while providing a better living and working environment. The remaining energy can then be supplied by renewables. The three-tier design is not a simple stack-up approach, but rather, an integrative and iterative process, which means that when designing Tie 1, Tie 2 and Tie 3 may also be considered simultaneously. One example is a high quality but expensive insulation material may not be an optimal solution when solely design Tie-1; however, if the use of this insulation largely reduces the cooling energy and results in a much smaller air-conditioner (AC) (or even no AC), this insulation will be a valid option as it leads to the smallest overall project cost. This conclusion cannot be reached if an integrative design is not implemented.

1.8   Two Case Studies

Rocky Mountain Institute Headquarter

- **Location**: Old Snowmass, CO (Figure 1.13)
- **Function**: Home/Workplace/Showcase
- **Main features**: passive solar heating and superinsulation – no heating system. Instead, relies on the greenhouse "furnace" and one million pounds of heat-storing thermal mass for heating needs.
- **Other features**: tracking PV, solar thermal water heater, daylighting, efficient appliances. Masonry taken from local area and helps building blend in with surroundings.
- **Economic cost and benefits**: Completed in 1984 for $500k, or $130/ft^2. Energy savings paid off extra cost within 10 months. Saves 99% space/water heating, 90% household electricity, 50% household water.

Fossil Ridge High School

- **Location**: Fort Collins, CO (Figure 1.14)
- **Function**: School
- Completed in August 2004, designed by RB+B Architects, Ft. Collins
- **Energy**:
  - east/west orientation
  - PV shading structure
  - daylighting
  - compact fluorescent (CF) lights and sensors
  - natural ventilation
  - very good envelope
  - ice-storage cooling

Figure 1.13   Rocky Mountain Institute (RMI) headquarter. *Source:* Roger Ressmeyer/Corbis Historical/Getty Images.

Figure 1.14 Fossil Ridge High School. *Source:* Photography by David Patterson Photography, designed by RB+B Architects.

- **Materials**:
  - 17% recycled products
  - 50%+ made locally
  - Preserved 1930s building
  - 70%+ debris recycled

- **Water**:
  - ○ rainwater storage
- **IAQ**: No VOC materials
  - ○ Fresh air flushing
- LEED certified

Homework Problems

1 Browse relevant websites and/or references to find a "green building" example, which is most attractive to you. Identify the "green" building technologies and design concepts employed in the design of the building. Prepare ONE-page power point file (PPT) which contains:

 (a) Title: location, name, function, complete year, cost, and designer (if available) of the building;

 (b) Body: representative "green" image(s) of the building;

 (c) Notes: brief descriptions of the "green" points of the building.

2 Compare small office building energy use (i.e., the total annual site energy, and the energy end use for heating, cooling and lighting, respectively, in MJ/m^2, called "Energy Use Intensity" [EUI]) for Baltimore, MD (see "scorecard" in the html format at https://www.energy.gov/sites/default/files/2013/12/f5/refbldg_4a_usa_md_baltimore_new2004_v1.3_5.0.zip) and Boulder, CO (https://www.energy.gov/sites/default/files/2013/12/f5/refbldg_5b_usa_co_boulder_new2004_v1.3_5.0.zip). Comment on the comparison. The benchmark building energy use (for the same climate and building type) can be used as the energy base for a design.

3 **Staged Project Assignment (Teamwork)**

 (a) Overall goal:

   - Design a commercial building using three containers of given sizes (e.g., 2.32 m (W) × 12.05 m (L) × 2.38 m (H)) for a given location (e.g., Washington, DC) at a specific site (e.g., National Mall, without immediately adjacent buildings). Designers may decide the orientation and land need.
   - Function: any commercial function (e.g., office/visitor center, restaurant/bar, shop/retail, etc.). Designers may decide the function zones, occupant numbers etc.
   - Judge criteria: meet codes [15%]; functions [10%]; design (shape, envelope, system) [30%]; performance (operation, comfort, health, energy) [30%]; creativity [10%]; constructability [5%].

 (b) Assignment:

   - Determine building function
   - Check existing container basic information: 3D sizes, materials, color, etc.
   - Find and study two to three case studies of interest

References

Dorgan, C.B., Dorgan, C.E., Kanarek, M.S., and Willman, A.J. (1998). Health and productivity benefits of improved indoor air quality. *ASHRAE Transactions* 104 (1): 4161.

Fisk, W.J. (2000). Health and productivity gains from better indoor environments and their relationship with building energy efficiency. *Annual Review Energy Environment* 25: 537–566.

McQuiston, F.C., Parker, J.D., and Spitler, J.D. (2005). *Heating, Ventilating and Air Conditioning Analysis and Design*, 6e. Wiley ISBN-13: 978-0471470151.

Miller, L.A., Ramaswami, A., and Ranjan, R. (2013). Contribution of water and wastewater infrastructures to urban energy metabolism and greenhouse gas emissions in cities in India. *Journal of Environmental Engineering* 139 (5): 738–745.

Reddy, T., Kreide, J.F., Curtiss, P.S., and Rabl, A. (2016). *Heating and Cooling of Buildings: Principles and Practice of Energy Efficient Design*, 3e. CRC Press ISBN-13: 978-1439899892.

Roodman, D.M. and Lenssen, N.K. (1995). *Worldwatch Paper #124: A Building Revolution: How Ecology and Health Concerns are Transforming Construction*. Washington D.C.: Worldwatch Institute.

Smith, K., Liu, S., and Chang, T. (2016). Contribution of urban water supply to greenhouse gas emissions in China. *Journal of Industrial Ecology* 20 (4): 792–802.

U.S. Environmental Protection Agency (1998). Characterization of Building-Related Construction and Demolition Debris in the United States. Report No. EPA530-R-98-010. Washington D.C.: U.S. Environmental Protection Agency.

U.S. Environmental Protection Agency (2005). *Municipal Solid Waste in the United States: 2005 Facts and Figures*. Washington D.C.: U.S. Environmental Protection Agency.

UNEP (United Nations environment Programme) (2004). Our increasing appetite for natural resources. https://www.coursehero.com/file/69878603/wastereport-full-ilovepdf-compressed-1-20pdf.

University of Michigan (2018). Center for Sustainable Systems. U.S. Material Use Fact Sheet. Center for Sustainable Systems, University of Michigan. "U.S. Material Use Factsheet." Pub. No. CSS05-18.

USGBC (2003). Building Momentum: National Trends and Prospects for High-Performance Green Buildings. Report prepared for the US Senate Committee on Environment and Public Works. https://www.usgbc.org/resources/building-momentum-national-trends-and-prospects-high-performance-green-buildings.

Wakeel, M., Chen, B., Hayat, T. et al. (2016). Energy consumption for water use cycles in different countries: A review. *Applied Energy* 178: 868–885.

World Commission on Environment and Development (1987). *Our Common Future*. New York: Oxford University Press.

2

Life Cycle Cost Analysis

2.1 Life Phases of a Building

A typical building undergoes various phases during its life span:

1) **Design** (typical duration is 1–2 years) which includes:
   a) *Conceptual design*, which brainstorms and analyzes the design strategies and concepts based on the provided building program (requirements and resources);
   b) *Schematic design* (SD), where a number of design alternatives will be analyzed and compared with the assistance of both visual and computational tools for various fundamental systems and their associated subsystems in the building;
   c) *Design development* (DD), which details the selected design solutions with more architectural and technical descriptions and quantitative analyses;
   d) *Construction document* (CD), which prepares and delivers the final code-compliant design documents to be used for construction.
2) **Construction** (typical duration is 1–3 years) which includes:
   a) Civil construction (foundation, structure, shell, etc.);
   b) Installation of various building systems (water, mechanical, electrical, lighting, safety, etc.);
   c) Start-up commissioning to assure systems installed properly and operating as intended.
3) **Operation and maintenance** (O&M) over its active life (typically 50–60 years for the shell, and 20–25 years for the mechanical, electrical, and plumbing [MEP] systems).
4) **Addition and retrofit**
   a) Addition that expands the building sizes and system capacities, and/or alters building functions and layouts;
   b) Retrofit the building and systems to meet the original operation conditions, typically, after a thorough retro-commissioning of the existing building and systems.
5) **Demolition**, which demolishes the entire building and systems, where some elements may be reused, recycled, or regenerated.

Energy Efficient Buildings: Fundamentals of Building Science and Thermal Systems, First Edition. Zhiqiang (John) Zhai.
© 2023 John Wiley & Sons, Inc. Published 2023 by John Wiley & Sons, Inc.

2.2 Design Process of a Building

The design process of a building includes the following steps:

1) **Planning process**: this phase starts with conceptualizing the project, selecting the site, determining the project scope and duration, and planning the budget, capital, and program. The project will also create an initial design management team and solicit interests of design and construction companies. The design program will define the building use, size, function, and any future expansion. It will also define specific standards/codes to be met, generally local codes. A sustainable building project will increase client awareness and goal setting by explicitly defining the green vision, project goals, green design criteria, and green standards such as Leadership in Energy and Environmental Design (LEED), CA title 24.

2) **Team development**: The management team will determine a short list of qualified design teams and select the main architect and engineering firm after the interview and negotiation. Some projects (especially government-funded projects) may require a public competition process to decide the final design team. A typical design fee ranges from 5 to 10% of the entire construction cost, which will be documented in the design contract.

3) **Design process**: this includes the SD, DD, and CD process (Figure 2.1). The entire green design process will focus on well-integrated design (among various disciplines) and conduct effective resource management and on-going performance goal evaluation.
   a) During the SD process, the building program will be studied, and the required codes will be analyzed. Existing utilities and infrastructures will be coordinated for connection and expansion. The design team will brainstorm and deliver multiple alternative solutions with a preliminary pros and cons analysis, to be reviewed by the management team. During the SD period, 30% project tasks are completed and 15% of the total time is spent.
   b) During the DD process, selected designs and features from SD will be further detailed and evaluated, mostly with the assistance of architectural and engineering tools. Solid details for the determined

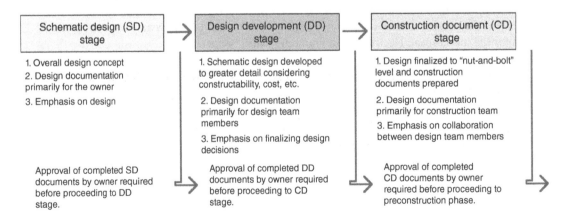

Figure 2.1 Building design process.

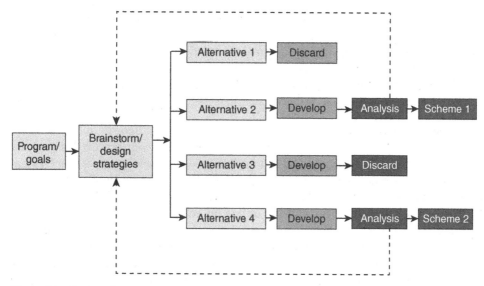

Figure 2.2 Design alternative development process.

components (e.g. window glasses and insulation materials) will be quantitatively finalized. Unsuitable options found during the evaluation will be dropped and alternatives will be proposed for analysis and confirmation (Figure 2.2). All the building systems and subsystems (e.g. mechanical and lighting systems) will be determined at this stage. This should include the entire floor and elevation plans of the architectural design, the specifications of selected systems, and the operation schemes of MEP systems, as well as the cost estimates. During the DD period, 50% project tasks are completed and 35% time is spent.

c) During the CD process, the final code-compliant design documents (aka blueprints) will be prepared and delivered for construction. For international projects, this stage is mostly conducted by the local companies who are familiar with the local building codes. The blueprints should have all the details for structural, mechanical, electrical, lighting, and plumbing elements. Building information modeling (BIM) has been popularly used during this stage to check the conflicts among the design aspects and elements, which also provides a nice 3D rendering for the entire project from the overview to the details. During the CD period, 100% of design tasks are completed and 90% time is spent. The remaining 10% time is budgeted for the on-site visit and explanation and correction, if needed, during construction.

4) **Bid process**: the final reviewed and approved design will be released for a public/open bid or a bid for a selected (preapproved) group of bidders (i.e. preferred contractors). Meetings will be arranged among the design team, the management team, and the potential construction companies to answer technical questions and discuss project goals, time, and cost expectations. The submitted bids will be reviewed by the design and management teams considering both technical and financial contents. A winner will be selected and announced thereafter. For some projects, a re-bid may be necessary if either the technical or financial goals are not met. This usually requires a value engineering to adjust the design scope and/or budget.

5) **Construction administration**: This starts with the groundbreaking ceremony and ends with the opening ceremony. During the construction, the design documents (blueprints) will be followed but may find conflicts and needed changes during the actual construction, which will call for re-evaluation and adjustment. Smaller decisions can be made by the field engineers with changes marked on the blueprints that form the as-built documents. Bigger changes will be evaluated by the design and management teams, which may have larger implications to the function, the systems, the cost, and the project duration.

For a sustainable building development, the preset green design goals and criteria during the planning and programing process should be revisited and rechecked during the entire design process. The goals and criteria may be adjusted according to technical and cost feasibilities. For instance, a net-zero energy goal may have to be adjusted if the renewables are not identified as the best option for the site economically. Another example is that the preset LEED goal may experience dynamic adjustment based on the design decisions made by the design and management teams during the entire process. This is especially true when life cycle cost analysis (LCAA) is applied during the design, to be discussed later in this chapter.

It is widely acknowledged that earlier design decisions have larger impacts on building performance, while these decisions may require a smaller degree of effort (Figure 2.3). Examples include the determination of building orientation (which directly impacts the energy demand for heating, cooling, and lighting, as well as photovoltaic [PV] placement), the insulation level, and the window-wall-rate (which directly impacts the energy demand for heating, cooling, and system sizes). When the design evolves into the DD and CD stages, many design opinions may be missed, hard to incorporate, or changed. Efforts are used to quantify the values for the decisions determined at DD and CD stages, which may have minor impacts on the building's overall performance. The two peaks at the end of the CD and construction management stages are attributed to the inspection at the end of each stage. These inspections may identify and correct inevitable problematic points during the design that can improve the building performance.

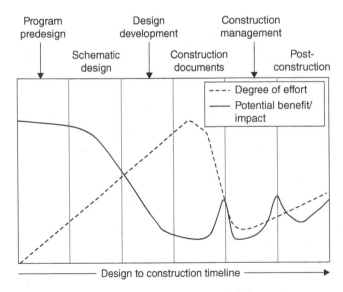

Figure 2.3 Impact of early design input on building performance.

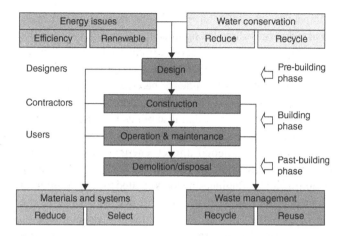

Figure 2.4   Impact of building life stage on energy, water, material, and waste performance.

Figure 2.4 indicates the building life stages that may impact the energy, material and water use, and efficiency. Most energy and water conservations are determined during the design phase. For instance, if a water-free urinal is not designed, this water saving potential is missed from the beginning. Material selection and use are closely related to almost every stage of a building's life, revealing the importance of construction and operation in addition to the design.

2.3   Integrated Design Process of a Sustainable Building

Buildings have been traditionally designed using a sequential design process (Figure 2.5) in which design elements are one after another. In this method, the architect will provide a design to the engineering consultants with each consultant sequentially applying their design. Collaboration between disciplines does not occur until the coordination phase, which is typically during the DD phase of a project. This is long

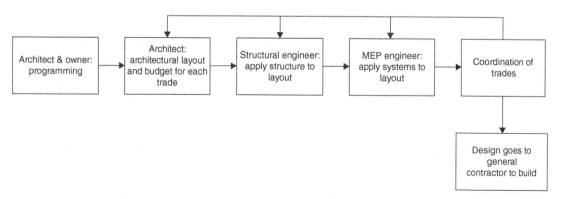

Figure 2.5   Sequential design process.

after many of the crucial design decisions that impact sustainability have already been made. The primary limitation of the sequential design process is that it does not allow for significant interaction between the design trades. Most sustainable technologies are implemented across several traditional building trades making the sequential design process unsuitable for use in sustainable building design.

Sustainable building design recognizes the high level of interdependence between numerous design elements and therefore uses an integrated design process (IDP). This holistic approach to the design of buildings allows for synergistic collaboration between various design disciplines in order to provide optimized design solutions to meet the project goals. The integrated building design (IBD) method balances the need for architectural expression, minimal environmental impact, energy efficiency, indoor environmental quality, and resource use by promoting collaboration early in the design process using simulation tools to provide feedback for design optimization. Engineering tools created for environmental assessment, LCCA, energy analysis, and indoor environmental quality inform decisions early in the development of a building's design in order to simplify the implementation of innovative solutions.

Studies have shown that innovative sustainable design solutions are more effective when collaboration among the various design consultants occurs in the early stages of the project development. This is due to the fact that sustainable strategies such as daylighting, rainwater catchment, and passive heating and cooling require the coordination of design elements that were traditionally across differing building trades. Therefore, sustainable building design requires a new approach to not only the individual building technologies but also the entire process of building design.

The IDP can be thought of as a series of iterative design decisions in which all team members offer valuable assistance. The majority of the work that will lead to the sustainable design objectives will be accomplished in the early stages of design. When applied to the traditional construction phases, the most important sustainable design work decisions often occur in the SD phase where multiple design alternatives are compared iteratively in order to optimize the design based on performance criteria. The general flow of the IDP is shown in Figure 2.6.

The IDP acknowledges the interdependency between the different disciplines of a project. Traditionally, building services engineers have added to an already developed conceptual layout, whereas the IDP requires all of the design team to work in tandem to develop a well-coordinated and optimal design. The collaboration between these disciplines is highlighted throughout all phases of a building design project, which is a significant difference from the traditional sequential design process. In addition, the majority of decisions are shifted from the DD phase to the SD phase. This method of design is critical for the proper design of sustainable buildings since it allows for the comparison of multiple design options. The design team can use this process to minimize the energy use, carbon emissions, or environmental impact while maximizing indoor environment quality.

The IDP requires the iteration of design concepts in order to refine the integrated performance of various building components. This requires a fairly complex interaction between design team members since many different members will be providing information related to different aspects of the design. Iteration is crucial to the proper implementation of sustainable technologies because of the large interdependence between the operation of different systems. One example of this interdependence is the relationship between heating, cooling, and lighting. Lighting power density, lamp types, and daylighting aperture reduce heating loads and increase cooling loads. Switching from incandescent to fluorescent lamps or adding a daylighting strategy can impact the size of heating, ventilation and air conditioning (HVAC) systems. An iterative design process allows the overall impact of these design strategies to be evaluated

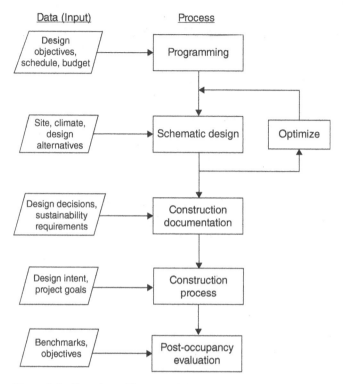

Figure 2.6 Overview of integrated design process.

and refined to determine a solution that provides the best balance of all trade-offs. In order to iteratively optimize these systems, the HVAC engineer, lighting engineer, and architect all must collaborate to evaluate and revise the lighting and HVAC systems. The nature of iterative design seems increasing complex and effortful at the beginning of a project, mostly due to diverse requirements and constraints from different perspectives of a design. However, this can accelerate and streamline the entire design process harmoniously with a consistent design goal and evaluation target (Figure 2.7). Charrettes with active participation of designers of all disciplines at critical stages are the important vehicle to assess and recalibrate the design goals and associated decisions.

One of the overall objectives of sustainable design is to maximize the use of energy that is available on-site. In order to reduce the size (or even eliminate the use) of mechanical systems, the building site, structure, and envelope should be evaluated and optimized first. This allows the building loads and resource consumption to be minimized. This optimization should be based on criteria and assessment methods that are defined in the Owner's Project Requirements. Second, the passive systems should be considered, designed, and optimized. This allows for the analysis of the passive systems and provides feedbacks to the building design at an early stage. Lastly, the mechanical systems should be optimally designed. At this point, the mechanical systems are designed based on building loads that are greatly reduced from the initial design. This type of design method (Figure 2.8) can be applied not only to HVAC systems but to civil, structural, lighting, and plumbing systems as well. By focusing the initial efforts on minimizing

Integrative process

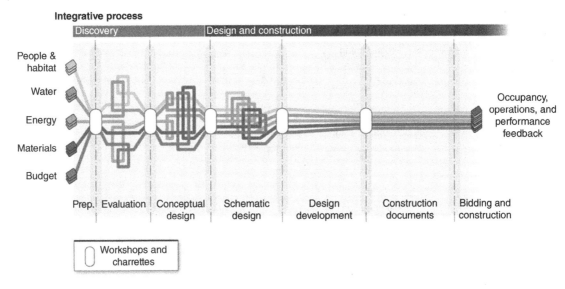

Figure 2.7 Iterative design process.

resource use, the passive and active systems that are achieved can be much smaller, freeing up more of the budget for high-performance building components and sustainable design technologies.

2.4 Basics of Cost and Economic Analysis

- **Present value (cost):** the amount of money that can be used or invested now, or the amount of money that would have to be invested now in an interest-bearing account in order for the amount to grow to a given future value.
- **Future value (cost):** the value of a present investment at some time in the future. Future value for discrete situations is calculated using the appropriate compound interest formula. The future value of a continuous income stream is the total accumulated value of the income stream and its earned interest.
- **Interest:** the cost of borrowing, which compensates lenders for the risk they take in making their money available to borrowers. Interest is usually expressed at an annual rate: the amount of interest that would be paid during a year divided by the amount of money loaned. For one dollar invested in 2020 ($\$_{2020}$) with an interest rate ($r_{interest}$) of 2%, the total value in 2021 becomes ($\$_{2021}$), whose relationship is:

$$\$_{2021}Y = \$_{2020} 1 \times (1 + r_{interest}) = \$_{2020} 1 \times (1 + 2\%) = \$_{2021} 1.02 \tag{2.1}$$

The general expression can be presented as:

$$\$_M Y = \$_N X \times (1 + r_{interest})^{(M-N)} \tag{2.2}$$

where M is a later year and N is the earlier year when the investment occurs.

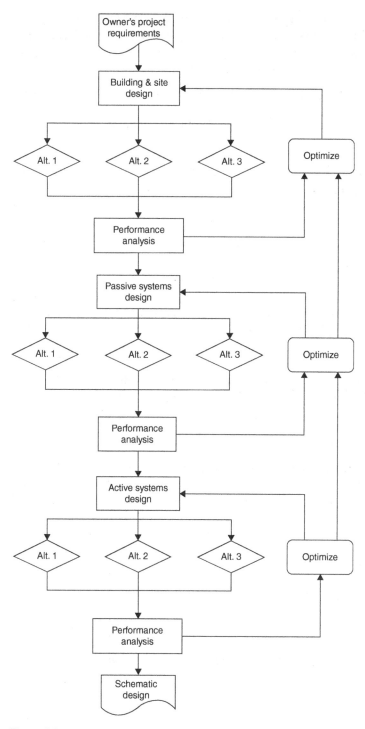

Figure 2.8 Example of an iterative design process for energy efficiency.

- **Inflation:** an ongoing rise in the average level of absolute prices. The annual inflation rate ($r_{inflation}$) for the United States is 5.0% for the 12 months that ended May 2021 after rising 4.2% previously, according to US Labor Department data published 10 June 2021. Because of inflation, one dollar in 2021 ($\$_{2021}$) is less valuable than one dollar in 2020 ($\$_{2020}$), whose relationship is:

$$\$_{2021} Y = \$_{2020} 1 \times 1/(1 + r_{inflation}) = \$_{2020} 1 \times 1/(1 + 5\%) = \$_{2021} 0.95 \tag{2.3}$$

In order to keep the same value (or purchasing power) of one dollar in 2021, the amount of money to be saved in 2000 should be:

$$\$_{2021} 1 = \$_{2000} X \times 1/(1 + r_{inflation})^{(2021 - 2000)} = \$_{2000} X \times 1/(1 + 5\%)^{21} \tag{2.4}$$

$$\$_{2000} X = \$_{2021} 1 \times (1 + 5\%)^{21} = \$_{2000} 2.79 \tag{2.5}$$

The general expression is thus:

$$\$_{N + t} Y = \$_N X \times 1/(1 + r_{inflation})^t \tag{2.6}$$

Or

$$\$_N X = \$_{N + t} Y \times (1 + r_{inflation})^t \tag{2.7}$$

where N is the current year. When considering both investment interest and inflation, one dollar invested in 2020 with an interest rate ($r_{interest}$) of 2% and an inflation rate ($r_{inflation}$) of 5%, will become, in 2021,

$$\$_{2021} Y = \$_{2020} 1 \times (1 + r_{interest})/(1 + r_{inflation}) = \$_{2020} 1 \times (1 + 2\%)/(1 + 5\%) = \$_{2021} 0.97 \tag{2.8}$$

Since the interest rate is less than the inflation rate, the dollar is still discounted from 2020 to 2021. The **real interest rate** ($r_{interest\ 0}$) of this investment can be defined and calculated as:

$$1 + r_{interest\ 0} = (1 + r_{interest})/(1 + r_{inflation}) \tag{2.9}$$

$$r_{interest\ 0} = (r_{interest} - r_{inflation})/(1 + r_{inflation}) \tag{2.10}$$

- **Constant dollars (currency):** dollar values that have been adjusted for inflation by means of price indexes to eliminate inflationary factors and allow direct comparison across years. To consider the real purchasing power and eliminate the inflation in comparison, converting to constant dollars for a given year (N + t) of interest is calculated as the current dollar value at year N multiplied by the purchasing power of the dollar based on a dollar value of $1 in year (N + t):

$$\$_{N + t} Y = \$_N X \times (1 + r_{inflation})^t \tag{2.11}$$

- **Growth rate:** the percentage change of a specific variable within a specific time period, usually expressed at an annual (nominal) growth rate (r_{growth}):

$$\$_{N + t} Y = \$_N X \times (1 + r_{growth})^t \tag{2.12}$$

- **Real growth rate:** the percentage change of a specific variable within a specific time period that considers the inflation. Similar to the real interest rate in Equations (2.9) and (2.10), the real growth rate ($r_{growth\ 0}$) is defined as:

$$r_{growth\ 0} = (r_{growth} - r_{inflation})/(1 + r_{inflation}) \tag{2.13}$$

For low inflation rates, Equation (2.13) can be approximated as:

$$r_{growth\ 0} = r_{growth} - r_{inflation} \qquad (2.14)$$

An analysis using constant currency and real rates is exactly equivalent to one with inflating currency and nominal rates if not considering loan, depreciation, etc. For some applications, the real growth rate is fairly well known, while the inflation rate is hard to predict. For some products or commodities, the constant currency is much more stable than the market or inflating currency. Therefore, it is instructive to think in terms of real currency and real rates.

Example 2.1

In 1980, the market electrical price is $0.08/kWh; in 2010, the market electrical price became $1.12/kWh. If the inflation rate is 3% between 1980 and 2015, what are the real prices of 1980 and 2010 prices in 2015? What are the nominal and real growth rate between 1980 and 2010?

Solution

From Equation (2.11):

$$\$_{2015}\ Y = \$_{1980}\ 0.08 \times (1 + 3\%)^{35} = \$_{2015}\ 0.23$$

$$\$_{2015}\ Y = \$_{2010}\ 1.12 \times (1 + 3\%)^{5} = \$_{2015}\ 1.30$$

From Equation (2.12):

$$r_{growth} = -1 + \sqrt[30]{1.12/0.08} = 0.09$$

From Equation (2.13):

$$r_{growth0} = \left(r_{growth} - r_{inflation}\right)/(1 + r_{inflation}) = (0.09 - 0.03)/(1 + 0.03) = 0.06$$

or

$$r_{growth0} = -1 + \sqrt[30]{1.30/0.23} = 0.06$$

- **Discount rate:** the discounting of future money back to the present value, which is similar to the growth rate but in the backward direction:

$$\$_N\ X = \$_{N+t}\ Y \times 1/(1 + r_{discount})^{t} \qquad (2.15)$$

Considering the inflation, the real discount rate is:

$$r_{discount\ 0} = \left(r_{discount} - r_{inflation}\right)/(1 + r_{inflation}) \qquad (2.16)$$

Both the growth rate and discount rate depend on the investment activities. For instance, if the money is hidden under the bed, both rates are zero; if it is invested in stock, the stock change rate is the growth/discount rate.

The present value (P) can be calculated by using either future market value (F_t) and market discount rate ($r_{discount}$) or future real value (F_{t0}) and real discount rate ($r_{discount\ 0}$), as proved below.

$$P = F_t/(1 + r_{discount})^t = \left[F_t \times (1 + r_{inflation})^t\right]/\left[(1 + r_{discount})^t \times (1 + r_{inflation})^t\right]$$

$$= \left[F_t/(1 + r_{inflation})^t\right]/\left[(1 + r_{discount})/(1 + r_{inflation})\right]^t = F_{t0}/(1 + r_{discount\,0})^t \qquad (2.17)$$

The ratio of the present and future value (P/F$_t$) is called the "present worth factor."

- **Equivalent cash flow:** the total present value of continuous equal future value (A) (e.g. payment or investment) in regular interval (e.g. yearly):

$$P = A/(1 + r_{discount})^1 + A/(1 + r_{discount})^2 + \dots + A/(1 + r_{discount})^t = A\left[1 - (1 + r_{discount})^{-t}\right]/r_{discount}$$

$$(2.18)$$

when $r_{discount} \neq 0$. If $r_{discount} = 0$, $P = A \times t$. This calculation process is called as levelizing.

Example 2.2

A heat pump has an end-life (salvage) value of $200 after 20 years. What is the equivalent levelized annual value if the discount rate is 5%?

Solution

Equation (2.18): $t = 20$, $r_{discount} = 5\%$;

$$P = A\left[1 - (1 + r_{discount})^{-t}\right]/r_{discount} \qquad \text{(value at year} - 0)$$

$$F = P \times (1 + r_{discount})^t = \$200 \qquad \text{(value at year} - 20)$$

$$\$200/1.05^{20} = A(1 - 1.05^{-20})/0.05$$

$$A = \$6/\text{year}$$

Example 2.3

A home buyer obtained a mortgage of $300 000 for an annual interest rate of 4% over 30 years. What is the total payment of this mortgage at year-30? What is the annual payment?

Solution

$$t = 30\,\text{years}$$

$$F = P \times (1 + r_{interest})^t = \$300\,000 \times 1.04^{30} = \$973\,019 \text{ (total value/payment at year} - 30)$$

$$P = A\left[1 - (1 + r_{interest})^{-t}\right]/r_{interest} = \$300\,000 = A(1 - 1.04^{-30})/0.04$$

$$A = \$17\,349/\text{year}$$

If the monthly interest rate is assumed to be $4\%/12 = 0.3333\%$, what is the total payment of this mortgage at year-30 and what is the monthly payment?

$$t = 30 \times 12 = 360\,\text{months}$$

$$F = P \times (1 + r_{interest})^t = \$300\,000 \times 1.003333^{360} = \$994\,049 \text{ (total value/payment at year} - 30)$$

$$P = A\left[1 - (1 + r_{interest})^{-t}\right]/r_{interest} = \$300\,000 = A(1 - 1.003333^{-360})/0.003333$$

$$A = \$1432/\text{month}$$

- **Compounding**: the process whereby interest is credited to an existing principal amount as well as to interest already paid. Compounding can thus be construed as interest on interest – the effect of which is to magnify returns to interest over time, the so-called "miracle of compounding." When banks or financial institutions credit compound interest, they will use a compounding period such as annual, monthly, or daily.
- **Discrete and continuous cash flow**: the discrete or continuous value change (e.g. interest or growth) with compounding intervals such as daily or monthly or annually. The discrete compounding interest ($r_{compound}$) can be expressed as:

$$1 + r_{compound} = (1 + r_{nominal}/m)^m \tag{2.19}$$

$r_{nominal}$ is the nominal interest rate without considering compounding; m is the discrete time interval. If m approaches infinite, Equation (2.19) becomes:

$$1 + r_{compound} = \exp(r_{nominal}) \tag{2.20}$$

This is the continuous cash flow, representing the largest growth factor.

Example 2.4

If the loan has a nominal annual interest rate of 4%, what is the annual interest rate with monthly, daily, and continuous compounding?

Solution

$$r_{compound, 12} = (1 + r_{nominal}/12)^{12} - 1 = 0.0407415$$

$$r_{compound, 365} = (1 + r_{nominal}/365)^{365} - 1 = 0.0408085$$

$$r_{compound, cont} = \exp(r_{nominal}) - 1 = 0.0408108$$

2.5 Life Cycle Cost Analysis

2.5.1 Terminologies

- **Life cycle cost**: the sum of the present values of all cost components through the life of a product, a device, a system, or a project/building.
- **Capital cost**: the total initial investment (e.g. for a product, a device, a system, or a project/building).
- **Operation cost**: the total cost to operate the product, device, system, or project/building. This usually refers to the cost of energy, water, etc. required to run the device, system, and building.
- **Maintenance, repair, and replace cost**: the cost to maintain, repair, and replace the product, device, system, or project/building. Sometimes, people may put this with the operation cost and call it the "Operation and Maintenance" (O&M) cost. However, since the operation cost may be readily predictable while maintenance is not, separation of the energy and water use cost from the maintenance cost may be a wise choice for some energy-focused construction and retrofit projects, especially when the energy use is dominant among the cost components.
- **Resale value**: the salvage value of the product, device, system, or project/building when it is replaced. The resale or remaining values vary largely by nature, age, and condition of goods, as well as location, economics, etc., and thus are somehow challenging to precisely determine. Remaining value is

important to consider for replacing products or systems that are not at the end of their lives. A fair market price can be a good estimate, and therefore this cost is called the "resale value."

- **Insurances**: various necessary insurances to assure the safety and value of the product, device, system, or project/building through their lives.
- **Taxes**: various required taxes, which may vary during the life.

2.5.2 Life Cycle Cost

The life cycle cost of a product, a system, or a project may include a number of expenditures from the start to the end of the life of the product, system, or project. This includes the initial obtaining cost (down payment), the cost of loan (if borrowed from a bank), the cost of operation (e.g. energy), the cost of maintenance, etc. This cost can be reduced if interest tax credit, incentive, depreciation, and salvage are considered. For simplicity, this chapter discards these savings, which are location, time, and condition dependent, to allow a common (or worst) scenario analysis.

- **Initial cost (down payment)**: the part of the capital cost (e.g. purchase, installation, construction, etc.) that is directly taken out of pocket. If there is no loan involved (i.e. full cash buying), this is the total capital cost for the product, the system, or the project. This cost can be put forward all at once (day one) or by an agreed payment schedule with or without interest.

$$C_{initial} = C_{cap} \times (1 - f_{loan}) \tag{2.21}$$

where f_{loan} is the fraction of the capital cost (C_{cap}) borrowed from a lender. $f_{loan} = 80\%$ is common for purchasing a residential dwelling to avoid the mortgage insurance and to acquire a good interest rate.

- **Cost of loan**: the total payback amount of a loan with both principal and interest in the present value.

$$C_{loan} = C_{cap} \times f_{loan} \times \left\{ r_{interest} / \left[1 - (1 + r_{interest})^{-N} \right] \right\} / \left\{ r_{discount} / \left[1 - (1 + r_{discount})^{-N} \right] \right\} \tag{2.22}$$

where

$$C_{cap} \times f_{loan} \times \left\{ r_{interest} / \left[1 - (1 + r_{interest})^{-N} \right] \right\} = C_{loan-payment} \tag{2.23}$$

$C_{loan-payment}$ is the scheduled regular loan payment amount (e.g. monthly, quarterly, or annually). N is the total installation of the payment. $r_{interest}$ is the interest for the payment period (e.g. monthly, quarterly, or annually).

- **Cost of operation**: the total cost for the operation of the product, the system, or the project. This may include both supply (e.g. energy and water) and labor costs. For some projects, such as buildings, this cost is much more dominant (over 90% of the whole life cycle cost) than the capital cost. Therefore, a green feature may seem overpriced within the capital cost but it may significantly reduce the operating cost as a return. Therefore, this feature might be missed if only considering capital cost during the design. The LCCA will ensure the best gain through the entire life of the project. The average operation cost during each period (e.g. year) is

$$C_{operation} = \sum_{i=1}^{S} Q_i R_i \tag{2.24}$$

where Q is the demand (in energy, water, labor, etc.) and R is the rate of cost (e.g. energy rate in $/kWh, water in $/ton, labor in $/hour).

$$C_{operation,present} = C_{operation} \times \left[1 - (1 + r_{discount})^{-N}\right] / r_{discount} \tag{2.25}$$

Equation (2.25) converts every period (year) operation cost to the present value throughout the entire life period. If additional cost $C_{addition}$ is used for a specific year t, this cost can be converted to the present value using Equation (2.15):

$$C_{addition,present} = C_{addition} / (1 + r_{discount})^{t} \tag{2.26}$$

For more detailed estimates, different discount rates for different categories of cost (e.g. energy, water, and labor) can be used, where $C_{operation}$ will be divided into each category to calculate individual discounts of the current cost.

- **Cost of maintenance:** the total cost of maintenance during the lifetime of a product, a system, or a project. This may include the repair and replacement cost. Equation (2.25) can be used to convert any regular maintenance cost to the present value, and Equation (2.26) can be applied to include any additional cost for the maintenance (e.g. unexpected repair and replacement).

2.5.3 Life Cycle Savings

The difference of the life cycle cost between the reference (base) case and its alternative is called "life cycle savings." For a new project (or building), the reference case is often referred to the standard one, which meets the minimum code requirements for the specific location. For a retrofit project (or building), the reference case is simply the existing one.

Assuming the same interest and discount rates, the comparison of two designs mainly differs in the capital cost and O&M costs. The most complicated, yet valuable, study using the life cycle cost concept would be the case, as described earlier, that has a larger capital cost while O&M costs are smaller. Some simple but widely used measures to quantify the saving are below.

- **Payback time (N_p):** the ratio of the extra capital cost $\Delta C_{capital}$ to the first-year saving $C_{saving, 0}$.

$$N_p = \Delta C_{capital} / C_{saving, 0} \tag{2.27}$$

The inverse of N_p is sometimes called the "return on investment" (ROI).

$$ROI = 1/N_p = C_{saving, 0} / \Delta C_{capital} \tag{2.28}$$

The shorter the N_p (or the higher the ROI), the higher the profitability. Both Equations (2.27) and (2.28) do not consider any time value of the saving (for the future years), which may increase the payback time and reduce ROI. Therefore, both represent the maximum benefits of a design. This N_p is hence called simple payback time. Although simple with fundamental inadequacy, N_p is intuitive in quickly comparing and determining appropriate alternatives with approximating ranks that show distinct disparities in cost performance.

Example 2.5

Compare two alternatives for the desk lamps at home: an incandescent bulb and an LED bulb, both giving light of the same quantity and quality. Make the following assumptions.

- First cost of incandescent bulb = $1; First cost of LED bulb = $15
- Power drawn by incandescent bulb = 60 W; Power drawn by LED bulb = 15 W
- Lifetime of incandescent bulb = 2000 hours; Lifetime of LED bulb = 10 000 hours
- Light is turned on 2000 h/year; Electricity price paid = 15¢/kWh

Solution

Difference in capital cost $\Delta C_{capital} = \$15 - \$1 = \$14$
Savings in electricity in the first year $C_{saving, 0} = [(60-15) \text{ W}] \times 2000 \text{ h} \times \$0.15/\text{kWh} = \$13.50$
From Equation (2.27), $N_p = \$14/\$13.5 = 1.037$ year = 12.4 months

- **Internal rate of return:** the value of the discount rate $r_{discount}$ at which the life cycle savings are zero. Similar to Equation (2.25), the lifetime saving due to the operation saving (e.g. lower energy consumption) can be calculated:

$$\Delta C_{operation, saving, present} = \Delta C_{operation, saving} \times \left[1 - (1 + r_{discount})^{-N} \right] / r_{discount} \qquad (2.29)$$

If the value is equal to the additional capital cost $\Delta C_{capital}$ that leads to this savings, such a $r_{discount}$ is called the internal rate of return. The higher the internal rate of return, the higher the profit of the investment.

Example 2.6

Continue on **Example 2.5** to calculate the internal rate of return of using the LED bulb rather than the incandescent bulb for the lifetime of the LED bulb.

Solution

$\Delta C_{operation, saving} = [(60-15) \text{ W}] \times 2000 \text{ h} \times \$0.15/\text{kWh} = \$13.50$ (annual savings)
$\Delta C_{capital} = \$15 - \$1 \times 5 = \$10$ (for the lifetime of the LED bulb – five years, assuming the incandescent bulb keeps the same price for the five years or five incandescent bulbs are purchased at year-1).
From Equation (2.29),

$$\$10 = \$13.50 \times \left[1 - (1 + r_{discount})^{-5} \right] / r_{discount}$$

Solving this yields: $r_{discount} = 133\%$

Example 2.7

Continue on **Example 2.5** to calculate the simple payback time of using the LED bulb rather than the incandescent bulb for the lifetime of the LED bulb.

Solution

$\Delta C_{capital} = \$15 - \$1 \times 5 = \$10$ (for the lifetime of the LED bulb – five years, assuming the incandescent bulb keeps the same price for the five years or five incandescent bulbs are purchased at year-1).

Savings in electricity in the first year $C_{saving, 0} = [(60-15) \text{ W}] \times 2000 \text{ h} \times \$0.15/\text{kWh} = \$13.50$
From Equation (2.27), $N_p = \$10/\$13.5 = 0.74$ year = 8.9 months
This is a very profitable investment. The rule of thumb of an excellent payback time is less than 1/3 of the equipment life (i.e. 20 months for the LED bulb).

From Equation (2.29),

$$\Delta C_{capital} = \Delta C_{operation, saving, present} = \Delta C_{operation, saving} \times \left[1 - (1 + r_{discount})^{-N}\right]/r_{discount} \qquad (2.30)$$

$$\Delta C_{operation, saving} = C_{saving, 0} \qquad (2.31)$$

Equation (2.27) can be written as:

$$N_p = \Delta C_{capital}/C_{saving, 0} = \left[1 - (1 + r_{discount})^{-N}\right]/r_{discount} \qquad (2.32)$$

Therefore, N_p can be determined by the internal rate of return or vice versa.
From Example 2.6, $N_p = [1 - (1 + 133\%)^{-N}]/133\% = 0.74$ year

Example 2.8
LCCA comparison of a green and conventional residential building

- 30-year term loan, 7% interest APR, 20% down payment
- Custom 4-bedroom house with moderate finishes on 2-acre lot in rural subdivision
- Purchasing homeowner's combined income: $85K/year

Solution

1. Conventionally-built house construction and cost calculation

- 2600 ft^2 house $\times$ \$130/ft^2 (including site, sitework, construction costs, and overhead/profit) = \$338K
- Architecture, engineering, and permit fees (12% $\times$ \$338K) = \$41K
- Total project cost: \$338K + \$41K = \$379K
- Finance: \$379K $-$ \$76K(=20% $\times$ \$379K; down payment) = \$303K = \$2022/month $\times$ 360 payments = \$728K over 30-years.

2. Green-built house construction and cost calculation

- 2200 ft^2 house (tightened design) $\times$ \$156/ft^2 (20% added for extra green features and costs) = \$343K
- Architecture, engineering, and permit fees (12% + 20% $\times$ 12% extra design/documentation cost = 14.4% $\times$ \$343K) = \$49K
- Total project cost: \$343K + \$49K = \$392K
- Finance: \$392K $-$ \$78K(=20% $\times$ \$392K; down payment) = \$314K = \$2089/month $\times$ 360 payments = \$752K over 30 years.

For extra \$67/month (or a 30-year investment totaling \$24K) + \$2K extra down payment, the homeowner could have a high-performance green home.

3. Conventionally-built house energy cost calculation

- Average combined electric and gas monthly bills = $250/month + 5%/year inflation = $199K over 30 years
- Combined energy payment would be $1029/month in year-30.
- Combined mortgage and energy costs = $2022/month + $250/month = $2272/month in year-1 and $3052/month in year-30.

4. Green-built house energy cost calculation

- Average combined electric and gas monthly bills = $250/month × 60% (40% efficiency saving assumed) + 3%/year inflation (a portion assumed to be produced by local renewables such as PV) = $86K over 30 years
- Combined energy payment would be $354/month in year-30.
- Combined mortgage and energy costs = $2089/month + $150/month = $2239/month in year-1 and $2443/month in year-30.

The green house is $33/month less expensive on the day the homeowner moves in. The green house will be $609/month less expensive in year-30 and saves $113K in energy costs over 30 years.

2.6 Life Cycle Cost Analysis Based Optimization

The concept of life cycle cost is important to optimize the building and system design. As explained earlier, considering the operation and maintenance costs can ensure no missing of feasible design alternatives due to their potentially high capital costs. Evaluating the inherent relations among building elements can also reveal the cost–performance interactions of these elements and therefore impact decisions of adopting these elements as a system. One example is that replacing single-pane windows with double-pane windows will increase the window budget but reduce the cost of the heating and air-conditioning systems. Selecting a proper window glass without considering its potential impacts on mechanical system selection may not provide the most performance and cost wise decision.

Figure 2.9 shows the conceptual path to a net-zero energy building, with energy savings on the x-axis and energy-related costs on the y-axis. The annual cost includes the mortgage payment (the increment to cover energy improvements) and utility bills, plus replacement costs (discounted) for measures with lifetimes less than the analysis period. This is essentially an adjusted year-1 energy-related cash flow. These terms are most "real" in the residential market. For the base-case building, the annual costs are entirely due to utility bills.

From the base case, there are some no-cost or low-cost options available for saving energy, so the cash flow curve follows the utility bill cost line. As the building design is further improved, utility bills continue to reduce, while there are increased mortgage payments to pay for the energy efficiency measures (EEM). Annual costs are at a minimum at Point 2 (Figure 2.10) which, in terms of the economic metric, is a global optimum for the homeowner. The slope of the cash flow curve represents the net marginal cost of saved energy, depending on the performance and cost of the adopted technologies. At Point 2, the cost of the last unit of saved energy just equals the cost of buying energy from the utility. However, one may desire to go

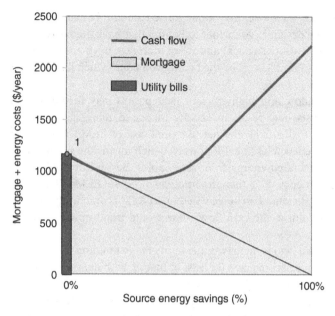

Figure 2.9 Conceptual path to a net-zero energy building.

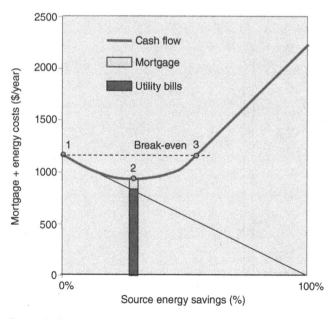

Figure 2.10 Energy efficiency measures to reach the global optimum and break-even design.

further, for noneconomic reasons. For instance, the owner may attain a fixed amount of loan or income that can be used for the house. This allows the design to go further to Point 3 that have the same annual energy related costs with more energy saving (utility savings) and more mortgage payment due to the extra investment on energy efficient measures. Point 3 is called the break-even point which has the same annual costs as the original design.

Advanced EEM (e.g. a superinsulated window and high-efficient heat pump) may further reduce building energy use. However, the high costs may result in rapidly increased mortgage payments (due to the diminishing ROI). At Point 4 in Figure 2.11, the net marginal cost of EEM (indicated by the slope of the curve) equals the net cost of renewables (e.g. PV power), which means the saved energy cannot economically compete against the generated energy from renewables. Beyond this point, it is more cost effective to invest in renewables (e.g. PV) than to implement more EEMs. The cost–performance slope at Point 4 can be used, as a threshold, to judge whether an EEM is practically feasible. If a design continues to add EEMs beyond Point 4, the cash flow curve would trend upward with an increasingly steep slope.

As the size of the PV system (or other renewables) is increased, net-zero energy building is eventually achieved at Point 5. At this point, the utility bills become zero (at least in theory, with net metering) and the annual energy-related costs comprise entirely of mortgage payments for the EEMs and PV. Note that Point 5 is much higher than Point 1 in the annual costs due to the high cost of PV, although this has been declining significantly in the past decade. It should also be noted that this trend curve may vary with future utility rates (Figure 2.12), which motivates the pursuit of net-zero energy designs.

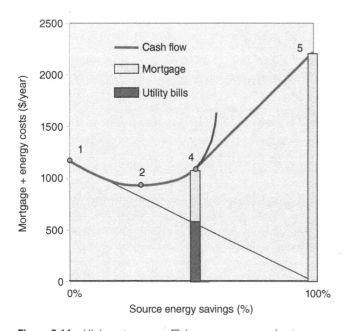

Figure 2.11 High-cost energy efficiency measures and net-zero energy design.

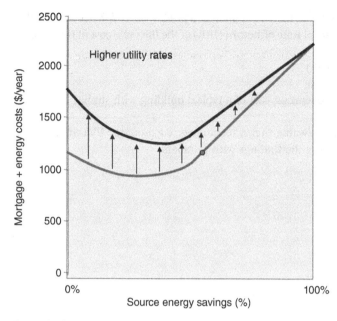

Figure 2.12 Impacts of utility rates on the path to a net-zero energy building.

Homework Problems

1 A high-quality insulation doubles the cost of a traditional insulation (assume $X), which is fully loaned for q years with an annual interest rate of m (in %). This results in an annual energy saving of n (in %) (assume the total annual energy cost is $Y). Assume the average monthly energy cost growth rate is p (in %). Please derive the equation to determine the month when the energy saving is over the extra loan payment due to the use of high-quality insulation. Calculate the number if X = 100 000, Y = 20 000, m = 8%, n = 3%, p = 0.4%, q = 30.

2 A cheap air conditioner (AC-1) has a lifetime of 20 years (no salvage value after that). The cost of this AC at year-1 is $2000. A high performance AC-2 has a lifetime of 30 years (no salvage value after that). The cost of this AC at year-1 is $3000. Assume the building lifetime is 60 years, AC-2 can save 300 kWh/year than AC-1. The currency inflation rate is 3%. The electricity cost is $0.8/kWh at year-1 with an annual growth rate of 5%. Please comment on the selection of the AC with calculation and numbers. What is the simple payback time for choosing the high-performance AC?

3 For 2020–2021, the cost of the solar PV system is $3/W. A single-family house plans to install a PV system of 3000 W, with 20% down payment and 80% for a 30-year-loan at an interest rate of 3.5%. Assume the PV system has a growth/discount rate is 5% and the average annual maintenance cost is $200. Please calculate the life cycle cost of the system in 30 years. If the system can generate

30 000 kWh electricity annually and sell this to the power plant at a price of $0.15/kWh, calculate the Return on Investment (ROI) and the Internal Rate of Return (IRR) of the life cycle cost of the system.

4 Staged Project Assignment (Teamwork)
  (a) Check and estimate the container costs.
  (b) Check the average construction and operation cost of a typical building with similar design function and size at the design location.
  (c) Check and analyze whether on-site renewable energy techniques (e.g., roof-top PV) would make economic sense for this building. Justify the findings with numbers.

3

Building Standards and Codes

3.1 Impacts of Building Codes

The building industry is fundamentally code-driven, from design to construction to operation to demolition. Residential and commercial building codes require increased building efficiency and help drive the transition to a more efficient economy. Recent, significant changes to **the American Society of Heating, Refrigerating, and Air-Conditioning Engineers** (ASHRAE) 90.1-2010 and the International Energy Conservation Code (IECC) 2012 require decreases of 9–38% and 12–15% in energy use from a 1975 baseline, respectively (Figure 3.1).

The process of sustainable building design has traditionally started with the use of rating systems such as the US Green Building Council (USGBC) LEED or Building Research Establishment's BREEAM programs. These systems have been instrumental in raising awareness about sustainable building methods among the architecture and engineering design professions and the general public.

3.2 Types of Design Regulations

A large variety of regulations, standards, guidelines, and rating systems can be encountered during the design and construction of a sustainable building. These can be created for one specific element of a building (e.g. heat pump and lighting sources), for one general design perspective (e.g. energy conservation and IAQ management), or even for the overall sustainability of a project. These are also developed for different building project phases, types, and scopes, as showed in Figure 3.2. Effectively integrating these into an actual project requires comprehensive knowledge on the regulations, standards, guidelines, and rating systems, and usually demands a team with in-depth knowledge on each of the key standards and practices. This chapter will first define the distinctions among regulations, standards, guidelines, and rating systems and will introduce some typical examples in each category. Then it will discuss how one can use these in an integrative fashion.

3.2.1 Federal Regulations

Entities who have the authority of managing the urban and house development can regulate the development and practice in the field. Regulations are enforced as laws that need to be complied with during

Energy Efficient Buildings: Fundamentals of Building Science and Thermal Systems, First Edition. Zhiqiang (John) Zhai.
© 2023 John Wiley & Sons, Inc. Published 2023 by John Wiley & Sons, Inc.

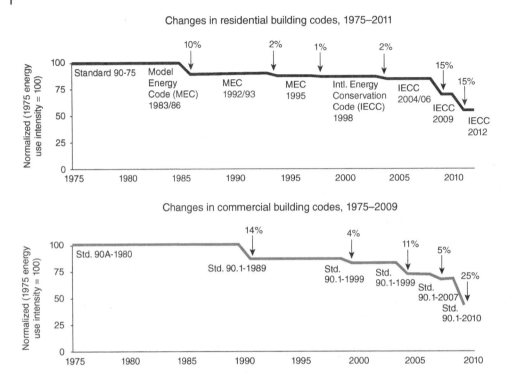

Changes in residential building codes, 1975–2011

Changes in commercial building codes, 1975–2009

Figure 3.1 ASHRAE 90.1-2010 and the International Energy Conservation Code (IECC) 2012 require decreases of 9–38% and 12–15% in energy use from a 1975 baseline.

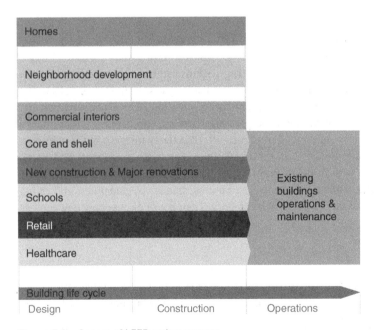

Figure 3.2 Scopes of LEED rating systems.

each step and aspect of a practice. Since different countries and regions have different authorizing models and control privileges, the creation, adoption, management, and enforcement of the regulations are somehow different. In the United States, the Code of Federal Regulations (CFR) is the codification of the general and permanent regulations published in the Federal Register by the executive departments and agencies of the federal government of the United States. The CFR is divided into 50 titles that represent broad areas subject to federal regulation, in which Title 10 is for energy and Title 24 is for Housing and Urban Development. In the United States, federal regulations are enforced for federal projects.

Specifically, Title 10 CFR Part 433 "Energy Efficiency Standards for the Design and Construction of New Federal Commercial and Multi-Family High-Rise Residential Buildings" establishes energy conservation performance standards for the design of new commercial and high-rise residential buildings. These standards are designed to achieve maximum practicable improvements in energy efficiency and increase the use of non-depletable sources of energy. These buildings must be designed to achieve energy consumption levels that are at least 30% below the levels established by the referenced baseline edition of ANSI/ASHRAE/IES Standard 90.1, which is the national model energy code for commercial buildings. The current federal standard for commercial buildings is based on Standard 90.1-2013, effective 5 January 2016.

Title 10 CFR Part 435 "Energy Efficiency Standards for the Design and Construction of New Federal Low-Rise Residential Buildings" establishes energy conservation performance standards for the design of new low-rise residential buildings. These buildings must be designed to achieve energy consumption levels that are at least 30% below the levels established by the referenced baseline edition of the IECC, which is the national model energy code for residential buildings. The current federal standard for residential buildings is based on the 2009 IECC, effective 10 August 2012.

The complete Electronic Code of Federal Regulations (eCFR) can be found at: https://www.ecfr.gov/cgi-bin/ECFR?page=browse. It is noted that regulations are often developed or based on other building codes and standards such as IECC and ANSI/ASHRAE/IES Standard, to be discussed next.

To show commitment to sustainable design, under the Energy Policy Act of 2005, the US government requires that all new or replacement federal buildings must be designed using sustainable design principles (U.S. Congress 2005). The General Services Administration now requires that new federal facilities be LEED certified and have an energy performance 20% better than ASHRAE Standard 90.1 (U.S. General Services Administration 2005). In the United States, regulation over the general public has been limited to the energy efficiency requirements of the Energy Policy Act of 2005, but the federal government has taken the initiative to lead by example with the adoption of the LEED program for their buildings.

The European Parliament and the Council of the European Union enacted the Energy Performance of Buildings Directive (EPBD), DIRECTIVE 2002/91/EC (Official Journal of the European Communities 2002), in December 2002 requiring member states to enact regulations of the energy performance of buildings by 4 January 2006. The EPBD is unique in several ways:

- A standardized framework is provided for developing an energy performance calculation methodology. The specific requirements for determining energy performance are specified by the individual nation-states.
- Energy performance requirements are set on a regional (nation-state) level.
- A certificate of energy performance is required every 10 years for all buildings that are constructed, sold, or rented. This certificate includes the benchmark performance, so consumers may make informed decisions related to the actual performance of each building.

- Heating and cooling system inspections are required regularly in order to identify operation and maintenance items that may aid in energy efficiency.
- All assessments of building performance, maintenance, and certification are conducted by independent experts. This maintains that the application of the EPBD is independent of any owner or stakeholder interests.

3.2.2 Building Codes

Building codes set up rules that specify the standards for constructed projects. Building code becomes law for a particular jurisdiction when formally enacted by the appropriate governmental or private authority. Different authorities (e.g. state, county, and city in the United States) may decide what building codes should be adopted and enforced as local laws for the jurisdiction area. In the United States, a lower authority (e.g. city) may override the adoption decision of an upper authority (e.g. state), while in other countries (e.g. China) a lower authority (e.g. city) must comply with the codes of an upper authority (e.g. province) with additional local rules if desired.

Building designers and builders must conform to the code to obtain planning permission, usually from a local council/committee. The most widely used/adopted building code is the IECC that sets out minimum efficiency standards for new construction for a structure's walls, floors, ceilings, lighting, windows, doors, duct leakage, and air leakage.

The IECC is referred to as America's model energy code because building codes are state laws; there is no national building energy code. The first IECC was created by the International Code Council in 2000. Every three years, officials from municipalities and states across the nation vote on proposed changes to the IECC to incorporate new building technologies and practices as they evolve over time and ensure that new American homes and commercial buildings meet modern-day minimum levels of safety, fire protection, and efficiency. From 2006 to 2021, the IECC increased its efficiency requirements by about 40%, or an average of 8% a cycle. The current version is 2021 IECC, which can be purchased and downloaded at: https://webstore.ansi.org/Standards/ICC/ICCIECC2021.

The IECC serves as the go-to source for states adopting an energy code; an ICC code is in use or adopted in all 50 states and beyond. It is important to get each three-year version right, so states can be confident that they are adopting an updated, well-vetted, and feasibly implementable building energy code. Attention should be paid to the specific version of the IECC adopted by individual states, counties, and cities. For instance, although the state of Colorado, in the United States, does not enforce energy codes at the state level, it requires the board of county commissioners, who have enacted a building code, to adopt and enforce a building energy code that meets or exceeds the standards in the 2003 IECC (https://www.energycodes.gov/adoption/states/colorado, accessed in June, 2021).

Other important building codes include:

- **International Mechanical Code (IMC-2021):** establishes minimum regulations for mechanical systems using prescriptive and performance-related provisions.
- **International Plumbing Code (IPC-2018):** provides minimum regulations for plumbing facilities in terms of both performance and prescriptive objectives, and provides for the acceptance of new and innovative products, materials, and systems.

- **National Electrical Code (NFPA 70-2020):** provides benchmarks for safe electrical design, installation, and inspection to protect people and property from electrical hazards. (NFPA stands for National Fire Protection Association)
- **Fire Code (NFPA 1-2021):** advances fire and life safety for the public and first responders as well as property protection by providing a comprehensive, integrated approach to fire code regulation and hazard management.

3.2.3 Building Standards

Building standards describe the best practice for use by building professionals. Building codes often refer to building standards for specific rules and regulations. For example, the IECC adopts the latest ASHRAE Standards, plus any addendums and new data. This implies that the IECC is ultimately a more comprehensive and stringent code than the ASHRAE Standards. Compliance with building standards is not required by law. However, in a lawsuit, the designer can be shown to be negligent if design standards (i.e. common/best practices) are not met. ASHRAE Standards focus on the best practice for the design and operation of building HVAC systems. The most relevant standards to the design of sustainable buildings are briefed below.

ANSI/ASHRAE/IES Standard 90.1-2019 "Energy Standard for Buildings Except Low-Rise Residential Buildings" is the most popular energy efficiency standard that provides minimum requirements for the energy efficient design of buildings except low-rise residential buildings. It offers, in detail, the minimum energy efficiency requirements for the design and construction of new buildings and their systems, new portions of buildings and their systems, and new systems and equipment in existing buildings, as well as criteria for determining compliance with these requirements. It is an indispensable reference for engineers and other professionals involved in the design of buildings and building systems. Standard 90.1 has been a benchmark for commercial building energy codes in the United States and a key basis for codes and standards around the world for more than 35 years.

ANSI/ASHRAE/IES Standard 90.2-2018 "Energy-Efficient Design of Low-Rise Residential Buildings" establishes the minimum whole-building energy performance requirements for energy efficient residential buildings. It provides the minimum design, construction, and verification requirements for new residential buildings and their systems and new portions of existing residential buildings and their systems that use renewable and non-renewable forms of energy. Both Standards 90.1 and 90.2 have been referenced in many other standards, guidelines, and rating systems.

ANSI/ASHRAE/ICC/USGBC/IES Standard 189.1-2020 "Standard for the Design of High-Performance Green Buildings Except Low-Rise Residential Buildings" provides guidance for designing, building, and operating high-performance green buildings. This standard refers to and integrates other ASHRAE code-intended standards, which often focus on just one aspect of the building. This standard presents provisions in six major categories: (i) Site sustainability, (ii) Water use efficiency, (iii) Energy efficiency, (iv) Indoor environmental quality, (v) Impact on the atmosphere, materials, and resources, and (vi) Construction and plans for operation. Similar to Standard 90.1, most sections include mandatory provisions, prescriptive, and performance-based options. The mandatory provisions must be met in all cases along with either the prescriptive options or corresponding performance options.

The other popular standards related to building sustainability include:

- **ANSI/ASHRAE Standard 62.1-2019 "Ventilation for Acceptable Indoor Air Quality":** outlines minimum ventilation rates and other measures intended to provide IAQ that is acceptable to human occupants and that minimizes adverse health effects in commercial buildings.
- **ANSI/ASHRAE Standard 62.2-2019 "Ventilation and Acceptable Indoor Air Quality in Residential Buildings":** defines the roles and minimum requirements for mechanical and natural ventilation systems and the building envelope to provide acceptable indoor air quality in residential buildings.
- **ANSI/ASHRAE Standard 55-2020 "Thermal Environmental Conditions for Human Occupancy":** specifies the combinations of indoor thermal environment factors and personal factors that will produce thermal environmental conditions acceptable to a majority of the occupants within the space.
- **ANSI/IES RP-1-20 "Recommended Practice: Lighting Office Spaces":** sets standards and guidelines for the lighting design and calculation of office spaces (https://store.ies.org/product/rp-1-20-recommended-practice-lighting-office-spaces/).
- **ANSI/ASA S12.2-2019 "American National Standard Criteria for Evaluating Room Noise":** defines the three primary methods for evaluating room noise and also contains one ancillary set of criteria curves for evaluating acoustically induced vibrations or rattles (https://webstore.ansi.org/Standards/ASA/ANSIASAS122019).

The current popular ASHRAE standards and guidelines can be viewed at: https://www.ashrae.org/technical-resources/standards-and-guidelines/read-only-versions-of-ashrae-standards.

American Society of Testing and Materials (ASTM) also develops a number of standards that provide a definition for sustainability in the context of buildings. They have also initiated several standards related to life cycle assessments and the presentation of material data. Some examples are: ASTM E2432-19 "Standard Guide for General Principles of Sustainability Relative to Buildings," ASTM E2114-19 "Standard Terminology for Sustainability Relative to the Performance of Buildings," and ASTM E2129-18 "Standard Practice for Data Collection for Sustainability Assessment of Building Products."

International Standardization Organization (ISO) has developed a suite of standards that provide general definitions and frameworks for the establishment of sustainable buildings. These documents include methods for comparing and standardizing environmental assessment methods, such as: ISO 15392:2019 "Sustainability in buildings and civil engineering works – General principles," ISO 21930:2017 "Sustainability in buildings and civil engineering works – Core rules for environmental product declarations of construction products and services," and ISO/DIS 21931-1 "Sustainability in buildings and civil engineering works – Framework for methods of assessment of the environmental, social and economic performance of construction works as a basis for sustainability assessment – Part 1: Buildings."

3.2.4 Building Guidelines

Guidelines provide the instruction and guidance on calculations, methods, and procedures for performing a task. Some building guideline examples from ASHRAE are:

- **ASHRAE Guideline 0-2019 "The Commissioning Process":** presents best practices for applying whole-building commissioning to all phases of new construction and renovation projects and provides a uniform, integrated, and consistent approach to commissioning.

- **ASHRAE Guideline 14-2014 "Measurement of Energy, Demand, and Water Savings":** provides procedures for using measured pre-retrofit and post-retrofit billing data (e.g. kWh, kW, Mcf, kgal) for the calculation of energy, demand, and water savings.

ASHRAE recently developed a series of advanced energy guides to provide recommendations for achieving zero energy buildings or building energy savings over the minimum code requirements of ANSI/ASHRAE/IES Standard 90.1. The recommendations in the guides allow those involved in designing or constructing the various building types (e.g. K-12 School Buildings, Small to Medium Office Buildings, Grocery Stores, and Large Hospitals) to easily achieve advanced levels of energy savings without having to resort to detailed calculations or analyses. Prescriptive energy-saving recommendations are contained in a single table for each of the eight US climate zones for the 30 and 50% guides. The zero-energy guides provide recommendations that are detailed in tables throughout the How-to Strategies. These guides can be used as design references but are not required. The guides can be freely downloaded at: https://www.ashrae.org/technical-resources/aedgs.

3.2.5 Building Assessment and Rating Systems

A large number of different green building assessment methods have been developed in the past 20 years in order to meet the challenge of designing buildings that reach specific environmental and energy performance objectives. There are many different methods used internationally, and many of these methods have been developed for specific purposes. Due to the specific application-driven nature of these assessment methods, several methods should be compared to determine the best fit for an individual project. The International Energy Agency Annex 31-2004 conducted a detailed study of many different assessment methods, and a few of the key methods will be mentioned in this chapter (Tables 3.1 and 3.2).

a) **BREEAM (BRE):** BREEAM is the oldest and most developed assessment system. The BREEAM system focuses on global, local, and indoor environmental issues and is used throughout the life cycle of the building. The BREEAM assessment method has been expanded to various applications such as offices, retail, schools, homes, multi-residential, and more. Numerous tools are available for use at different stages of the building life cycle such as the Envest 2 web-based design tool which calculates the whole-building life cycle environmental impacts and costs for buildings.

b) **LEED (USGBC):** The LEED program was first publicly released in March 2000 as LEED Version 2.0. As of January 2008, there were over 9500 registered projects and over 1200 certified projects in over 40 countries (U.S. Green Building Council 2008). The LEED system has expanded to include various applications such as new-construction, existing buildings, commercial interiors, core and shell, schools, homes, and more. The LEED system is widely used in the United States and is based on a checklist system that provides design guidance. More details are provided below for recent developments and applications of LEED.

c) **SBC08 (IIBSE):** The Sustainable Building Challenge (SBC) is an extension of the Green Building Challenge, which is an international framework that has gone through several incarnations in the past 10 years. The SBC framework has been developed as a flexible framework that can be crafted by regional organizations for use in particular locations. It also allows for the assessment of up to 125 criteria that are critical to sustainable building design. Most notable about the SBC framework is its use of a flexible spreadsheet assessment tool, SBTool.

Table 3.1 Comparison of environmental assessment methods.

Organization	Program	Year Began	Country of origin	Number of projects
US Green Building Council (USGBC)	Leadership in Energy and Environment Design (LEED)	2000	United States	1200 projects in over 40 countries
Building Research Establishment (BRE)	BRE Environmental Assessment Method (BREEAM)	1990	United Kingdom	100 000 certified in the United Kingdom
International Initiative for a Sustainable Built Environment (iiSBE)	Sustainable Building Challenge 2008 (formerly Green Building Challenge)	1996	Canada	20 countries
Japan Sustainable Building Consortium (JSBC)	Comprehensive Assessment System for Building Environmental Efficiency (CASBEE)	2005	Japan	23 projects
HK-BEAM society	Hong Kong Building Environmental Assessment Methods (HK-BEAM)	1996	China	80 projects
Arup	Sustainable Project Appraisal Routine (SPeAR)	2002	United Kingdom	Available as consulting service

(Note that: project numbers are cited by 2007).

d) **CASBEE (JSBC):** The unique part of the CASBEE method is that energy efficiency, resource efficiency, local environment, and indoor environment are re-categorized into environmental loadings (L) and environmental quality (Q). A building environmental efficiency is then quantified as the ratio of environmental quality to environmental loading. Therefore, the CASBEE method rates buildings by comparing the environmental quality benefits of the building against the environmental impact associated with providing these benefits.

e) **HK-BEAM (HK-BEAM Society):** HK-BEAM uses the global, local, and indoor format of BREEAM but has been modified for use in Hong Kong. It includes standards for a wide range of building types. HK-BEAM standards are based on a checklist of best design practices that a project is awarded points for implementing. This method is a good example of the ability for the BREEAM method to be adapted to different regional applications.

f) **SPeAR (Arup):** SPeAR is a four-quadrant graphical assessment tool that quantifies building performance for the environmental, societal, economic, and natural resources sectors. This tool is currently a proprietary assessment tool that is available through Arup's sustainability consulting services. The main strength of the SPeAR assessment is that it conveys information in an intuitive format to an audience with both a technical and non-technical background.

The LEED (Leadership in Energy and Environmental Design) Green Building Rating System™ is a voluntary, consensus-based national standard for developing high-performance, sustainable buildings. LEED was developed by the USGBC in 2000. USGBC is the premier non-profit organization for advancing green building. This coalition of about 600 leading building-industry companies and organizations

Table 3.2 Comparison of environmental assessment phases.

Region phase	North America		Europe			Asia		
	USA	Canada	United Kingdom	Germany	Netherlands	Japan	Hong Kong	Korea
Programming phase	LEED ASHRAE GreenGuide	C-2000 IDP and CBIP GBTool	BREEAM	Guideline for sustainable construction		CASBEE	HK-BEAM	
Design phase	LEED ASHRAE GreenGuide	C-2000 IDP and CBIP GBTool	BREEAM	Guideline for sustainable construction	GreenCalc	CASBEE	HK-BEAM	GBRS
Building construction	LEED ASHRAE GreenGuide	C-2000 IDP and CBIP GBTool	BREEAM	Guideline for sustainable construction			HK-BEAM	GBRS
Building operation	LEED ASHRAE GreenGuide	C-2000 IDP and CBIP GBTool	BREEAM	Guideline for sustainable construction		CASBEE	HK-BEAM	GBRS
Building demolition		C-2000 IDP and CBIP GBTool		Guideline for sustainable construction				

connects through the USGBC the leading architects, environmentalists, engineers, utilities, product manufacturers, universities, building owners, and federal, state, local governments. As one of the most complete rating systems for green buildings, LEED accelerates the mainstreaming of green building programs, educates the next generation of students and professionals, and gets the word out.

The number of LEED-certified projects in the United States rose from 296 certifications in 2006 up to over 67 200 certifications in 2018. Figure 3.3 shows the cumulative number of LEED registration in the United States from 2000 to 2019 (https://www.statista.com/statistics/323383/leed-registered-projects-in-the-united-states/). More than 203 000 professionals have earned a LEED credential to help advance their careers, which requires passing a written exam to obtain the title of "LEED Green Associate." Professionals can then earn the title of "LEED AP with specialty" by passing additional specialty exams with LEED-project working experience preferred.

Although the organization of prerequisites and credits varies slightly depending on the building type and associated rating system, LEED is generally organized by the following broad concepts:

- **Location and transportation.** LEED emphasizes location and transportation issues by rewarding development that preserves environmentally sensitive places and takes advantage of existing infrastructure, community resources, and transit. It encourages access to open space for walking, physical activity, and time spent outdoors. Credits also encourage smart transportation choices and access to a diversity of uses.
- **Sustainable sites.** Choosing a building's site and managing that site during construction are important considerations for a project's sustainability. LEED credits addressing sustainable sites discourage development of previously undeveloped land and damage to ecosystems and waterways; they encourage regionally appropriate landscaping, and control of rainwater runoff, as well as reduced erosion, light pollution, heat island effect, and construction-related pollution.

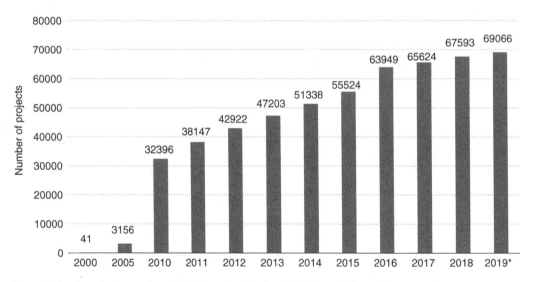

Figure 3.3 Cumulative number of LEED registration in the US from 2000 to 2019.

- **Water efficiency.** Buildings are major users of our potable water supply. The goal of credits addressing water efficiency is to encourage smarter water usage, inside and outside. Water reduction is typically achieved through more efficient appliances, fixtures, and fittings inside and water-wise landscaping outside.

- **Energy and atmosphere.** LEED encourages a wide variety of strategies to address energy consumption, including commissioning; energy use monitoring; efficient design and construction; efficient appliances, systems, and lighting; demand response; and the use of renewable and clean sources of energy, generated on-site or off-site.

- **Materials and resources.** During both construction and operations, buildings generate large amounts of waste and use tremendous volumes of materials and resources. These credits encourage the selection of sustainably grown, harvested, produced, and transported products and materials. They promote the use of life cycle assessments to holistically evaluate materials and the disclosure and optimization of a material's chemical ingredients.

- **Indoor environmental quality.** The average American spends about 90% of their day indoors, where pollutant concentrations may be 2 to 100 times higher than outdoor levels. Thus, indoor air quality can be significantly worse than outside. LEED credits promote strategies that can improve indoor air, provide access to natural daylight and views, and improve acoustics.

- **Innovation.** LEED promotes innovation by offering points for improving a building's performance well beyond what is required by the credits or for incorporating green building ideas that are not specifically addressed elsewhere in the rating system. This credit category also rewards the inclusion of a LEED Accredited Professional on the project team.

- **Regional priority.** USGBC's regional councils, chapters, and affiliates have identified the environmental concerns that are most important for every region of the country, and LEED credits that address those local priorities have been selected for each region. A project team that earns a regional priority credit earns one bonus point in addition to any points awarded for that credit.

Although the new LEED rating systems (released in 2009) include regional priority points, the potential constraint of "one size fits all" exists due to the large diversity in building function, need, micro-climate, etc. The risk of buying points due to unequal values and costs may still exist. In addition, most of the design decisions that require sophisticated simulations with many assumptions (including costs) could use more transparency of cost-benefits when performing strategy/technology ranking and alternating. Finally, the certification process is not free (typically 0.2–0.55% of construction costs), and can go from 0.66% for a basic certification to 6.8% for a platinum certification based on a statistics of 40 LEED buildings at different certification levels.

The LEED rating system can be a great starting point to brainstorm effective design strategies for a specific project. The checklist as shown in Figure 3.4 can be used to pre-check and pre-calculate the Yes/No/Maybe points so as to estimate the distance between pre-set green/LEED goals and reality so that a value engineering may be performed to justify or adjust the goals. For each category, there are detailed credits (points) to be claimed that require comprehensive calculations and documentations, in addition to several prerequisites. It is important to weigh the systematic values of points related to their capital and operation costs, as well as societal and environmental benefits.

LEED v4 for BD+C: New Construction and Major Renovation
Project Checklist

Project Name:
Date:

Y ? N

	Credit	Integrative Process	1

Location and Transportation		16
Credit	LEED for Neighborhood Development Location	16
Credit	Sensitive Land Protection	1
Credit	High Priority Site	2
Credit	Surrounding Density and Diverse Uses	5
Credit	Access to Quality Transit	5
Credit	Bicycle Facilities	1
Credit	Reduced Parking Footprint	1
Credit	Green Vehicles	1

Sustainable Sites		10
Prereq	Construction Activity Pollution Prevention	Required
Credit	Site Assessment	1
Credit	Site Development - Protect or Restore Habitat	2
Credit	Open Space	1
Credit	Rainwater Management	3
Credit	Heat Island Reduction	2
Credit	Light Pollution Reduction	1

Water Efficiency		11
Prereq	Outdoor Water Use Reduction	Required
Prereq	Indoor Water Use Reduction	Required
Prereq	Building-Level Water Metering	Required
Credit	Outdoor Water Use Reduction	2
Credit	Indoor Water Use Reduction	6
Credit	Cooling Tower Water Use	2
Credit	Water Metering	1

Energy and Atmosphere		33
Prereq	Fundamental Commissioning and Verification	Required
Prereq	Minimum Energy Performance	Required
Prereq	Building-Level Energy Metering	Required
Prereq	Fundamental Refrigerant Management	Required
Credit	Enhanced Commissioning	6
Credit	Optimize Energy Performance	18
Credit	Advanced Energy Metering	1
Credit	Demand Response	2
Credit	Renewable Energy Production	3
Credit	Enhanced Refrigerant Management	1
Credit	Green Power and Carbon Offsets	2

Materials and Resources		13
Prereq	Storage and Collection of Recyclables	Required
Prereq	Construction and Demolition Waste Management Planning	Required
Credit	Building Life-Cycle Impact Reduction	5
Credit	Building Product Disclosure and Optimization - Environmental Product Declarations	2
Credit	Building Product Disclosure and Optimization - Sourcing of Raw Materials	2
Credit	Building Product Disclosure and Optimization - Material Ingredients	2
Credit	Construction and Demolition Waste Management	2

Indoor Environmental Quality		16
Prereq	Minimum Indoor Air Quality Performance	Required
Prereq	Environmental Tobacco Smoke Control	Required
Credit	Enhanced Indoor Air Quality Strategies	2
Credit	Low-Emitting Materials	3
Credit	Construction Indoor Air Quality Management Plan	1
Credit	Indoor Air Quality Assessment	2
Credit	Thermal Comfort	1
Credit	Interior Lighting	2
Credit	Daylight	3
Credit	Quality Views	1
Credit	Acoustic Performance	1

Innovation		6
Credit	Innovation	5
Credit	LEED Accredited Professional	1

Regional Priority		4
Credit	Regional Priority: Specific Credit	1
Credit	Regional Priority: Specific Credit	1
Credit	Regional Priority: Specific Credit	1
Credit	Regional Priority: Specific Credit	1

TOTALS	Possible Points:	110

Certified: 40 to 49 points, Silver: 50 to 59 points, Gold: 60 to 79 points, Platinum: 80 to 110

Figure 3.4 LEED v4 for BD+C checklist.

3.3 Integrative Use of All

3.3.1 Integrated Design

Integrated design is the key (approach) throughout a sustainable project, which pulls together all the stakeholders, architects, engineers, consultants, builders for goal setting to strategy determination to detail design to blueprints to construction to commissioning (Figure 3.5). This can bring visions, regulations, standards, codes, and rating systems from the very beginning of the project and from all aspects of the project. Conflicts can be found and sorted out from the start. Harmonious solutions can be proposed that comply with pre-set goals. More effort provided at an earlier stage can reduce efforts needed at a later stage. Iteration during the design process will ensure that the right direction is taken by addressing feedback received from the later stages.

Integrated design often is implemented via the format of a "charrette," which is a hands-on, collaborative, fast-paced, and intensive working session typically organized by sustainable consulting firms. A charrette can be organized throughout the project with different purposes such as stakeholder motivation, goal setting, technical solutions, industry-wide collaboration, etc. Figure 3.6 illustrates the five key elements to a successful charrette, which starts from desired outcomes, to determine needed participants, suitable approach and facilitation, and the follow-up.

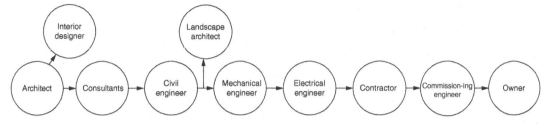

Traditional design process by professions

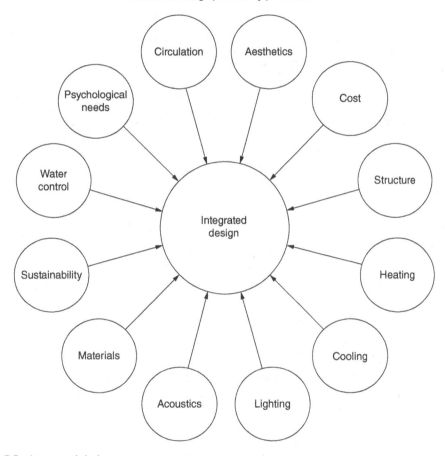

Figure 3.5 Integrated design.

3.3.2 Life Cycle Cost Analysis Based Design

Life cycle cost analysis (LCCA) is fundamental in determining proper design solutions in order to meet codes while promising return on investment. LCCA is not just important for economic reasons, but it is also crucial for achieving the inherent principles of sustainability, e.g. to reduce resource uses (material, energy, and water) and to reduce waste and pollutant generation, as well as enhancing quality of life.

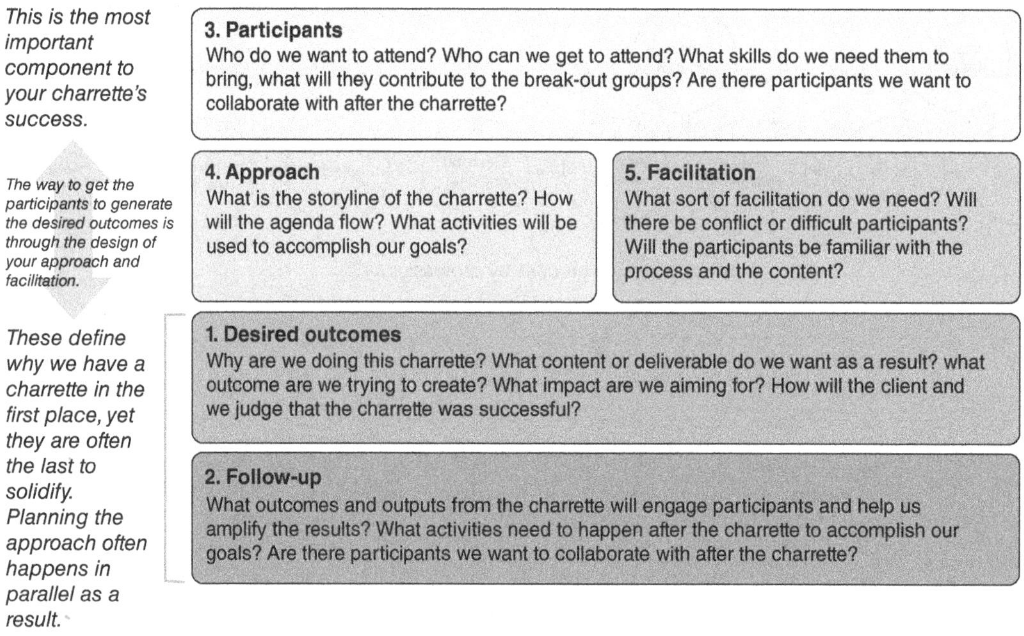

This is the most important component to your charrette's success.

3. Participants
Who do we want to attend? Who can we get to attend? What skills do we need them to bring, what will they contribute to the break-out groups? Are there participants we want to collaborate with after the charrette?

The way to get the participants to generate the desired outcomes is through the design of your approach and facilitation.

4. Approach
What is the storyline of the charrette? How will the agenda flow? What activities will be used to accomplish our goals?

5. Facilitation
What sort of facilitation do we need? Will there be conflict or difficult participants? Will the participants be familiar with the process and the content?

These define why we have a charrette in the first place, yet they are often the last to solidify. Planning the approach often happens in parallel as a result.

1. Desired outcomes
Why are we doing this charrette? What content or deliverable do we want as a result? what outcome are we trying to create? What impact are we aiming for? How will the client and we judge that the charrette was successful?

2. Follow-up
What outcomes and outputs from the charrette will engage participants and help us amplify the results? What activities need to happen after the charrette to accomplish our goals? Are there participants we want to collaborate with after the charrette?

Figure 3.6 Five key elements to a successful charrette.

Technology-optimization without LCCA will lead to the use of the best products and systems available. However, the use of these products and systems may not provide the best overall performance for a specific project. For some projects, e.g. a pilot project for advanced technologies or a special-purpose building (e.g. a military facility), LCCA may not be necessary as the operation performance is the main focus of the demonstration. LCCA is a must for most, if not all, commercial and residential building developments, upon which optimization can be carried out to produce a best and long-lasting design.

3.3.3 Building Information Modeling

An area that offers considerable potential for aiding in the transfer of integrated building design to professional practice is building information modeling (BIM). Sophisticated BIM tools have several important benefits for sustainable building design.

- Coordination of building trades during early design
- Project documents are better coordinated with less mistakes
- Integration of analysis with design
- Ability to do online cost estimating and take-offs
- Automatic code or standard checking

The future of BIM is being extended as far as the imagination can go. The industry is focusing heavily on the interoperability among BIM software and other analysis tools. Schemes such as IFC, XML, dxf, and pdf are being evaluated for use in transferring geometric and property data between software. The integration of analysis tools into the design process will allow for a greater number of design options to be evaluated quickly using advanced analysis tools such as computer energy simulation, computational fluid dynamics, and lighting rendering. BIM is also being evaluated for post-occupancy evaluation of buildings. The BIM database can store maintenance and facility condition information, energy and water consumption data, and occupant satisfaction data. The importance of BIM in sustainable building design is that it offers an integrated database of building information that begins at the conceptual stage and progresses into construction and operation.

Homework Problems

1   Refer to ASHRAE 90.1-2019 (Table 5.5) (https://www.ashrae.org/technical-resources/standards-and-guidelines/read-only-versions-of-ashrae-standards) and compare the building envelope requirements for Miami, FL and Washington DC (i.e., two different climate zones), and comment on the findings.

2   **Staged Project Assignment (Teamwork)**
    (a) Check DOE benchmarks (http://energy.gov/eere/buildings/new-construction-commercial-reference-buildings), EIA statistics (http://www.eia.gov/consumption/commercial/), EPA portfolio manager (http://www.energystar.gov/buildings/tools-and-resources/portfolio-manager-quick-start-guide), or German Passive House (http://passivehouse.com/) (or others) to identify and justify your **energy design goal** (in EUI) (including overall energy demand, HVAC energy demand, etc.).
    (b) Identify the specific **energy codes** (e.g., ASHRAE 90.1 and IECC) required for the design location (e.g., Washington DC) (indicate the code title-year while reading the relevant requirements to be used later).
    (c) Check **the LEED rating system** (https://www.usgbc.org/leed/v41#bdc) to brainstorm the design strategies and pre-estimate possible LEED credits and certification level.

References

Official Journal of the European Communities (2002). *Directive 2002/91/EC of the European Parliament and of the Council of 16 December 2002*. European Parliament and of the Council of the European Union.

U.S. Congress. 2005. Energy Policy Act of 2005. 08/08/2005 Became Public Law No: 109-58.

U.S. General Services Administration (2005). *Facilities Standards for the Public Buildings Service. PBS-P100*. Washington, DC: U.S. General Services Administration, Office of the Chief Architect.

U.S. Green Building Council (2008). *Green Building Facts: Green Building by the Numbers*. March 2008. Washington, DC: U.S. Green Building Council.

4

Air Properties and Psychrometric Chart

As stated in Chapter 1, creating thermally comfortable and healthy indoor environments is one of the primary goals of designing and operating a sustainable building. To achieve this, a high-quality building envelope (including the roof, walls, windows, doors, and ground) will be crucial, which confines and forms the indoor space and connects the indoor and outdoor environments. A variety of approaches, dependent on climate, can be applied to reach the design goal, in which the most energy efficient solution with reliable controls is the best choice. This may include using various passive strategies such as natural ventilation, evaporative cooling, passive solar heating along with active systems that include traditional and advanced heating and cooling systems. These passive and active measures regulate room air properties to stay within a comfortable range.

Chapter 4 introduces various properties of air and their inherent connections. These properties and their relationships can be plotted on an engineering tool – a psychrometric chart – upon which various heating, cooling, humidification, and dehumidification processes can be presented and calculated.

4.1 Air Composition

Air in the real environment is a mixture of various components. Air in most environments contains, by volume:

- **Nitrogen**: 78.084%
- **Oxygen**: 20.948%
- **Argon**: 0.934%
- **Carbon dioxide**: 0.031%
- **Minor gases**: 0.003%

where oxygen is what humans need in their living environments, through proper ventilation.

Air in the real environment is not fully dry but a mixture of dry air and water vapor, which determines how wet (humid) the air (or environment) is. Beside air temperature, air humidity is an extremely important factor in determining thermal comfort. With the same air temperature, Phoenix, AZ, may provide a much different experience from Miami, FL, in the summer.

Energy Efficient Buildings: Fundamentals of Building Science and Thermal Systems, First Edition. Zhiqiang (John) Zhai.
© 2023 John Wiley & Sons, Inc. Published 2023 by John Wiley & Sons, Inc.

Air in the real environment is also not pure. It contains various pollutants such as dust, fog, and microbes. Therefore, outdoor air needs filtering (or purification) before it can be conditioned and supplied to indoor applications. This becomes more a serious issue when the surrounding air (or microenvironment) is highly polluted, such as near highways or in the downtown area of large cities. This chapter will focus on the fundamental properties of air. Air pollutants will be discussed in Chapter 7.

4.2 Moist Air and Its Properties

4.2.1 Ideal Gas Law

Air and its components can be treated as an ideal gas, following the ideal gas law:

$$Pv = RT \tag{4.1}$$

where P is the air pressure (Pa), $v = 1/\rho$ is the specific volume of air (m^3/kg), ρ is the air density (kg/m^3), R is the gas constant (J/kg−K), and T is the air temperature (K). The reference or molar gas constant is:

$$R_{ref} = pV_m/T = \left(101\,325\,\text{N/m}^2 \times 22.4\,\text{m}^3/\text{Kmol}\right)/273.15\,\text{K} = 8314.66\,\text{J/Kmol-K} \tag{4.2}$$

The gas specific constant is thus:

$$R = R_{ref}/M \tag{4.3}$$

where M is the molecular mass of the gas. For water vapor, $M_v = 18$; and for dry air, $M_{dry\text{-}air} = 29$.

4.2.2 Properties

4.2.2.1 Pressure: P (Unit: Pa)

Air pressure is caused by the weight of the air molecules above the measurement location. This is similar to water pressure, which depends on the height of the water body above the measuring point ($P = \rho gh$). Figure 4.1 shows various terminologies of pressure. The atmospheric pressure $P_{atm} = 101\,325$ Pa at sea level. Pressure below P_{atm} is called a vacuum environment. A full vacuum environment (zero pressure) is hard to achieve in engineering. Absolute pressure is pressure measured related to zero pressure, and gage pressure is pressure measured related to atmospheric pressure.

Air pressure is often measured using a water (or mercury) U-tube as shown in Figure 4.2. The height difference in the U-tube can be used to directly represent the pressure difference. For instance, $H = 4''$ WG (Water Gage) is typical in HVAC systems of buildings. The actual pressure is calculated as:

$$\Delta P = \rho_{fluid} \times g \times H \tag{4.4}$$

This is the pressure friction (loss) to be overcome by the fan. The static pressure in Figure 4.2 is the gage pressure of still air, while the total pressure includes the dynamic pressure caused by air velocity (and elevation pressure, if any).

$$P_{total} = P_{static} + P_{dynamic} + P_{elevation} = P_{static} + 1/2 \times \rho_{air} \times V^2 + \rho_{air}\,gh_{elevation} \tag{4.5}$$

Relations of absolute, gage, and vacuum pressures.

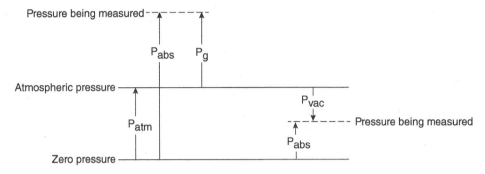

Figure 4.1 Terminologies of air pressure.

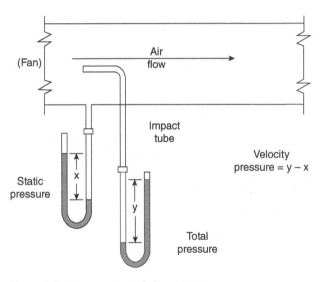

Figure 4.2 Measurement of air pressure.

where ρ_{air} is the air density, V is the air flow velocity, $h_{elevation}$ is the elevation difference between the two openings of the U-tube (usually too small to ignore).

Air pressure is additive, which means the total pressure of air is the sum of the pressure of each component in the air mixture:

$$P = P_1 + P_2 + P_3 + ... + P_n \text{ (Unit : Pa)} \tag{4.6}$$

Injecting another gas into the air space will thus increase the total pressure of the space. Certainly, when a space has a high air pressure, it will need more effort (more mechanical work) to inject additional gas.

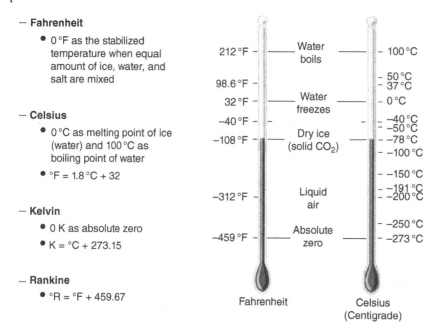

— **Fahrenheit**

- 0 °F as the stabilized temperature when equal amount of ice, water, and salt are mixed

— **Celsius**

- 0 °C as melting point of ice (water) and 100 °C as boiling point of water

- °F = 1.8 °C + 32

— **Kelvin**

- 0 K as absolute zero

- K = °C + 273.15

— **Rankine**

- °R = °F + 459.67

212 °F — Water boils — 100 °C

98.6 °F — 50 °C / 37 °C

32 °F — Water freezes — 0 °C

−40 °F — −40 °C / −50 °C

−108 °F — Dry ice (solid CO_2) — −78 °C / −100 °C

−150 °C

−312 °F — Liquid air — −191 °C / −200 °C

−250 °C

−459 °F — Absolute zero — −273 °C

Fahrenheit

Celsius (Centigrade)

Figure 4.3 Temperature measurement and units.

4.2.2.2 Temperature: T (Unit: K, C, F, R)

Air temperature represents the internal energy of the air (i.e. molecular activity). Temperature can be measured and presented in different units as shown in Figure 4.3. Temperature is the most important variable that affects the quality of living environments, and it also impacts other properties such as air velocity, relative humidity (RH), etc. All of the components in the air mixture share the same temperature. Adding another stream of hot gas into the air will increase the overall temperature of the mixture and thus every single component's temperature will rise by the same amount.

4.2.2.3 Humidity Ratio: W (Unit: Kg/Kg$_{dry-air}$)

Humidity ratio defines the amount of water vapor contained in the air:

$$W = m_v / m_a \tag{4.7}$$

where m_v is the mass of water vapor in the air and m_a is the mass of the air containing the vapor. Using the ideal gas law for both dry air and water vapor provides:

$$P_a V_a = m_a R_a T_a \tag{4.8}$$

$$P_v V_v = m_v R_v T_v \tag{4.9}$$

Since the dry air and water vapor share the same volume and the same temperature,

$$V_a = V_v \tag{4.10}$$

$$T_a = T_v \tag{4.11}$$

This yields,

$$W = m_v/m_a = (P_v R_a)/(P_a R_v) = 0.622\, P_v/(P - P_v) \tag{4.12}$$

where $R_a/R_v = M_v/M_{\text{dry-air}} = 18/29 = 0.622$, and $P_a = P - P_v$ is the pressure of the dry air in the air mixture.

The humidity ratio represents the actual water vapor content in the air; however, this value does not directly relate to the dampness one feels in the air, which is represented by the property of "relative humidity" as defined below.

4.2.2.4 Relative Humidity: ϕ (Unit: %)

Relative humidity (RH), expressed as a percentage, indicates a present state of absolute humidity relative to the maximum humidity capacity of the air at the same pressure and temperature condition. RH is defined as:

$$\phi = P_v/P_{v,s} \times 100\% \tag{4.13}$$

where P_v is the current water vapor pressure in the air; $P_{v,s}$ is the saturated (maximum) water vapor pressure that can be contained in the air at the same air conditions. The value of $P_{v,s}$ can be found in water vapor/steam property table (as attached in Tables 4.1 and 4.2), as a function of temperature.

For dry air, $\phi = 0\%$, while for saturated air, $\phi = 100\%$. The relations between the humidity ratio (W) and the RH can be found below:

$$W = 0.622 P_v/(P - P_v) = 0.622 \times \phi \times P_{v,s}/(P - \phi \times P_{v,s}) \tag{4.14}$$

$$W_{v,s} = 0.622 \times P_{v,s}/(P - P_{v,s}) \tag{4.15}$$

Table 4.1 Saturation properties for steam – temperature table.

Temp °C	Pressure MPa	Temp °C	Pressure MPa	Temp °C	Pressure MPa	Temp °C	Pressure MPa
0.01	0.0006117	60	0.01995	140	0.3615	270	5.5030
5	0.0008726	65	0.02504	150	0.4762	280	6.4166
10	0.001228	70	0.03120	160	0.6182	290	7.4418
15	0.001706	75	0.03860	170	0.7922	300	8.5879
20	0.002339	80	0.04741	180	1.0028	310	9.8651
25	0.003170	85	0.05787	200	1.5549	320	11.284
30	0.004247	90	0.07018	210	1.9077	330	12.858
35	0.005629	95	0.08461	220	2.3196	340	14.601
40	0.007385	100	0.1014	230	2.7971	350	16.529
45	0.009595	110	0.1434	240	3.3469	360	18.666
50	0.01235	120	0.1987	250	3.9762	370	21.044
55	0.01576	130	0.2703	260	4.6923	373.95	22.064

Table 4.2 Saturation properties for steam – pressure table.

Pressure MPa	Temp °C	Pressure MPa	Temp °C	Pressure MPa	Temp °C	Pressure MPa	Temp °C
0.001	7.0	0.014	52.5	0.18	116.9	3	233.9
0.0012	9.7	0.016	55.3	0.2	120.2	4	250.4
0.0014	12.0	0.018	57.8	0.3	133.5	6	275.6
0.0016	14.0	0.02	60.1	0.4	143.6	8	295.0
0.0018	15.8	0.03	69.1	0.6	158.8	10	311.0
0.002	17.5	0.04	75.9	0.8	170.4	12	324.7
0.003	24.1	0.06	85.9	1	179.9	14	336.7
0.004	29.0	0.08	93.5	1.2	188.0	16	347.4
0.006	36.2	0.1	99.6	1.4	195.0	18	357.0
0.008	41.5	0.12	104.8	1.6	201.4	20	365.8
0.01	45.8	0.14	109.3	1.8	207.1	22.064	373.95
0.012	49.4	0.16	113.3	2	212.4		

As $P_{v,s} \ll P$, $(P - \phi \times P_{v,s}) \approx (P - P_{v,s}) \approx P$,

$$\phi \approx [(0.622 \times \phi \times P_{v,s})/(P - \phi \times P_{v,s})]/[(0.622 \times P_{v,s})/(P - P_{v,s})] = W/W_{v,s} \times 100\% \qquad (4.16)$$

It is important to note that a higher W does not imply a higher ϕ. It is more about the maximum capability of air to hold vapor. Air at a higher temperature can hold more water vapor than air at a lower temperature. Therefore, the air will feel dryer in the summer than in the winter if the air contains the same amount of water vapor in each season.

4.2.2.5 Dewpoint Temperature: T_{dew} (Unit: K, C, F, R)

Dewpoint temperature is the saturated temperature of a given air mixture at the same pressure and humidity ratio. This is the lowest temperature that air can reach before condensation occurs. The RH at the dewpoint temperature is 100%. Any temperature below the dewpoint temperature of air will lead to the condensation of water vapor and thus the reduction of water vapor content in the air (i.e. humidity ratio). During condensation, the RH will maintain a value of 100%.

4.2.2.6 Wet-Bulb Temperature: T_{wet} (Unit: K, C, F, R)

"Wet-bulb" temperature is related to "dry-bulb" temperature which is the common air temperature as mentioned earlier. The names "wet-bulb" and "dry-bulb" stem from the measurement process as discussed in Example 4.2. Wet-bulb temperature represents the temperature at which the moist air becomes saturated if water is adiabatically evaporated into the moist air. Unlike the dewpoint temperature, which reaches saturation by reducing the temperature while keeping the same water content (i.e. humidity ratio), wet-bulb temperature reaches saturation by adding more water vapor to the air and thus the humidity ratio is increased.

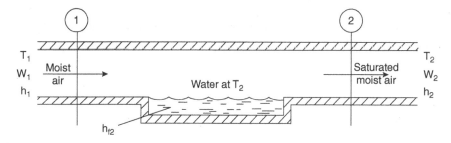

Figure 4.4 Principle of the process to reach saturation of moist air by evaporation.

Figure 4.4 shows the principle of the process. Moist air is moving through a large reservoir in a fully insulated container (i.e. no external heat loss and gain during the process). Moist air becomes fully saturated at State-2. The humidity ratio is W2 > W1 due to the additional water vapor received from the body of water via evaporation. The air temperature is T2 < T1 because evaporation takes heat from the moist air. T2 is called wet-bulb temperature as it reaches saturation. The total energy (h) of the air (to be introduced later) is constant during the process as there is no additional heat gain or loss from State-1 to State-2. Water becomes vapor, which takes energy from the air (thereby reducing the air temperature). The same amount of energy is added back into the air mixture, which can be attributed to the additional vapor received. This is a process of converting sensible heat (related to temperature change) to latent heat (related to material phase change) while the total energy is unchanged (to be introduced later). Strictly speaking, h2 is slightly larger than h1 due to the enthalpy of additional water from the reservoir, but this difference is very small.

4.2.2.7 Enthalpy: h (Unit: kJ/kg$_{dry\text{-}air}$, Btu/lb$_{dry\text{-}air}$)

Enthalpy describes the energy contained in an air mixture, which can be divided into two parts: dry air and water vapor:

$$h = h_a + W \cdot h_v \tag{4.17}$$

where h is the total enthalpy, W is the humidity ratio, and h_a is the enthalpy of dry air:

$$h_a = C_{p,a} T \tag{4.18}$$

where T is the air temperature and $C_{p,a}$ is the specific heat of dry air. The enthalpy of water vapor h_v is calculated as:

$$h_v = h_g + C_{p,v} T \tag{4.19}$$

where h_g is the enthalpy of saturated water vapor at 0 °C and $h_g = 2501.3$ kJ/kg (at 0 °C) = 1061.2 Btu/lb (at 32 °F); $C_{p,v}$ is the specific heat of water vapor. Equation (4.17) can then be expressed as:

$$h = C_{p,a} T + W(h_g + C_{p,v} T) \tag{4.20}$$

$$h = 1.01T + W(2501.3 + 1.86T) \quad (kJ/kg_{dry-air}) \text{ [SI unit]} \tag{4.21}$$

$$h = 0.24T + W(1061.2 + 0.444T) \left(Btu/lb_{dry-air}\right) \text{ [IP unit]} \tag{4.22}$$

Example 4.1

Determine the humidity ratio of moist air at a temperature of 24 °C and a relative humidity of 50% at a standard pressure 1 atm; find T_{dew} of this air mixture; find the enthalpy of the air mixture.

Solution

1) $W = 0.622 \times \phi \times P_{v,s}/(P - \phi \times P_{v,s})$

   $P = P_{atm} = 1.013 \times 10^5$ Pa

   Check Table 4.1, find $P_{v,s} = 0.03 \times 10^5$ Pa at T = 24 °C,

   $W = 0.622 \times 50\% \times 0.03/(1.013 - 50\% \times 0.03) = 0.0093$ kg$_v$/kg$_{dry\text{-}air}$
   $= 9.3$ g$_v$/kg$_{dry\text{-}air}$

2) At dewpoint temperature, $\phi_{dew} = 100\%$ while W is constant as 9.3 g$_v$/kg$_{dry\text{-}air}$.

   $W = 0.622 \times \phi_{dew} \times P_{v,s,dew}/(P - \phi_{dew} \times P_{v,s,dew})$
   $= 0.622 \times 1 \times P_{v,s,dew}/(P - 1 \times P_{v,s,dew}) = 9.3$ g$_v$/kg$_{dry-air}$

   $P_{v,s,dew} = \phi \times P_{v,s} = 0.5 \times 0.03 \times 10^5 = 0.015 \times 10^5$ Pa

   Check Table 4.2, find $T_{dew} = 12.7$ °C at $P_{v,s,dew} = 0.015 \times 10^5$ Pa

3) $h = C_{p,a}T + W(h_g + C_{p,v}T) = 1.01T + W(2501.3 + 1.86T)$
   $= 1.01 \times 24 + 0.0093 \times (2501.3 + 1.86 \times 24) = 48$ kJ/kg$_{dry\text{-}air}$

Example 4.2

Use a sling psychrometer to measure the relative humidity. If the temperature difference between the dry-bulb and wet-blub tips is 10 °C, and the dry-bulb temperature is 25 °C, determine the relative humidity of the air.

Solution

The principle of a psychrometer (Figure 4.5) is to evaporate air around one tip in a wet cloth sock (thus called "wet-bulb") and measure the saturated (thus cooler) air temperature (wet-bulb temperature)

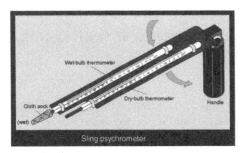

Figure 4.5 Use sling psychrometer to measure relative humidity. *Source:* Joe Haupt/Flickr.

around this tip. The other tip is to measure the original air temperature (dry-bulb temperature). The temperature difference can then be used to calculate the relative humidity of the air. In order to accelerate the evaporation, the user needs to spin the sling psychrometer. For better accuracy, a local air velocity $v > 2.5$ m/s is desired.

At the wet-bulb, the evaporation energy needed for the wet-bulb is

$$Q_1 = m \times r \tag{4.23}$$

where m is the mass of the water evaporated and r is the latent heat of water evaporation. This heat is provided by the air around the wet-bulb:

$$Q_2 = \alpha(T_{dry} - T_{wet})A \tag{4.24}$$

where α is the heat transfer coefficient and A is the area of the wet-bulb.

According to the mass transfer principle,

$$m = \beta(p'_{v,s} - p_v)A \tag{4.25}$$

where β is the mass transfer coefficient, $p'_{v,s}$ is the saturated water vapor pressure under T_{wet}, and p_v is the present water vapor pressure. Since $Q_1 = Q_2$, this leads to

$$\beta(p'_{v,s} - p_v)Ar = \alpha(T_{dry} - T_{wet})A \tag{4.26}$$

$$p_v = p'_{v,s} - \frac{\alpha}{\beta r}(T_{dry} - T_{wet}) \tag{4.27}$$

where α and β are a function of local air velocity (v) around the wet-bulb:

$$\frac{\alpha}{\beta r} = 65 + \frac{6.75}{v} \quad \text{(SI unit)} \tag{4.28}$$

Equation (4.28) works for sea level and may change for other elevations. As a result,

$$\phi = \frac{p_v}{p_{v,s}} = \frac{p'_{v,s} - \frac{\alpha}{\beta r}(T_{dry} - T_{wet})}{p_{v,s}} \tag{4.29}$$

where $P_{v,s}$ is the saturated water vapor pressure under T_{dry}.

For the problem, $T_{dry} = 25\,°C$, $T_{wet} = T_{dry} - 10\,°C = 15\,°C$. Checking Table 4.1 provides, $P_{v,s} = 0.003170 \times 10^6$, $p'_{v,s} = 0.001706 \times 10^6$. Using Equation (4.29) and assuming $v = 2.5$ m/s,

$$\phi = \frac{p_v}{p_{v,s}} = \frac{p'_{v,s} - \frac{\alpha}{\beta r}(T_{dry} - T_{wet})}{p_{v,s}} = \frac{0.001706 \times 10^6 - (65 + (6.75/2.5)) \times 10}{0.003170 \times 10^6} = 0.3246 = 32.46\%$$

4.3 Construction of a Psychrometric Chart

Principally, any two air properties can determine all other properties. Free online calculators, Excel and MATLAB based tools can be found or built to conduct these conversions. For engineering design purposes, sometimes it is more convenient and intuitive to use graphs to identify air properties, especially during an air treatment process. As a result, a psychrometric chart is introduced and widely used in building and system design. A psychrometric chart is basically a graphical representation of the equations discussed earlier. The following paragraphs introduce the construction of a psychrometric chart, using the above equations and knowledge.

4.3.1 Construction of Air Saturation Line as a Function of Temperature

Using the dry-bulb temperature as the X-coordinate (Figure 4.6), the maximum (saturated) humidity ratio ($W_{v,s}$), as the Y-coordinate, can be determined, which forms the saturation line.

$$W_{v,s} = 0.622 P_{v,s}/(P - P_{v,s}) \tag{4.23}$$

where $P_{v,s}$ is the function of T (Table 4.1), and P is the local total air pressure (e.g. P_{atm}). Therefore, a psychrometric chart is location dependent (usually categorized by elevation). It is important to use the right psychrometric chart for a specific location/evaluation (e.g. sea level vs 1500 m).

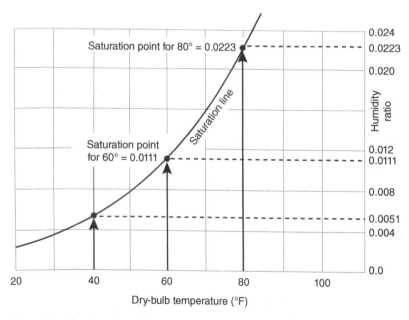

Figure 4.6 Construction of air saturation as a function of temperature.

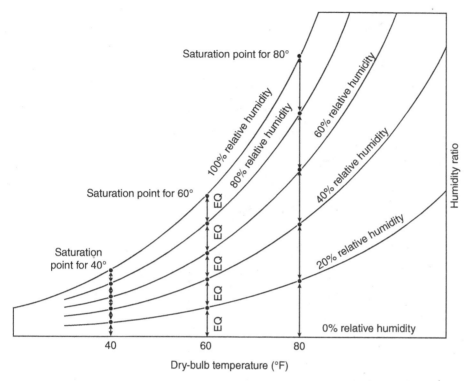

Figure 4.7 Construction of relative humidity lines.

4.3.2 Construction of Relative Humidity Lines

The RH on the air saturation line is 100% (fully wet). At a given dry-bulb temperature T (thus a fixed $W_{v,s}$), the RH is proportional to the humidity ratio W, according to Equation (4.16). Connecting the same percentage of RH for different temperatures produces the RH lines (Figure 4.7), with 0% on the X-coordinator and 100% on the saturation line.

4.3.3 Construction of Enthalpy Lines

The enthalpy lines can be built using Equation (4.21) or (4.22), as a function of dry-bulb temperature (T) and humidity ratio (W). These straight lines represent the iso-enthalpy line with variable temperatures and humidity ratios (Figure 4.8). This implies that different combinations of T and W can achieve the same enthalpy. This will be important in order to understand the conversion between sensible heat and latent heat during the heating and cooling processes to be introduced in Chapter 12.

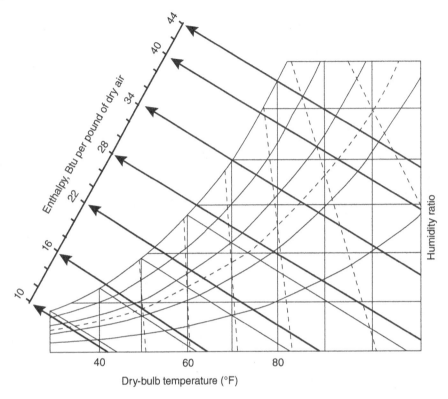

Figure 4.8 Construction of enthalpy lines.

4.3.4 Construction of Wet-Bulb Temperature Lines

The wet-bulb temperature lines (Figure 4.9) can also be created, which almost follow the iso-enthalpy lines. This is because any air (point) with a property of T and W can go through an iso-enthalpy evaporation process (as shown in Figure 4.4) to reach saturation (fully wet, i.e. $\phi = 100\%$). Wet-bulb temperature lines end at the saturation lines ($\phi = 100\%$). The X-coordinate values of the intersections of the wet-bulb temperature lines and the saturation lines are the wet-bulb temperatures for all the dots on the same wet-bulb temperature lines.

Figure 4.10 presents the relationships among various temperature definitions. For a given air point with a known dry-bulb temperature, the dewpoint temperature can be found when the point moves horizontally (with the constant humidity ratio W) to meet the saturation line ($\phi = 100\%$); the wet-bulb temperature can be determined by moving the point along the wet-bulb temperature lines (or the iso-enthalpy lines) to meet the saturation line ($\phi = 100\%$). Among the three temperatures, the dewpoint temperature is the lowest and the dry-bulb temperature is the highest while the wet-bulb temperature is in between.

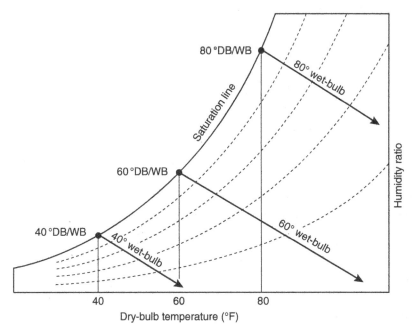

Figure 4.9 Construction of wet-bulb temperature lines.

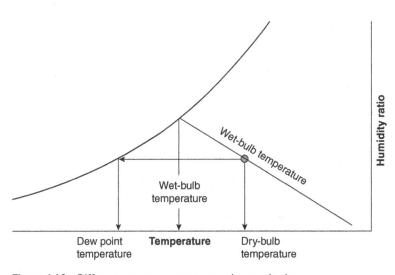

Figure 4.10 Different temperatures on a psychrometric chart.

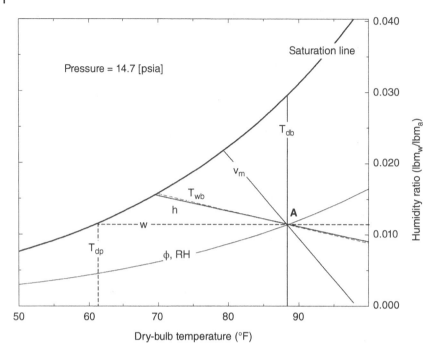

Figure 4.11 Illustration of the final format of a psychrometric chart.

4.3.5 The Final Format of a Psychrometric Chart

Figure 4.11 illustrates the final format of a psychrometric chart, in which air with two given properties can be used to determine the other air properties. Note that this chart is for the atmospheric pressure of 14.7 Psia (lbf/inch^2) (i.e. 1 atm). This chart can also determine the volume of 1 kg dry air plus water vapor in the air by using the following equation:

$$V_m = (0.082 \times T_{dry} + 22.4) \times (1/29 + W/18) \tag{4.24}$$

Figures 4.12 and 4.13 show the professional format of a psychrometric chart for sea level and a 5000 ft elevation (1500 m), respectively, made by ASHRAE. It should be noted that different countries or regions may present a psychrometric chart in different formats while using the same principles and inherent equations. As an example, Figure 4.14 shows the psychrometric chart commonly used in China.

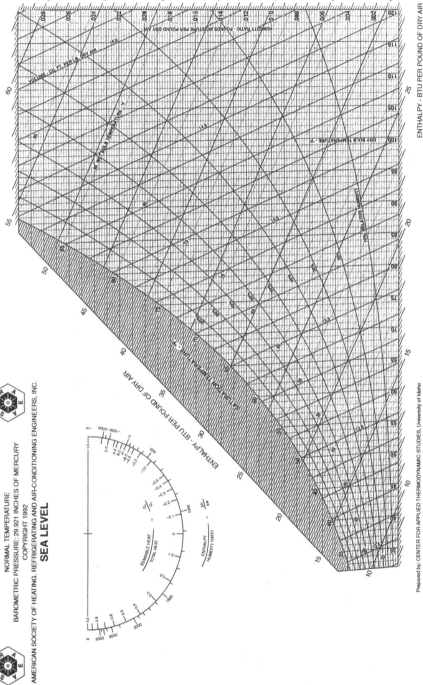

Figure 4.12 Psychrometric chart at the sea level. *Source:* Image by ASHRAE.

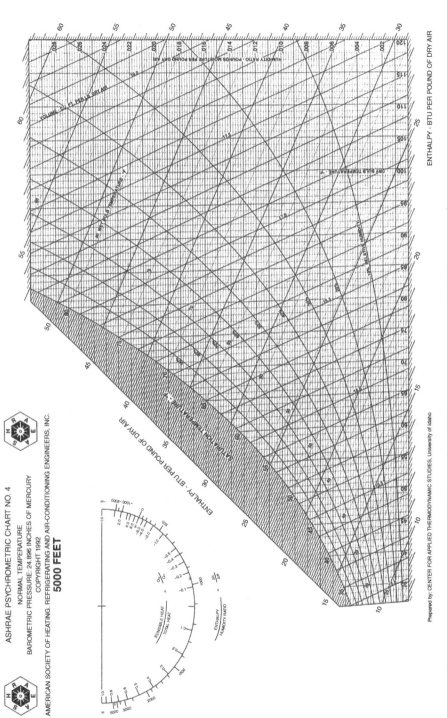

Figure 4.13 Psychrometric chart at the 5000 ft elevation (1500 m). *Source:* Image by ASHRAE.

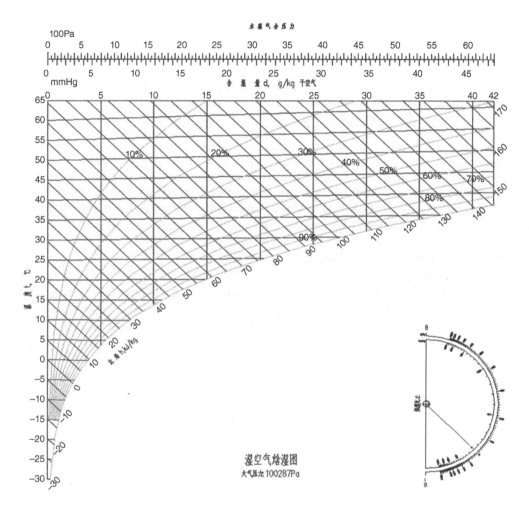

Figure 4.14 Psychrometric chart used in China. *Source:* Mandarin.

Homework Problems

1 (*Do not use a psychrometric chart for this problem.*)
  (a) The dry- and wet-bulb temperature in a room at Boston (sea level) are measured to be 27 °C and 15 °C, respectively. Please calculate the air humidity ratio and relative humidity.
  (b) Atmospheric air at 29 °C, 60% RH is compressed to 400 kPa and then cooled in an intercooler before entering a second stage of compression. What is the minimum temperature to which the air can be cooled without condensation?

2 Complete the following table for air conditions at an elevation of 5000 ft. (*You can use the psychrometric chart for this problem.*)

Dry-bulb temperature	Relative humidity	Humidity ratio	Enthalpy	Wet-bulb temperature	Dewpoint temperature
T	ϕ	W	i	T_{wb}	T_d
95 °F	30%				
80 °F				67 °F	

3 The air in a room at sea level is 68 °F and 50% relative humidity. What is the wet-bulb temperature of the air? Will moisture condense on a window whose surface is at 45 °F? Please use both the psychrometric chart and hand calculation to obtain your results.

4 **Staged Project Assignment (Teamwork)**
   (a) Locate the psychrometric chart that can be used for the design location (e.g., Washington, DC).
   (b) Analyze and comment on the typical summer and winter outdoor environments using the psychrometric chart (e.g., hot/cold, wet/dry, etc.).
   (c) Check and analyze the conventional heating and cooling strategies (both passive and active) for the location and for this type of building.

5

Climate and Site Analysis

5.1 Climate Analysis

5.1.1 Meteorological Year Data

Building design and performance are highly dependent on location/weather. Therefore, climate analysis is critical to help select proper building strategies and features for energy efficiency, such as whether to focus on winter (heating) or summer (cooling) or whether to enable the use of passive designs (e.g. natural ventilation, passive heating, evaporative cooling, etc.). Climate analysis requires the analysis of both the local/regional weather data and the microenvironment around the building site.

Typical meteorological year (TMY) data is the most widely used weather data, for prediction purposes, to determine the best building design, and to predict building performance. TMY data is a collation of selected weather data for a specific location, listing hourly values of solar radiation and meteorological elements for a one-year period. TMY files were first created by the US Department of Energy's (DOE) National Renewable Energy Laboratory (NREL) in 1981, based on the weather data of 1952–1975. The second-generation TMY2 files, that use 30 years of data, replaced the initial TMY file in around 1990, using an enhanced weighted average selection method. Approximately 1000 TMY2 files were created from weather station data from 1961 to 1990 mostly from airport-based stations. These served as the default weather files for the building energy simulation software EnergyPlus until late 2010. The third-generation TMY3 files that use 15 years of data (1991–2005) were introduced in 2005 with a greater emphasis on solar radiation variables and also included precipitation as a variable. While statistically stable files require 30 years of data, the TMY3 utilized only 15 because that was the period in which adequate satellite input was available. This effectively increased the availability and accuracy of solar radiation data. In late 2010, the DOE EnergyPlus converted to this file type as the default. TMY3 files are available for about 2500 sites primarily in the United States and Europe. The National Solar Radiation Data Base (NSRDB) (https://nsrdb.nrel.gov/data-sets/tmy) provides details and a downloadable form of the data.

TMY files are presented in a readable comma separated value (CSV) format. The TMY3 data files are named according to the US Air Force (USAF) site identifier as 999999TY.CSV, where 999999 represents the six-digit USAF station identifier. The TMY3 data format has 2 file header lines (Tables 5.1 and 5.2) and

Energy Efficient Buildings: Fundamentals of Building Science and Thermal Systems, First Edition. Zhiqiang (John) Zhai.
© 2023 John Wiley & Sons, Inc. Published 2023 by John Wiley & Sons, Inc.

Table 5.1 TMY3 data header (line 1).

Field	Element	Unit or description
1	Site identifier code	USAF number
2	Station name	Quote delimited
3	Station state	Two-letter US Postal abbreviation
4	Site time zone	Hours from Greenwich, negative west
5	Site latitude	Decimal degree
6	Site longitude	Decimal degree
7	Site elevation	Meter

Table 5.2 TMY3 data header (line 2).

Field	Element
1–68	Data field name and units (abbreviation or mnemonic)

8760 lines (hours) of data, each with 68 data fields (Table 5.3). It should be noted that each hour of data is neither an average nor a mean value of the 30 or 15 years of weather data; rather, it is typical (or representative) data of these years selected by the weighting algorithm. For example, the data for 1 a.m. on 1 January of the TMY3 file may be from 1993 while 2 a.m. on 1 January may be from 1995.

As a comparison, actual meteorological year (AMY) data is actual hourly data over the last year or time period when energy use data is available but put into the same formats as a TMY file. AMY is commonly used for existing building energy simulation (e.g. for building retrofit) to compare and/or calibrate the predicted results against the actual energy consumption. It should also be noted that AMY data is collected at weather stations (usually airports) and thus may be somehow different from the microenvironment at a building site. Figure 5.1 shows one example displaying the air temperature difference between the site and the weather station.

Because of the readability/editability of both TMY and AMY data, some extreme conditions of interest (e.g. without solar) could be assessed by manipulating specific rows and/or columns in the weather file. It can be interesting to investigate the impacts of particular weather factors.

5.1.2 Typical Meteorological Year (TMY) Data on Psychrometric Chart

Plotting TMY data on a psychrometric chart is a common practice to analyze weather conditions and further determine climate-specific design strategies. Figure 5.2 shows one example of analyzing Beijing weather, where points in the "Comfort" zone (dark green) require neither heating nor cooling, those in "Comfort 2m/s" (light green) require some ventilation to achieve thermal comfort, those in "Hot and/or humid" (red) are either too hot or too humidity that would require cooling (and dehumidification), and those in "Below comfort" (blue) require some kind of heating.

Table 5.3 TMY3 data field.

Field	Element	Unit or range	Resolution	Description
1	Date	MM/DD/YYYY	—	Date of data record
2	Time	HH:MM	—	Time of data record (local standard time)
3	Hourly extraterrestrial radiation on a horizontal surface	Watt-hour per square meter	1 Wh/m^2	Amount of solar radiation received on a horizontal surface at the top of the atmosphere during the 60-min period ending at the timestamp
4	Hourly extraterrestrial radiation normal to the sun	Watt-hour per square meter	1 Wh/m^2	Amount of solar radiation received on a surface normal to the sun at the top of the atmosphere during the 60-min period ending at the timestamp
5	Global horizontal irradiance	Watt-hour per square meter	1 Wh/m^2	Total amount of direct and diffuse solar radiation received on a horizontal surface during the 60-min period ending at the timestamp
6	Global horizontal irradiance source flag	1–2	—	
7	Global horizontal irradiance uncertainty	Percent	1%	Uncertainty based on random and bias error estimates – see NSRDB User's Manual
8	Direct normal irradiance	Watt-hour per square meter	1 Wh/m^2	Amount of solar radiation (modeled) received in a collimated beam on a surface normal to the sun during the 60-min period ending at the timestamp
9	Direct normal irradiance source flag	1–2	—	
10	Direct normal irradiance uncertainty	Percent	1%	Uncertainty based on random and bias error estimates – see NSRDB User's Manual (Wilcox, 2007b)
11	Diffuse horizontal irradiance	Watt-hour per square meter	1 Wh/m^2	Amount of solar radiation received from the sky (excluding the solar disk) on a horizontal surface during the 60-min period ending at the timestamp
12	Diffuse horizontal irradiance source flag	1–2	—	
13	Diffuse horizontal irradiance uncertainty	Percent	1%	Uncertainty based on random and bias error estimates – see NSRDB User's Manual
14	Global horizontal illuminance	Lux	100 lx	Average total amount of direct and diffuse illuminance received on a horizontal surface during the 60-min period ending at the timestamp

(Continued)

Table 5.3 (Continued)

Field	Element	Unit or range	Resolution	Description
15	Global horizontal illuminance source flag	1–2	—	
16	Global horizontal illuminance uncertainty	Percent	1%	Uncertainty based on random and bias error estimates – see Section 2.10
17	Direct normal illuminance	Lux	100 lx	Average amount of direct normal illuminance received within a 5.7° field of view centered on the sun during 60-min period ending at the timestamp
18	Direct normal illuminance source flag	1–2	—	
19	Direct normal illuminance uncertainty	Percent	1%	Uncertainty based on random and bias error estimates – see Section 2.10
20	Diffuse horizontal illuminance	Lux	100 lx	Average amount of illuminance received from the sky (excluding the solar disk) on a horizontal surface during the 60-min period ending at the timestamp
21	Diffuse horizontal illuminance source flag	1–2	—	
22	Diffuse horizontal illuminance uncertainty	Percent	1%	Uncertainty based on random and bias error estimates – see Section 2.10
23	Zenith luminance	Candela per square meter	10 cd/m^2	Average amount of luminance at the sky's zenith during the 60-min period ending at the timestamp
24	Zenith luminance source flag	1–2	—	
25	Zenith luminance uncertainty	Percent	1%	Uncertainty based on random and bias error estimates – see Section 2.10
26	Total sky cover	Tenths of sky	1 tenth	Amount of sky dome covered by clouds or obscuring phenomena at the time indicated
27	Total sky cover flag (source)			
28	Total sky cover flag (uncertainty)			
29	Opaque sky cover	Tenths of sky	1 tenth	Amount of sky dome covered by clouds or obscuring phenomena that prevent observing the sky or higher cloud layers at the time indicated
30	Opaque sky cover flag (source)			
31	Opaque sky cover flag (uncertainty)			

Table 5.3 (Continued)

Field	Element	Unit or range	Resolution	Description
32	Dry-bulb temperature	Degree Celsius	0.1°	Dry-bulb temperature at the time indicated
33	Dry-bulb temperature flag (source)			
34	Dry-bulb temperature flag (uncertainty)			
35	Dewpoint temperature	Degree C	0.1°	Dewpoint temperature at the time indicated
36	Dewpoint temperature flag (source)			
37	Dewpoint temperature flag (uncertainty)			
38	Relative humidity	Percent	1%	Relative humidity at the time indicated
39	Relative humidity flag (source)			
40	Relative humidity flag (uncertainty)			
41	Station pressure	Millibar	1 mbar	Station pressure at the time indicated
42	Station pressure flag (source)			
43	Station pressure flag (uncertainty)			
44	Wind direction	Degrees from north (360° = north; 0° = undefined, calm)	10°	Wind direction at the time indicated
45	Wind direction flag (source)			
46	Wind direction flag (uncertainty)			
47	Wind speed	Meter/second	0.1 m/s	Wind speed at the time indicated
48	Wind speed flag (source)			
49	Wind speed flag (uncertainty)			
50	Horizontal visibility	Meter[a]	1 m	Distance to discernable remote objects at the time indicated (7777 = unlimited)
51	Horizontal visibility flag (source)			

(Continued)

Table 5.3 (Continued)

Field	Element	Unit or range	Resolution	Description
52	Horizontal visibility flag (uncertainty)			
53	Ceiling height	Meter[a]	1 m	Height of the cloud base above local terrain (77777 = unlimited)
54	Ceiling height flag (source)			
55	Ceiling height flag (uncertainty)			
56	Precipitable water	Centimeter	0.1 cm	The total precipitable water contained in a column of unit cross section extending from the earth's surface to the top of the atmosphere
57	Precipitable water flag (source)			
58	Precipitable water flag (uncertainty)			
59	Aerosol optical depth, broadband	[unitless]	0.001	The broadband aerosol optical depth per unit of air mass due to extinction by the aerosol component of the atmosphere
60	Aerosol optical depth, broadband flag (source)			
61	Aerosol optical depth, broadband flag (uncertainty)			
62	Albedo	[unitless]	0.01	The ratio of reflected solar irradiance to global horizontal irradiance
63	Albedo flag (source)			
64	Albedo flag (uncertainty)			
65	Liquid precipitation depth	Millimeter[a]	1 mm	The amount of liquid precipitation observed at the indicated time for the period indicated in the liquid precipitation quantity field
66	Liquid precipitation quantity	Hour[a]	1 h	The period of accumulation for the liquid precipitation depth field
67	Liquid precipitation depth flag (source)			
68	Liquid precipitation depth flag (uncertainty)			

[a] Value of −9900 indicates the measurement is missing.

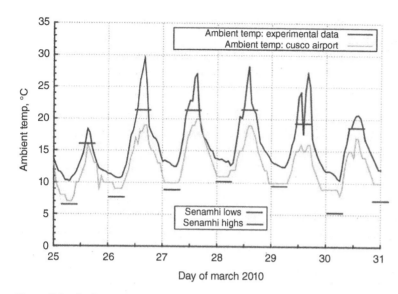

Figure 5.1 Onsite measured air temperature vs those from the airport during a five-day study period.

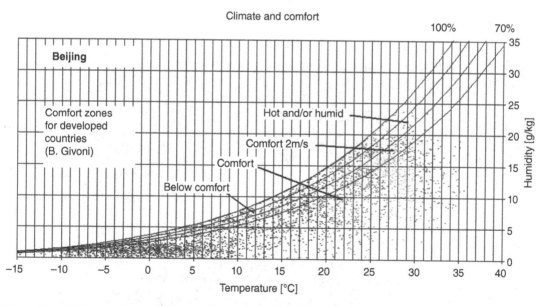

Figure 5.2 Beijing weather analysis on psychrometric chart.

A more in-depth and quantitative analysis can be conducted by using climate analysis software, e.g. the Ecotect Weather Tool (which was discontinued by Autodesk in 2015 but integrated into the Revit product family) or Climate Consultant (http://www.energy-design-tools.aud.ucla.edu/climate-consultant/request-climate-consultant.php). Figure 5.3 shows a series of weather analyses for Denver using the Ecotect Weather Tool:

a) For the monthly temperatures and solar radiation (Figure 5.3a). Other environmental conditions (such as relative humidity, wind speed, and cloud cover) can also be presented as needed.
b) For the seasonal wind conditions (direction, speed, and frequency/hours; Figure 5.3b). The detailed daily and monthly wind conditions can also be analyzed, as well as their corresponding air temperature. Note that wind speed is typically higher at the airport weather station than the project site.
c) For the hourly outdoor air temperature and humidity conditions plotted on a psychrometric chart (Figure 5.3c). The comfort zone in the yellow box is plotted as well as the extended comfort zones if using selected passive design strategies (i.e. passive solar heating, thermal mass effects, direct evaporative cooling).
d) For the estimated monthly benefits/savings of using different passive design strategies by counting hours/dots in the extended comfort zones for each passive design strategy (Figure 5.3d). For instance, passive solar heating may save about 30% energy in fall and spring and about 13% in winter; natural ventilation may save about 21% in summer; and thermal mass may save 2–6% in fall and spring and 61% in summer. A combined effect of savings can also be presented. These saving estimates are solely based on climate analysis, therefore only indicating potential values. The actual performance of a building will vary with its design.

(a)

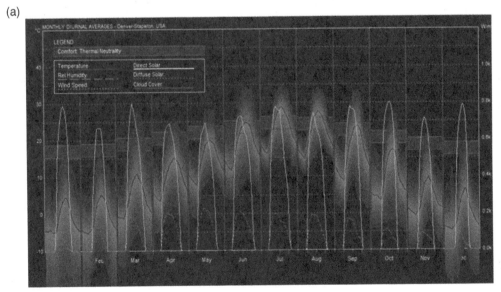

Figure 5.3 Denver weather analysis using Ecotect. (a) Monthly temperatures and solar radiations; (b) seasonal wind conditions; (c) hourly outdoor air conditions on a psychrometric chart; (d) monthly saving potentials of using different passive design strategies; and (e) optimal building orientation.

(b)

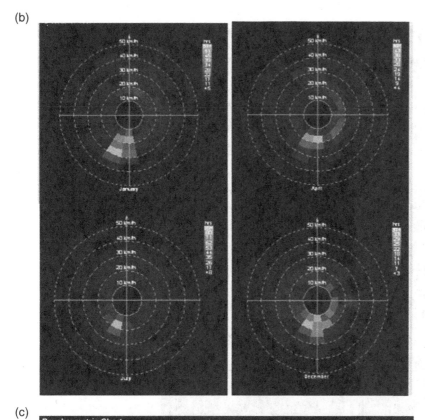

(c)

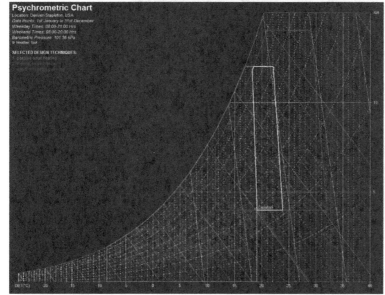

Figure 5.3 (Continued)

(d)

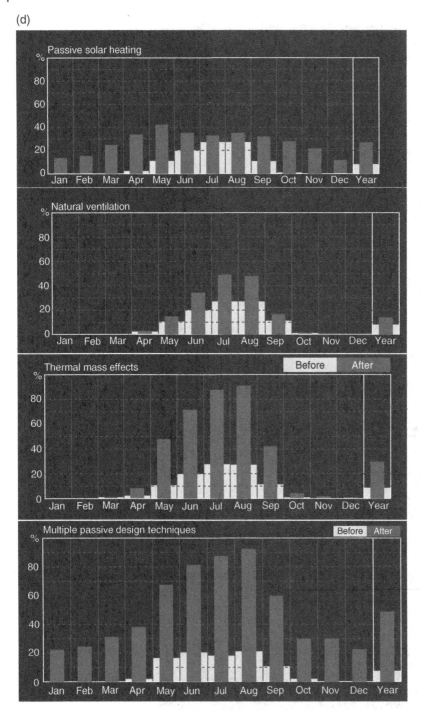

Figure 5.3 (Continued)

(e)

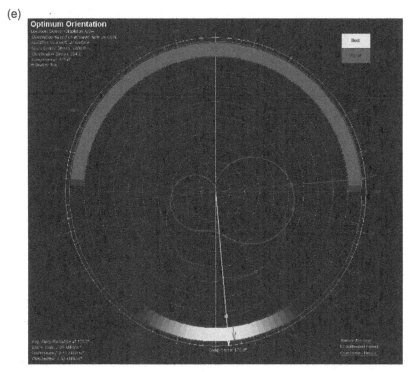

Figure 5.3 (Continued)

e) For the optimal building orientation (Figure 5.3e). This considers both summer and winter solar gains (in W/m² on a vertical surface at varying orientation angles) and demands (for cooling and heating). Users can define their own summer and winter time periods (months).

Climate Consultant was developed by the University of California at Los Angeles in the 1980s and has been continuously improved since then. This free tool, using TMY data in EnergyPlus weather (EPW) format, graphically displays climate data in dozens of ways that are useful to architects including temperature data, wind velocity data, sky cover data, percent sunshine data, psychrometric charts, timetables of bioclimatic needs, sun charts, and sun dials showing hours when solar heating is needed and when shading is required. The psychrometric analysis recommends the most appropriate passive design strategy. The software requires only basic familiarity with computers and architectural vocabulary. Figure 5.4 demonstrates the weather analyses for Washington DC, United States, using Climate Consultant.

A user can also create their own plot and chart according to the instruction of the software. Climate Consultant is under continuous development at UCLA and can be used with both a PC and Mac. Weather data analysis is the first step in brainstorming and evaluating proper regional design strategies to take full advantage of local climate characteristics to then determine the best systematic design solutions.

(a)

WEATHER DATA SUMMARY									LOCATION: Washington Dc Reagan Ap, VA, USA				
									Latitude/Longitude: 38.87° North, 77.03° West, Time Zone from Greenwich 5				
									Data Source: TMY3 724050 WMO Station Number, Elevation 3 m				

MONTHLY MEANS	JAN	FEB	MAR	APR	MAY	JUN	JUL	AUG	SEP	OCT	NOV	DEC	
Global Horiz Radiation (Avg Hourly)	199	251	321	402	352	416	407	399	349	333	210	195	Wh/sq.m
Direct Normal Radiation (Avg Hourly)	274	292	333	361	259	307	303	312	313	413	270	308	Wh/sq.m
Diffuse Radiation (Avg Hourly)	97	119	133	164	167	194	185	186	158	125	100	84	Wh/sq.m
Global Horiz Radiation (Max Hourly)	539	649	804	961	996	951	934	895	818	743	572	483	Wh/sq.m
Direct Normal Radiation (Max Hourly)	931	951	979	899	886	798	782	786	812	866	825	899	Wh/sq.m
Diffuse Radiation (Max Hourly)	323	351	458	435	559	447	432	431	399	361	316	249	Wh/sq.m
Global Horiz Radiation (Avg Daily Total)	1915	2627	3779	5273	4994	6116	5878	5386	4287	3650	2082	1819	Wh/sq.m
Direct Normal Radiation (Avg Daily Total)	2624	3050	3905	4743	3678	4515	4381	4226	3822	4511	2677	2871	Wh/sq.m
Diffuse Radiation (Avg Daily Total)	937	1251	1576	2148	2372	2855	2664	2503	1948	1370	996	785	Wh/sq.m
Global Horiz Illumination (Avg Hourly)	21475	27085	34847	43063	38429	44976	43955	43001	37520	35343	22438	20875	Lux
Direct Normal Illumination (Avg Hourly)	24771	27491	32160	35359	25824	30805	30522	31077	30863	39566	24850	27918	Lux
Dry Bulb Temperature (Avg Monthly)	2	1	6	14	17	24	27	25	19	15	8	6	degrees C
Dew Point Temperature (Avg Monthly)	-4	-6	-2	5	11	17	18	17	15	9	4	0	degrees C
Relative Humidity (Avg Monthly)	59	57	56	61	72	67	62	63	80	72	77	66	percent
Wind Direction (Monthly Mode)	330	310	330	180	170	180	210	170	170	330	200	340	degrees
Wind Speed (Avg Monthly)	4	4	5	4	4	3	3	3	3	3	4	3	m/s
Ground Temperature (Avg Monthly of 3 Depths)	10	6	5	5	8	12	16	20	22	21	18	14	degrees C

Back   Next

Figure 5.4 Washington DC weather analysis using Climate Consultant. (a) The monthly key weather data summary; (b) The comfort model selection and criteria specification (to be introduced in Chapter 6) (Ecotect also allows selecting and specifying particular comfort criteria); (c) The monthly and annual dry-bulb temperature range with comfort zone marked; (d) The monthly diurnal averages of temperature (dry-bulb and wet-bulb) and solar radiation (global horizonal, direct normal, and diffuse); (e) The monthly and annual radiation range (global horizonal, direct normal, and total surface), where the surface can be tilted and the ground reflectance can be altered; (f) The monthly and annual illumination range (direct normal and global horizonal); (g) The monthly and annual sky cover range; (h) The monthly and annual wind velocity range; (i) The monthly and annual ground temperature; (j) The daily dry-bulb temperature and relative humidity with comfort zones marked; (k) The daily dry-bulb temperature and dewpoint temperature with comfort zones marked; (l) The sun shading chart indicating when the shading is needed and when the sun is needed (to be introduced in Chapter 9); (m) The timetable (2D) plot of various weather variables (e.g. temperature, humidity, solar radiation, illumination, sky cover, etc.); (n) The 3D charts of various weather variables (e.g. temperature, humidity, solar radiation, illumination, sky cover, etc.); (o) The weather data plotted on a psychrometric chart with the comfort zone (in the blue box) and the extended comfort zones by using selected passive and active design strategies, along with the quantitative estimates of appliable hours (in both hours and percentile) for each selected design strategy; (p) The recommended design strategies (design guidelines) with graphic illustration for each strategy; and (q) The wind wheel that shows detailed wind conditions (monthly, daily, and hourly wind direction, temperature, frequency/ hours) with both charts and animation.

(b)

COMFORT MODEL

LOCATION: **Washington Dc Reagan Ap, VA, USA**
Latitude/Longitude: 38.87° North, 77.03° West, **Time Zone from Greenwich** 5
Data Source: TMY3 724050 WMO Station Number, **Elevation** 3 m

COMFORT MODELS:

Human Thermal comfort can be defined primarily by dry bulb temperature and humidity, although different sources have slightly different definitions. Select the model you wish to use:

○ **California Energy Code Comfort Model, 2013 (DEFAULT)**

For the purpose of sizing residential heating and cooling systems the indoor Dry Bulb Design Conditions should be between 68°F (20°C) to 75°F (23.9°C). No Humidity limits are specified in the Code, so 80% Relative Humidity and 66°F (18.9°C) Wet Bulb is used for the upper limit and 27°F (-2.8°C) Dew Point is used for the lower limit (but these can be changed on the Criteria screen).

⦿ **ASHRAE Standard 55 and Current Handbook of Fundamentals Model**

Thermal comfort is based on dry bulb temperature, clothing level (clo), metabolic activity (met), air velocity, humidity, and mean radiant temperature. Indoors it is assumed that mean radiant temperature is close to dry bulb temperature. The zone in which most people are comfortable is calculated using the PMV (Predicted Mean Vote) model. In residential settings people adapt clothing to match the season and feel comfortable in higher air velocities and so have wider comfort range than in buildings with centralized HVAC systems.

○ **ASHRAE Handbook of Fundamentals Comfort Model up through 2005**

For people dressed in normal winter clothes, Effective Temperatures of 68°F (20°C) to 74°F (23.3°C) (measured at 50% relative humidity), which means the temperatures decrease slightly as humidity rises. The upper humidity limit is 64°F (17.8°C) Wet Bulb and a lower Dew Point of 36F (2.2°C). If people are dressed in light weight summer clothes then this comfort zone shifts 5°F (2.8°C) warmer.

○ **Adaptive Comfort Model in ASHRAE Standard 55-2010**

In naturally ventilated spaces where occupants can open and close windows, their thermal response will depend in part on the outdoor climate, and may have a wider comfort range than in buildings with centralized HVAC systems. This model assumes occupants adapt their clothing to thermal conditions, and are sedentary (1.0 to 1.3 met). There must be no mechanical Cooling System, but this method does not apply if a Mechanical Heating System is in operation.

Back   Next

(c)

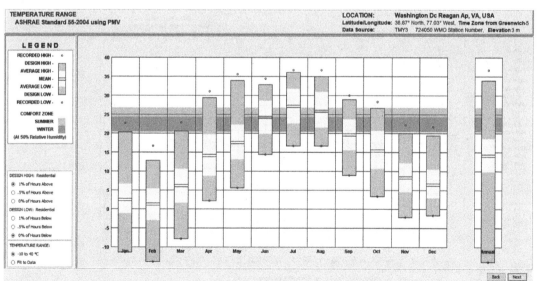

Figure 5.4 (Continued)

(d)

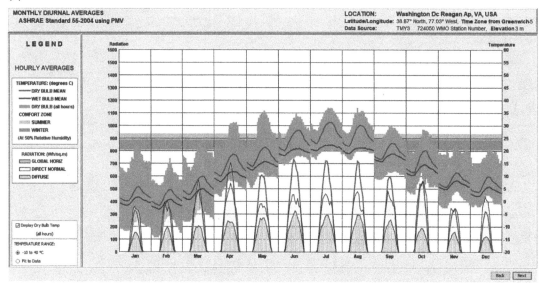

(e)

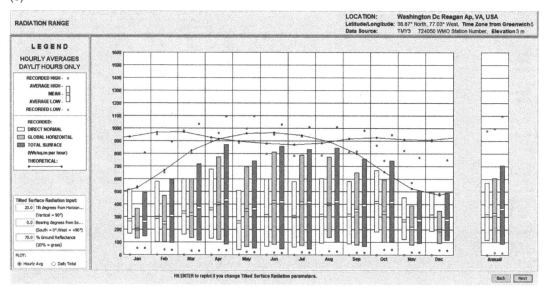

Figure 5.4 (Continued)

(f)

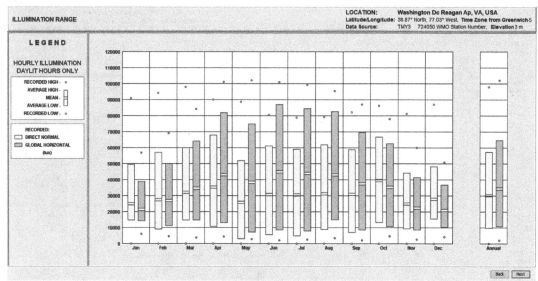

(g)

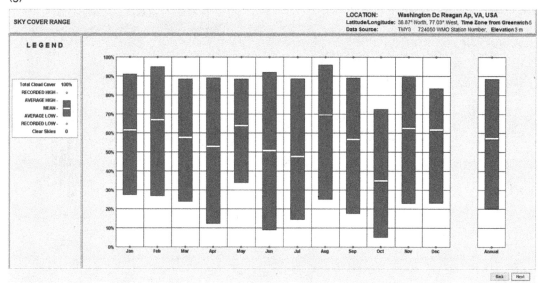

Figure 5.4 (Continued)

(h)

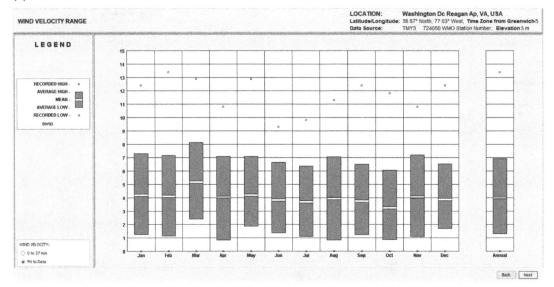

(i)

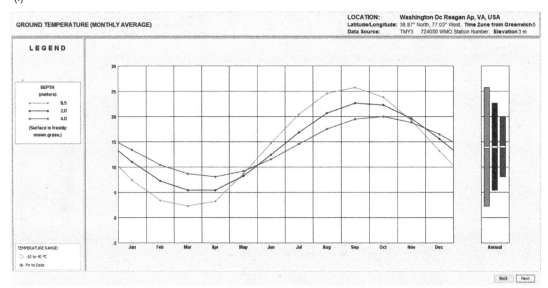

Figure 5.4 (Continued)

(j)

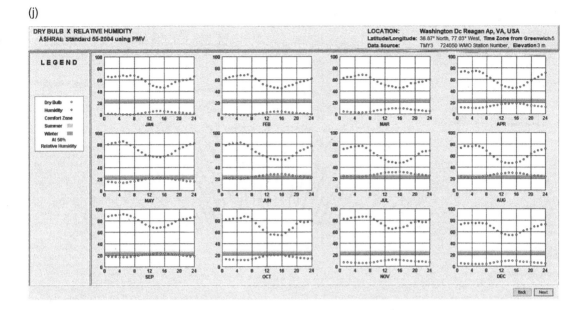

(k)

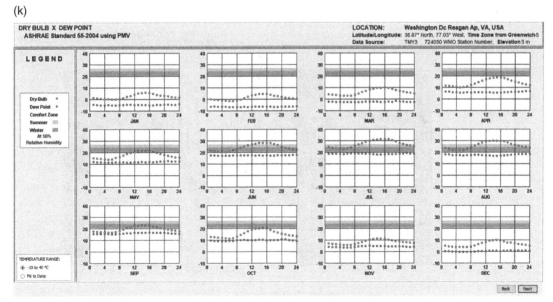

Figure 5.4 (Continued)

(l)

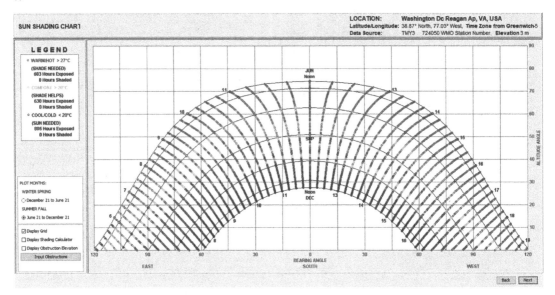

(m)

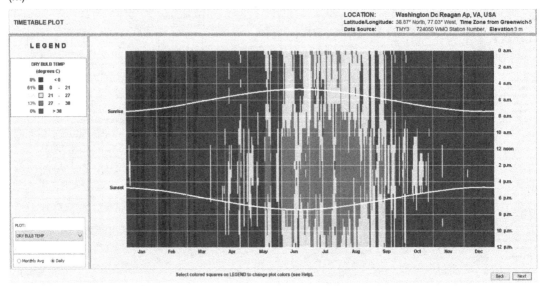

Figure 5.4 (Continued)

(n)

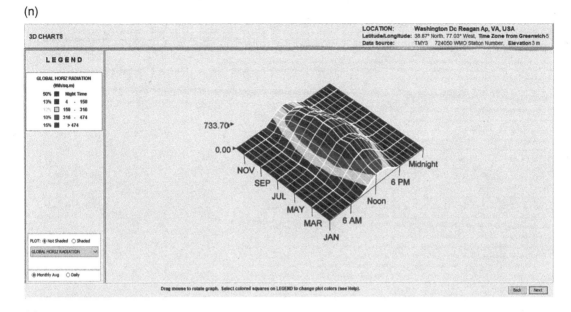

(o)

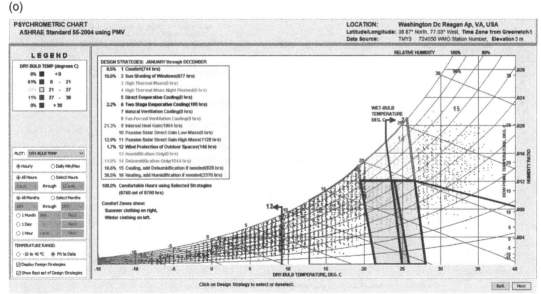

Figure 5.4 (Continued)

(p)

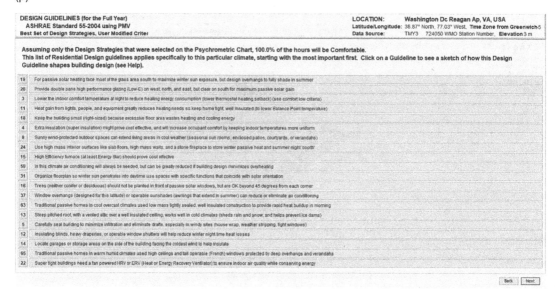

(q)

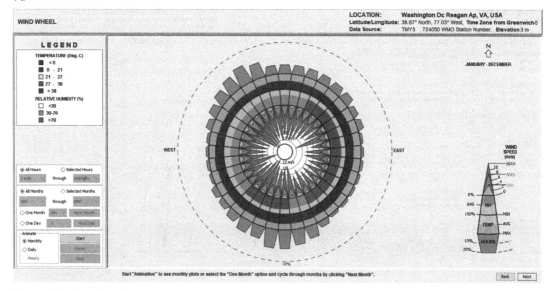

Figure 5.4 (Continued)

5.2   Heating and Cooling Design Climatic Data

Hourly climatic data, either typical or actual, is useful to predict detailed monthly, seasonal, and annual energy costs of a building, upon which energy efficient solutions can be tested and compared. Selecting and sizing a HVAC system (e.g. air conditioner, furnace, or heat pump) requires specific outdoor conditions (i.e. air temperature and relative humidity), and indoor design/preferred conditions to calculate the amount of energy (load) needed to compensate for the heat loss (winter) and heat gain (summer). These load calculations can then be used to determine the appropriate system sizes (depending on the system efficiency).

Ideally, the selected systems would be expected to cover a wide range of climatic conditions from the hottest to the coldest temperatures in the location's history. This, however, may lead to an oversized HVAC system designed for extreme or rare weather conditions that might occur once every hundred years or only occur in a short time span (e.g. coldest early morning hours or hottest later afternoon hours). Systems designed based on these extreme or rare weather conditions will have much larger capacities and sizes, and thus require more spaces and capital costs. For regular operating conditions, the system is not able to run at full (or close to full) capacity. Running on partial capacity often implies a lower efficiency, resulting in greater operating costs. On the contrary, if mild climatic conditions were used when designing the system, the preferred/designed indoor comfort conditions would not be met for enough time each day, which contradicts the purpose of the system design. Therefore, selecting appropriate outdoor design conditions are critical in designing and sizing proper building systems.

To help with choosing appropriate outdoor conditions for system design, weather statistics corresponding to several levels of probability have been determined and adopted. These are the weather conditions that are exceeded at the site in question during a specified portion of time (in percentile) and are determined based on historic climatic data over a long period of time. The annual percentiles, widely adopted by designers, are:

- **For heating:** 99.6 or 99%, which means that 0.4 or 1% of the 8760 hours during a year (i.e. 35 or 88 hours) may not meet the designed indoor conditions if adopting the heating systems that were sized using the associated heating (winter) design conditions.
- **For cooling:** 0.4, 1, or 2%, which means 0.4 or 1 or 2 of the 8760 hours during a year (i.e. 35 or 88 or 175 hours) might not meet the designed indoor conditions if adopting the cooling systems that were sized using the associated cooling (summer) design conditions.

It should be noted that those non-met hours may not imply an uncomfortable indoor environment but rather an indoor condition that has deviated from the original design condition. For instance, if a designed or preferred summer indoor condition is 25 °C, a "non-met hour" may result in an indoor temperature of 27 °C, using the designed cooling system, due to the fact that the actual outdoor temperature is higher than the outdoor design temperature for the hour in question. In some cases, these non-met hours might not even be noticed because of the small deviation, short time period, and/or specific moment in time (e.g. early morning when people sleep). Therefore, it is important to balance design needs and energy conservation goals.

The ASHRAE fundamentals handbook provides the design climatic data for major cities. Table 5.4 shows the winter/heating and summer/cooling design conditions for Denver, CO. These are used

Table 5.4 Annual heating and cooling design conditions (in IP unit) for Denver, CO.

Design conditions for DENVER/CENTENNIAL, CO, USA Station information

Station name	WMO#	Lat	Long	Elev	StdP	Hours +/- UTC	Time zone code	Period
1a	1b	1c	1d	1e	1f	1g	1h	1i
DENVER/CENTENNIAL	724666	39.57 N	104.85 W	5883	11.829	−7.00	NAM	8401

Annual heating and humidification design conditions

Coldest month	Heating DB 99.6%	99%	Humidification DP/MCDB and HR 99.6% DP	HR	MCDB	99% DP	HR	MCDB	Coldest month WS/MCDB 0.4% WS	MCDB	1% WS	MCDB	MCWS/PCWD to 99.6% DB MCWS	PCWD
2	3a	3b	4a	4b	4c	4d	4e	4f	5a	5b	5c	5d	6a	6b
12	−1.5	4.2	−6.5	4.8	8.4	−1.2	6.4	15.8	26.4	40.2	24.2	39.2	6.6	140

Annual cooling, dehumidification, and enthalpy design conditions

Hottest month	Hottest month DB range	Cooling DB/MCWB 0.4% DB	MCWB	1% DB	MCWB	2% DB	MCWB	Evaporation WB/MCDB 0.4% WB	MCDB	1% WB	MCDB	2% WB	MCDB	MCWS/PCWD to 0.4% DB MCWS	PCWD
7	8	9a	9b	9c	9d	9e	9f	10a	10b	10c	10d	10e	10f	11a	11b
7	27.3	91.2	60.6	88.9	60.1	86.1	59.7	65.4	81.3	63.9	79.6	62.4	78.0	11.2	350

Dehumidification DP/MCDB and HR

	0.4%			1%			2%	
DP	HR	MCDB	DP	HR	MCDB	DP	HR	MCDB
12a	12b	12c	12d	12e	12f	12g	12h	12i
61.2	100.8	68.7	59.2	93.7	68.1	57.3	87.4	67.6

Enthalpy/MCDB

0.4%		1%		2%	
Enth	MCBD	Enth	MCDB	Enth	MCDB
13a	13b	13c	13d	13e	13f
33.8	81.5	32.4	79.5	31.2	78.2

Extreme annual design conditions

Extreme annual WS

1%	2.5%	5%	Extreme Max WB
14a	14b	14c	15
24.8	21.9	18.8	73.2

Extreme Annual DB

mean		Standard deviation	
Max	Min	Max	Min
16a	16b	16c	16d
96.6	-8.5	2.0	7.9

n-year Return period values of extreme DB

n = 5 yr		n = 10 yr		n = 20 yr		n = 50 yr	
Max	Min	Max	Min	Max	Min	Max	Min
17a	17b	17c	17d	17e	17f	17g	17h
98.0	-14.2	99.2	-18.8	100.3	-23.2	101.8	-29.0

WMO#: World Meteorological Organization number; Lat: Latitude,°; , Long: Longitude,°; Elev: Elevation, ft; StdP: Standard pressure at station elevation, psi; DB: Dry-bulb temperature, °F; DP: Dewpoint temperature, °F; WB: Wet-bulb temperature, °F; WS: Wind speed, mph; Enth: Enthalpy, Btu/lb; HR: Humidity ratio, grains of moisture per lb of dry air MCDB: Mean coincident dry-bulb temperature, °F; MCWB: Mean coincident wet-bulb temperature, °F; MCWS: Mean coincident wind speed, mph; PCWD: Prevailing coincident wind direction, °, 0 = North, 90 = East
Source: Data from ASHRAE.

for annual calculations and design. It also includes the extreme weather conditions for comparison. If specific months are of particular interest, ASHRAE also provides monthly outdoor design conditions as shown in Table 5.5. Table 5.6 lists the design climatic data for some major cities in the United States (in SI unit).

Table 5.5 Monthly heating and cooling design conditions (in IP unit) for Denver, CO.

	Monthly design dry-bulb and mean coincident wet-bulb temperatures											
	Jan		Feb		Mar		Apr		May		Jun	
%	DB	MCWB	DB	MCWB	DB	MCWB	DB	MCWB	DB	MCWB	DB	MCWB
	18a	*18b*	*18c*	*18d*	*18e*	*18f*	*18g*	*18h*	*18i*	*18j*	*18k*	*18l*
0.4	63.8	42.7	65.4	44.2	72.6	49.7	77.3	51.8	85.7	56.2	93.4	59.9
1	60.4	41.5	63.1	42.5	69.9	47.2	75.1	50.6	82.5	55.7	91.2	59.9
2	56.9	39.8	60.6	41.7	67.4	46.3	73.0	49.9	80.7	55.1	89.7	59.3

	Jul		Aug		Sep		Oct		Nov		Dec	
%	DB	MCWB	DB	MCWB	DB	MCWB	DB	MCWB	DB	MCWB	DB	MCWB
	18m	*18n*	*18o*	*18p*	*18q*	*18r*	*18s*	*18t*	*18u*	*18v*	*18w*	*18x*
0.4	96.3	61.6	92.9	60.2	90.1	59.2	81.0	53.8	72.2	46.9	63.9	43.2
1	93.9	61.3	91.1	60.2	88.1	58.3	79.1	52.3	69.9	46.1	61.0	42.4
2	92.1	61.2	89.9	60.3	85.9	57.6	76.8	51.5	66.2	45.4	57.5	40.3

	Monthly design wet-bulb and mean coincident dry-bulb temperatures											
	Jan		Feb		Mar		Apr		May		Jun	
(%)	WB	MCDB	WB	MCDB	WB	MCDB	WB	MCDB	WB	MCDB	WB	MCDB
	19a	*19b*	*19c*	*19d*	*19e*	*19f*	*19g*	*19h*	*19i*	*19j*	*19k*	*19l*
0.4	44.6	60.1	45.7	63.4	52.5	67.8	54.1	72.0	60.3	76.6	65.2	83.8
1	42.5	58.0	43.9	60.8	49.8	67.9	52.9	71.2	58.9	75.4	64.1	82.2
2	40.9	55.6	42.2	58.4	47.0	64.1	51.6	69.4	57.5	74.0	62.9	80.7

	Jul		Aug		Sep		Oct		Nov		Dec	
%	WB	MCDB	WB	MCDB	WB	MCDB	WB	MCDB	WB	MCDB	WB	MCDB
	19m	*19n*	*19o*	*19p*	*19q*	*19r*	*19s*	*19t*	*19u*	*19v*	*19w*	*19x*
0.4	67.9	84.8	67.4	82.0	64.0	80.5	56.5	74.0	49.4	66.7	44.9	60.7
1	66.6	83.4	66.4	81.0	62.4	78.9	54.5	72.0	47.5	65.2	43.2	58.8
2	65.6	82.2	65.5	80.2	61.2	77.4	53.1	71.2	46.4	64.1	41.4	56.5

Table 5.5 (Continued)

					Monthly mean daily temperature range						
Jan	Feb	Mar	Apr	May	Jun	Jul	Aug	Sep	Oct	Nov	Dec
20a	*20b*	*20c*	*20d*	*20e*	*20f*	*20g*	*20h*	*20i*	*20j*	*20k*	*20l*
23.2	23.5	24.7	24.4	25.3	27.4	27.3	25.9	27.0	26.6	24.5	23.0

Source: Data from ASHRAE.

Table 5.6 Annual heating and cooling design conditions (in SI unit) for some major cities in the United States.

		Atlanta, GA	Chicago, IL	Denver, CO	New York, NY	Phoenix, AZ
Latitude		33.64 °N	41.99 °N	39.83 °N	40.66 °N	33.42 °N
Elevation		313 m / 1027 ft	205 m / 673 ft	1655 m / 5430 ft	7 m / 23 ft	339 m / 1106 ft
SI units						
Heating	Coldest month	1	1	12	1	12
	DBT (°C):					
	99.6 %	−5.8	−18.6	−17.5	−10.1	3.7
	99 %	−3.1	−15.7	−14.1	−7.9	5.3
	MCWS (m/s) at 99.6 % DBT	5.3	4.9	3.5	7.4	1.6
Cooling	Hottest month	7	7	7	7	7
	Hottest month daily range (°C)	9.4	10.2	15.3	7.4	11.8
	DBT /MCWBT (°C):					
	0.4%	34.4/ 23.4	33.0/23.5 31.5/22.9	34.6/15.5 33.3/15.5	32.1/22.7 30.3/22.1	43.5/20.9 42.4/20.8
	1%	33.1/ 23.3	30.0/22.1	31.8/15.4	28.7/21.7	41.3/20.7
	2%	32.1/ 23.0				
	MCWS (m/s) at 0.4 % DBT	3.9	5.1	4.1	5.8	4.2

Source: Data from ASHRAE.

5.3   Site Analysis

Site analysis is important in the design of buildings. Site analysis has a wide scope of tasks that can include the analyses of soil conditions, transportation circulations, utility links, view connections, solar accesses, wind obstructions, building setbacks, neighborhood contexts, etc. This section is focused on energy-related site analysis – the microclimate around a building that may impact its energy performance.

As stated earlier, either TMY or AMY data is obtained from the weather stations usually located at local airports. These weather data, although representative of the general region and location, may deviate from the conditions at the building site, e.g. in the downtown area of a big city. Hence, it is crucial to account for the microclimate around the project site. This may include, but is not limited to, the following impacts:

1) **Terrain:** where the top of the hill may experience more wind and the south slope is exposed more solar for a longer period of time (Figure 5.5). Designers may consider suitable project sites, if given the opportunity, according to different climatic conditions. For mountainous areas, the local wind direction can be altered due to significant diurnal temperature swings. Air tends to move down to

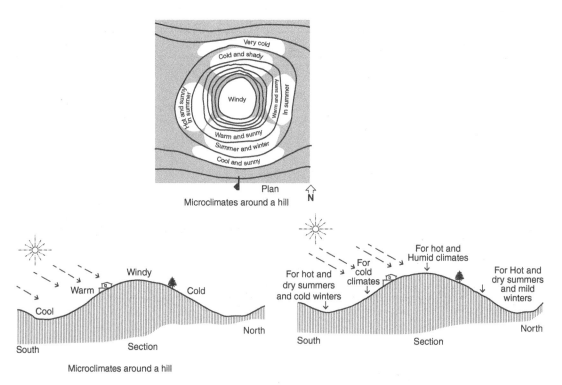

Figure 5.5   Microclimate: elevation impacts.

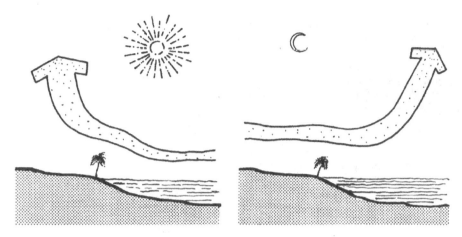

Figure 5.6   Microclimate: water impacts.

the valley at night and rise up to the peak during the day. Bodies of water (e.g. an ocean or lake) may also change the wind direction and speed, which can be attributed to the different heat capacities of sand/earth and water. Water changes temperature more slowly than sand/earth, resulting in a slower temperature change of the air above a body of water (Figure 5.6).

2) **Neighboring context:** This includes the building setback requirements – the minimum distance that a building must be set back from a street or road, neighboring objects (e.g. trees), and structures, as well as the layouts and heights of these objects (Figure 5.7). This information is critical for both the access and obstruction design of solar and wind. Information on the materials and colors of these objects is also important, which determines the amount of reflected solar, daylight and glare.

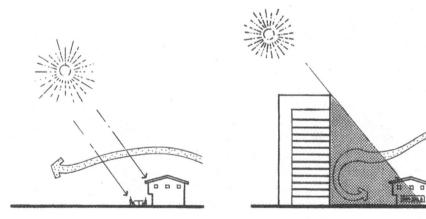

Figure 5.7   Microclimate: neighbor impacts.

(a)

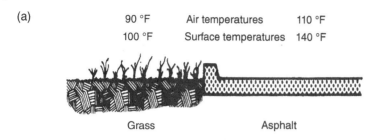

(b)

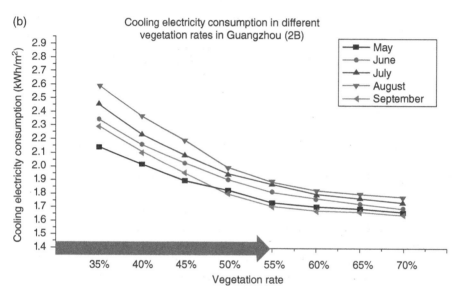

Figure 5.8 Microclimate: grass impacts. (a) Grass impact on surface and surrounding air temperature; and (b) grass impact on building cooling energy in a hot and humid city.

3) **Landscape:** It is widely recognized that landscape has a significant impact on microclimate. Greenery and bodies of water can change the local air temperature, humidity, and velocity and further influence the building's thermal and energy performances. In comparison to asphalt, grass can better reduce both surface and air temperatures (Figure 5.8a). Studies show that increasing the vegetation area coverage density from 35 to 55% can save 20% of summer cooling energy use for a hot and humid city (Guangzhou, China; Figure 5.8b). Plants around the project site can also affect the building solar and wind access and obstruction, as partially illustrated in Figure 5.9.

4) **Outdoor environment quality:** Factors such as air pollution and noise conditions can play important roles in determining appropriate building design strategies. For a noisy and polluted outdoor environment (Figure 5.10), natural ventilation may not be an acceptable solution even if weather data shows a great potential in utilizing this strategy. Special strategies, techniques, and products need to be designed and implemented to accommodate these nonideal outdoor microclimates.

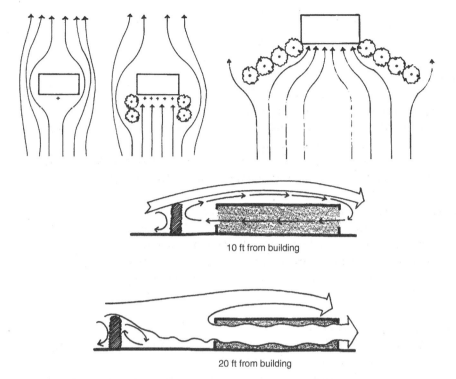

10 ft from building

20 ft from building

Figure 5.9 Microclimate: plant impacts. *Source:* ThKatz/Adobe Stock.

(b)

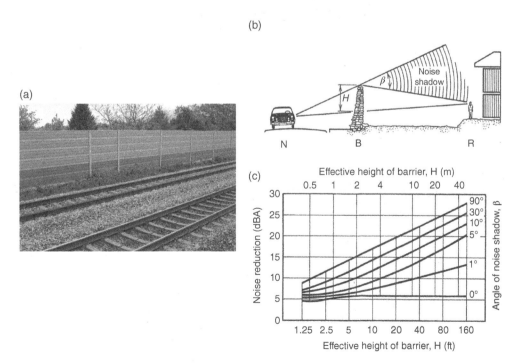

(a)

(c)

Figure 5.10 Microclimate: noise impacts. (a) Noise shadow created by barrier; (b) noise reduction related to barrier height; and (c) impact of effective height of barrier to noise reduction. *Source:* ThKatz/Adobe Stock.

Homework Problems

1 Study the TMY3 weather data for the design site (e.g., Washington, DC).

   (a) Find the total amount of direct and diffuse solar radiation received on the horizontal roof during the winter design day (21 December) (in $W\text{-}h/m^2$);

   (b) Calculate the heating and cooling degree days (assume the outdoor balance point temperature is 18 °C – when the indoor needs neither heating nor cooling – refer to Chapter 13).

2 Use a weather analysis tool (e.g., Ecotect or Climate Consultant) to

   (a) Identify the appropriate passive design strategies/technologies for four seasons (Jan, April, July, and Oct) for Boulder, CO (or your local city), as well as the potential savings;

   (b) Find the wind conditions (speed, frequency, direction, etc.) for four seasons (Jan, April, July, and Oct) for Boulder, CO (or your local city), and comment on relevant wind-blocking and natural ventilation approaches.

3 Staged Project Assignment (Teamwork)

   (a) Check the ASHRAE Fundamentals Handbook to determine and justify the outdoor design conditions for the design location (e.g., Washington, DC).

   (b) Use Climate Consultant (or similar weather tools) to describe the primary climate characteristics (e.g., temperature, humidity, wind, and solar for four seasons) and identify appropriate design strategies (and potential savings in each month) for the site.

   (c) Discuss the potential microclimate impacts and solutions.

6

Indoor Thermal Comfort

6.1 Indoor Environment Quality

Indoor environment quality (IEQ) is composed of four elements:

- **Thermal comfort**: mostly related to air and enclosure temperature, relative humidity, air speed and turbulence, clothing level, and occupant activities.
- **Indoor air quality (IAQ)**: closely related to indoor pollutants (including those from the outdoors; e.g. toxic gases, volatile organic compounds – VOCs, particulate matters, and their indoor distributions and controls).
- **Visual comfort (lighting)**: including both daylighting and artificial lights and their designs and controls.
- **Acoustic comfort (acoustics)**: including indoor and outdoor noise source controls (e.g. from traffic and devices), communication privacy, special environment sound quality (e.g. stadium and theater), as well as background noise management (e.g. from HVAC systems).

Building mechanical systems mostly affect indoor thermal comfort and indoor air quality. Therefore, this book focuses on these two aspects. Lighting and acoustics are important topics for most commercial and public buildings, and are extremely important for designing the indoor environments of buildings with special functions such as a museum, performance center, theater, etc.

6.2 Indoor Thermal Comfort

6.2.1 Heat and Mass Transfer Mechanisms

Thermal comfort describes the comfort level of the human body within the surrounding environment. In principle, this reflects the energy conservation capabilities of the human body as illustrated in Figure 6.1. Energy inputs via food and drink (fuels) are exhausted by various heat and mass transfer mechanisms (e.g. conduction, convection, radiation, evaporation) through the human body. As seen in Figure 6.2, the majority of heat (72%) is released through conduction, convection, and radiation. This fraction, however, varies with the surrounding air temperature (Figure 6.3). When the air temperature increases, the sensible heat transfer (driven by the temperature difference between the body and the surrounding air)

Energy Efficient Buildings: Fundamentals of Building Science and Thermal Systems, First Edition. Zhiqiang (John) Zhai.
© 2023 John Wiley & Sons, Inc. Published 2023 by John Wiley & Sons, Inc.

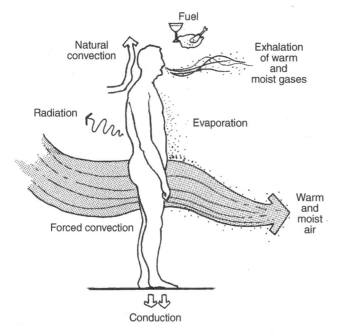

Figure 6.1   Energy conservation at human body.

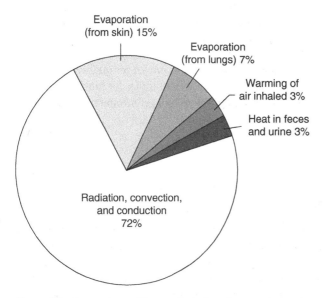

Figure 6.2   Proportions of human body heat loss at air temperature of 70 °F.

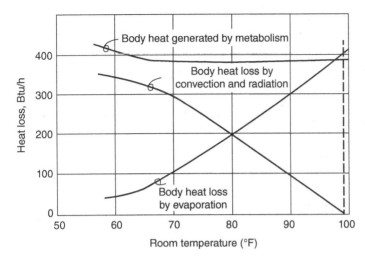

Figure 6.3 Variation of human body heat loss proportion with surrounding air temperature.

will decrease, while the latent heat transfer (due to phase changes such as evaporation) will increase. When the surrounding air temperature is above the body temperature (98 °F), the evaporation of body sweat will be the primary mechanism to remove the heat. Evaporation will be more difficult in humid locations (e.g. Miami, FL) than arid locations (e.g. Phoenix, AZ). Air-conditioning will ultimately be required for an extremely hot environment.

6.2.2 Energy Conservation Equation

The human body obeys the first law of thermodynamics (energy conservation law):

$$M - W = Q_{sk} + Q_{res} = (C_{sk} + R_{sk} + E_{sk}) + (C_{res} + E_{res}) \tag{6.1}$$

where M is the rate of metabolic heat production of a person (W/m^2, Btu/h·ft^2) (see Table 6.1, where 1 met = 58.2 W/m^2 body area); W is the rate of mechanical work conducted by a person (W/m^2, Btu/h·ft^2); Q is the heat losses; C is the convective heat losses; R is the radiative heat losses; E is the evaporative heat losses; the subscript "sk" stands for skin and "res" for respiration.

All of the convection (C), radiation (R), and evaporation (E) for both skin and respiration are the functions of air properties, such as air temperature, humidity, speed, enclosure surface temperature, etc. The detailed correlations were developed by Fanger in 1982 as shown below (in SI unit):

$$C_{sk} = h_c \times (T_{cloth} - T_{air}) \times A_{cloth}/A_{body} \tag{6.2}$$

$$h_c = 2.38 \times (T_{cloth} - T_{air})^{0.25} \quad \text{when } 2.38 \times (T_{cloth} - T_{air})^{0.25} > 12.1 \times V^{0.5} \tag{6.3}$$

$$h_c = 12.1 \times V^{0.5} \quad \text{when } 2.38 \times (T_{cloth} - T_{air})^{0.25} < 12.1 \times V^{0.5} \tag{6.4}$$

$$T_{cloth} = 35.7 - 0.0275 \times (M - W) - R_{cloth} \times \{(M - W) - 3.05 \times [5.73 - 0.007 \times (M - W) - p_v]$$

$$- 0.42 \times [(M - W) - 58.15] - 0.0173 \times M \times (5.87 - p_v) - 0.0014 \times M \times (34 - T_{air})\}$$

$$\tag{6.5}$$

Table 6.1 Metabolic rates for typical activities.

Activity	Metabolic rate		
	met	W/m^2	Btu/h·ft^2
Resting			
Sleeping	0.7	40	13
Reclining	0.8	45	15
Seated, quiet	1.0	60	18
Standing, relaxed	1.2	70	22
Walking (on level surface)			
0.9 m/s, 3.2 km/h, 2.0 mph	2.0	115	37
1.2 m/s, 4.3 km/h, 2.7 mph	2.6	150	48
1.8 m/s, 6.8 km/h, 4.2 mph	3.8	220	70
Office activities			
Reading, seated	1.0	55	18
Writing	1.0	60	18
Typing	1.1	65	20
Filing, seated	1.2	70	22
Filing, standing	1.4	80	26
Walking about	1.7	100	31
Lifting/packing	2.1	120	39
Driving/flying			
Automobile	1.0–2.0	60–115	18–37
Aircraft, routine	1.2	70	22
Aircraft, instrument landing	1.8	105	33
Aircraft, combat	2.4	140	44
Heavy vehicle	3.2	185	59
Miscellaneous occupational activities			
Cooking	1.6–2.0	95–115	29–37
House cleaning	2.0–3.4	115–200	37–63
Seated, heavy limb movement	2.2	130	41
Machine work			
Sawing (table saw)	1.8	105	33
Light (electrical industry)	2.0–2.4	115–140	37–44
Heavy	4.0	235	74
Handling 50 kg (100 lb) bags	4.0	235	74
Pick and shovel work	4.0–4.3	235–230	74–88

Table 6.1 (Continued)

Activity	Metabolic rate		
	met	W/m²	Btu/h·ft²
Miscellaneous leisure activities			
Dancing, social	2.4–4.4	140–255	44–81
Calisthenics/exercise	3.0–4.0	175–235	55–74
Tennis, single	3.6–4.0	210–270	66–74
Basketball	5.0–7.6	290–440	90–140
Wrestling, competitive	7.0–8.7	410–505	130–160

Source: ASHRAE 55-2020.

where p_v is the water vapor pressure (kPa).

$$A_{body} = 0.202 \times m^{0.425} \times l^{0.725} \tag{6.6}$$

where m is the body weight (kg) and l is the height (m)

$$A_{cloth}/A_{body} = f = 1.0 + 0.3 \times I_{cl} \tag{6.7}$$

where I_{cl} is the garment insulation value (Tables 6.2 and 6.3). The insulation value of clothing is measured in the unit clo (1.0 clo is equivalent to the typical American Man's Business suit in 1941). 1 clo = 0.88 ft²·h·°F/Btu = 0.155 m²·K/W. Table 6.2 lists the clothing insulation I_{cl} values for typical ensembles. Table 6.3 presents the garment insulation. Note that the actual I_{cl} can be calculated by adding the insulations of individual garments. R_{cloth} is the effective thermal resistance (R-value) of clothes (m²·K/W):

$$R_{cloth} = 0.155 \times I_{cl} \tag{6.8}$$

$$R_{sk} = \sigma \times \varepsilon_{cloth} \times \varepsilon_{ednclosure} \times \left[[T_{cloth} + 273]^4 - [T_{enclosure} + 273]^4 \right] \times A_{cloth}/A_{body} \tag{6.9}$$

Equation (6.9) can be simplified as:

$$R_{sk} = 3.96 \times 10^{-8} \times \left[(T_{cloth} + 273)^4 - (T_{enclosure} + 273)^4 \right] \times A_{cloth}/A_{body} \tag{6.10}$$

where $T_{enclosure}$ is the mean radiant temperature (T_{mrt}):

$$T_{enclosure} = \sum A_i T_i / \sum A_i \tag{6.11}$$

where T_i is the surface temperature of enclosure i, and A_i is the area of surface i.

$$E_{sk} = m_{sk} \times i_{fg} = 3.05 \times [5.73 - 0.007 \times (M - W) - p_v] + 0.42 \times [(M - W) - 58.15] \tag{6.12}$$

$$C_{res} = m_{res} \times C_{p,a} \times (T_{res} - T_{air}) = 0.0014 \times M \times (24 - T_{air}) \tag{6.13}$$

$$E_{res} = m_{res} \times i_{fg} = 0.0173 \times M \times (5.87 - p_v) \tag{6.14}$$

Table 6.2 Clothing insulation I$_{cl}$ values for typical ensembles.

Clothing description	Garments included[a]	I$_{cl}$, clo
Trousers	(1) Trousers, short-sleeve shirt	0.57
	(2) Trousers, long-sleeve shirt	0.61
	(3) #2 plus suit jacket	0.96
	(4) #2 plus suit jacket, vest, t-shirt	1.14
	(5) #2 plus long-sleeve sweater, t-shirt	1.01
	(6) #5 plus suit jacket, long underwear bottoms	1.30
Skirts/dresses	(7) Knee-length skirt, short-sleeve shirt (sandals)	0.54
	(8) Knee-length skirt, long-sleeve shirt, full slip	0.67
	(9) Knee-length skirt, long-sleeve shirt, half slip, long-sleeve sweater	1.10
	(10) Knee-length skirt, long-sleeve shirt, half slip, suit jacket	1.04
	(11) Ankle-length skirt, long-sleeve shirt, suit jacket	1.10
Shorts	(12) Walking shorts, short-sleeve shirt	0.36
overalls/coveralls	(13) Long-sleeve coveralls, t-shirt	0.72
	(14) Overalls, long-sleeve shirt, t-shirt	0.89
	(15) Insulated coveralls, long-sleeve thermal underwear tops and bottoms	1.37
Athletic	(16) Sweat pants, long-sleeve sweatshirt	0.74
sleepwear	(17) Long-sleeve pajama tops, long pajama trousers, short ¾-length robe (slippers, no socks)	0.96

[a] All clothing ensembles, except where otherwise indicated in parentheses, include shoes, socks, and briefs or panties. All skirt/dress clothing ensembles include pantyhose and no additional socks.
Source: ASHRAE 55-2020.

6.2.3 Predicted Mean Vote (PMV) and Predicted Percentage Dissatisfied (PPD) due to Thermal Comfort

The imbalance (thermal load on the body) of Equation (6.1) results in the discomfort.

$$L = M - W - (C_{sk} + R_{sk} + E_{sk}) - (C_{res} + E_{res}) \qquad (6.15)$$

When L=0, energy is balanced, and thus one feels thermally neutral or comfortable. When L > 0, more heat is generated than dissipated, and one feels warm or hot; when L < 0, more heat is lost than produced, therefore one feels cool or cold.

Fanger first introduced the concept of predicted mean vote (PMV) to statistically estimate the thermal condition in a built environment by subjectively voting on the thermal sensation using the following digital ranking:

Table 6.3 Garment insulation I_{cl}.

Garment description[a]	I_{clu}, clo	Garment description[a]	I_{clu}, clo
Underwear		Dress and skirts[b]	
Bra	0.01	Skirt (thin) mm	0.14
Panties	0.03	Skirt (thick)	0.23
Men's briefs	0.04	Sleeveless, scoop neck (thin)	0.23
T-shirt	0.08	Sleeveless, scoop neck (thick), i.e. jumper	0.27
Half slip	0.14	Short-sleeve shirtdress (thin)	0.29
Long underwear bottoms	0.15	Long-sleeve shirtdress (thin)	0.33
Full slip	0.16	Long-sleeve shirtdress (thick)	0.47
Long underwear top	0.20	Sweaters	
Footwear		Sleeveless vest (thin)	0.13
Ankle-length athletic socks	0.02	Sleeveless vest (thick)	0.22
Panty hose/stockings	0.02	Long-sleeve (thin)	0.25
Sandals/thongs	0.02	Long-sleeve (thick)	0.36
Shoes	0.02	Suit jackets and vests[c]	
Slippers (quilted, pile lined)	0.03	Sleeveless vest (thin)	0.10
Calf-length socks	0.03	Sleeveless vest (thick)	0.17
Knee socks (thick)	0.06	Single-breasted (thin)	0.36
Boots	0.10	Single-breasted (thick)	0.44
Shirts and blouses		Double-breasted (thin)	0.42
Sleeveless/scoop-neck blouse	0.12	Double-breasted (thick)	0.48
Short-sleeve knit sport shirt	0.17	Sleepwear and robes	
Short-sleeve dress shirt	0.19	Sleeveless short gown (thin)	0.18
Long-sleeve dress shirt	0.25	Sleeveless long gown (thin)	0.20
Long-sleeve flannel shirt	0.34	Short-sleeve hospital gown	0.31
Long-sleeve sweatshirt	0.34	Short-sleeve short robe (thin)	0.34
Trousers and coveralls		Short-sleeve pajamas (thin)	0.42
Short shorts	0.06	Long-sleeve long gown (thick)	0.46
Walking shorts	0.08	Long-sleeve short wrap robe (thick)	0.48
Straight trousers (thin)	0.15	Long-sleeve pajamas (thick)	0.57
Straight trousers (thick)	0.24	Long-sleeve long wrap robe (thick)	0.69
Sweatpants	0.28		
Overalls	0.30		
Coveralls	0.49		

[a] "Thin" refers to garments made of lightweight, thin fabrics often worn in the summer, "thick" refers to garments made of heavyweight, thick fabrics often worn in the winter.
[b] Knee-length dresses and skirts.
[c] Lined vests.
Source: ASHRAE 55-2020.

$$+\,3 \quad \text{hot}$$
$$+\,2 \quad \text{warm}$$
$$+\,1 \quad \text{slightly warm}$$
$$\text{PMV} = 0 \quad \text{neutral}$$
$$-\,1 \quad \text{slightly cool}$$
$$-\,2 \quad \text{cool}$$
$$-\,3 \quad \text{cold}$$

Example 6.1

A classroom has 50 students. The votes on the thermal comfort are: 5 hot, 12 warm, 15 slightly warm, 8 neutral, 5 slightly cool, 4 cool, and 1 cold. What is the overall PMV level in the classroom?

Solution

$\text{PMV} = [5 \times (+3) + 12 \times (+2) + 15 \times (+1) + 8 \times 0 + 5 \times (-1) + 4 \times (-2) + 1 \times (-3)]/50 = +0.72$ (this is between neutral and slightly warm)

The statistic of the vote is more accurate when the number of votes is bigger, which requires significant efforts to collect valid votes. In addition, a vote would only work for an existing indoor environment where people can sense their thermal comfort and provide a meaningful vote. For new construction where such an environment does not even exist, the voting approach would be impossible. Fanger thus correlated the PMV to the imbalanced heat L in Equation (6.15) to provide a mathematical approach to predicting the PMV for any indoor environment with given air properties, occupant activities, and clothing.

$$\text{PMV} = \left[0.0303 \times e^{-0.036 \times M} + 0.028\right] \times L \tag{6.16}$$

Test chamber tests with actual subjects were conducted by Fanger in1970. Studies on 1600 college-age students revealed certain interesting trends between comfort level, temperature, humidity, sex, and length of exposure. An empirical correlation has been subsequently developed:

$$\text{PMV} = a \times T_a + b \times p_v - c \tag{6.17}$$

where the numerical values of the coefficients a, b, and c are given in Table 6.4. The temperature T_a is in °C or °F, and the partial pressure of water vapor in the air (p_v) is in kPa or psi. In general, a change of 3 °C (5.4 °F) in temperature or a 3 kPa (0.44 psi) change in water vapor pressure is necessary to change a thermal sensation vote by one category.

Since women prefer slightly higher temperatures than men (some attribute this to metabolism rate while others to differences in clothing), separate correlations are provided for both sexes and also for a typical combined set of people. Note that duration of occupancy (expressed as "exposure period") is an additional factor.

Example 6.2

Calculate the difference in PMV for a combined group of men and women when exposed to their environment for 1 hour and 3 hours. Given: $T_a = 21$ °C, $T_{dew} = 12$ °C; Find PMV

Table 6.4 Coefficients for Equation (6.17).

Exposure period	Sex	SI units			IP units		
		a	b	c	a	b	c
1.0	Male	0.220	0.233	5.673	0.122	1.61	9.584
	Female	0.272	0.248	7.245	0.151	1.71	12.080
	Combined	0.245	0.248	6.475	0.136	1.71	10.880
3.0	Male	0.212	0.293	5.949	0.118	2.02	9.718
	Female	0.275	0.255	8.622	0.153	1.76	13.511
	Combined	0.243	0.278	6.802	0.135	1.92	11.122

Solution

Lookup values: The partial pressure corresponding to $T_{dew} = 12\,°C$ is $p_v = 1.402\,kPa$
Using Eq. (6.17) with the appropriate coefficients from Table 6.4, yields:

1-hour exposure: PMV $= 0.245 \times 21 + 0.248 \times 1.402 - 6.475 = -0.98$ (slightly cool)
3-hour exposure: PMV $= 0.243 \times 21 + 0.278 \times 1.402 - 6.802 = -1.31$ (slightly cool to cool)

Thus, subjects tend to feel cooler as their exposure period is increased.

While PMV is quantitative and useful for fundamental research in thermal comfort, the concept is not straightforward for designers, operators, and occupants to understand. Thus, another concept of predicted percentage dissatisfied (PPD) due to thermal comfort was introduced. The correlations between PMV and PPD are presented in Figure 6.4. In particular,

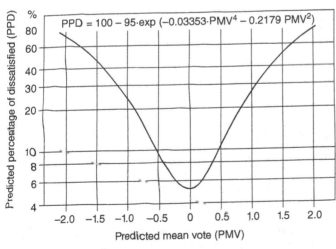

Figure 6.4 The correlations between PMV and PPD.

- PMV = 0 is equival ~~PPD~~ PPD = 5%, which means even with a neutral mean vote, there may be 5% of population who ~~ent~~ feel discomfort due to the diversity of personal sensation and preference.
- PMV = ±0.5 is ~~cial~~ ent to PPD = 10%, which is commonly used as a design threshold for most res-idential and c~~alent~~ buildings.
- PMV = ±1 is ~~alent~~ to PPD = 25%, which can be used for indoor spaces with less-strict thermal requiremen~~ch~~ as those with passive design features.

An empir~~formula~~ formula was created to quantitatively correlate PMV and PDD as follows:

$$PPD = 100 - 95 \times e^{\left(-0.03353 \times PMV^4 - 0.2179 \times PMV^2\right)} \tag{6.18}$$

Equation~~s~~ (6.1)–(6.18) can be (and have been) coded in various simulation software (e.g. building energy simulation and computational fluid dynamics programs) that can predict the indoor air properties. The results project the thermal comfort level in the space of interest under various weather, room, and system conditions.

6.3 Comfort Zone

Thermal comfort is the condition of mind that expresses satisfaction with the thermal environment. The task of air-conditioning systems is to maintain a thermally comfortable environment by simultaneous control of temperature, humidity, cleanliness, and air circulation. ASHRAE Standard 55 (Figure 6.5)

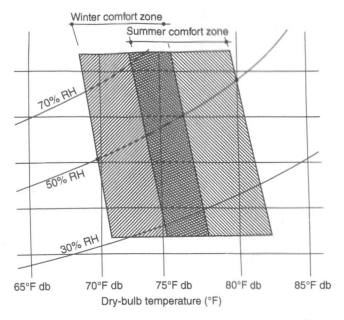

Figure 6.5 Winter and summer comfort zones for light activities in seasonal clothing (winter I_{cl} = 0.9 and summer I_{cl} = 0.5) with minimum air movement (V < 0.15 m/s for winter and V < 0.25 m/s for summer) and where dry-bulb air temperature and mean radiant temperature are equal. *Source:* ASHRAE (1981).

"Thermal Environmental Conditions for Human Occupancy" defines specified conditions for acceptable thermal environments and is intended for use in the design, operation, and commissioning of buildings and other occupied spaces.

The ASHRAE Standard 55 was first published in 1966 and is updated every 3–7 years based on current research, practical experience, and recommendations from designers, manufacturers, and end users. The most notable, as well as most recent, iterations of the standard are the 2004, 2010, and 2017 updated versions. Figures 6.5–6.8 show some historical evolutions of the thermal zones defined by ASHRAE as plotted on psychrometric charts. Figure 6.5 shows slightly shifted winter and summer comfort zones for light activities in seasonal clothing (winter $I_{cl} = 0.9$ and summer $I_{cl} = 0.5$) with minimum air movement ($V < 0.15$ m/s for winter and $V < 0.25$ m/s for summer) and where dry-bulb air temperature and mean radiant temperature are equal. Figure 6.6 shows the combined winter and summer comfort zones using the concept of effective temperature (ET). ET combines temperature and humidity in one index. It is the operative temperature (OT) of an enclosure at relative humidity $\varphi = 50\%$ that would cause the same sensible plus latent heat exchange from a person as would the actual environment. OT is commonly used to define the comfort zones, which combines convection and radiation in one index. Figures 6.7 and 6.8 show the comfort zones in 1997/2001 and 2004 using OT. OT (T_o) is what humans experience thermally in a space under the influence of the mean radiant temperature (T_{mr}) and air temperature (T_{db}). Usually, the radiative heat transfer coefficient (h_r) is at the same magnitude as the convective heat transfer coefficient (h_c), therefore,

$$T_o = (h_r T_{mr} + h_c T_{db})/(h_r + h_c) \approx (T_{mr} + T_{db})/2 \tag{6.19}$$

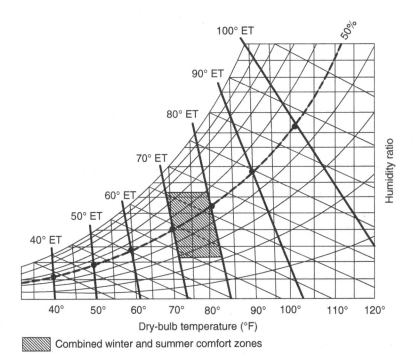

Combined winter and summer comfort zones

Figure 6.6 Combined winter and summer comfort zones using effective temperature. *Source:* ASHRAE (1989).

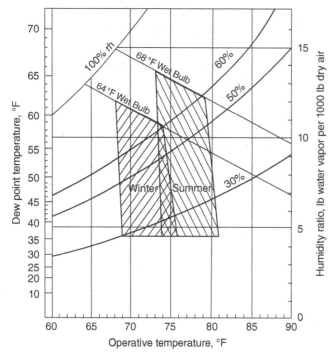

Figure 6.7 Winter and summer comfort zones using operative temperature. *Source:* ASHRAE (1997/2001).

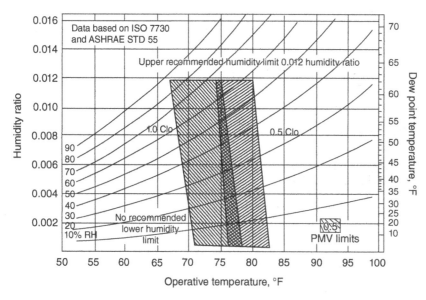

Figure 6.8 Winter and summer comfort zones using operative temperature. *Source:* ASHRAE (2004).

Table 6.5 shows the ASHRAE 55-2013 suggested optimal OT and acceptable range for light sedentary activity at 50% relative humidity and at mean velocity ≤0.15 m/s (30 ft/min).

The most recent ASHRAE 55-2020 determines the acceptable thermal environments in occupied spaces, according to average air speed in the spaces, with 0.20 m/s (40 fpm) as the threshold (Table 6.6). Figure 6.9 shows the comfort zones in SI and IP units, respectively, using the OT. The figures assume a metabolic rate of 1.3 met and clothing values of 1.0 and 0.6 (for two zones), as well as an average air speed of 20 fpm. There is no upper or lower limit on humidity. For other environmental and personal conditions, verification using a computer tool or alternative methods is required to achieve $-0.5 \leq PMV \leq +0.5$. The Center for the Built Environment (CBE) thermal comfort tool developed by the University of California, Berkeley (https://comfort.cbe.berkeley.edu/) is permitted by ASHRAE for compliance checks. Figure 6.10 shows a snapshot of the CBE thermal comfort tool.

The Elevated Air Speed Comfort Zone Method uses the Analytical Comfort Zone Method combined with the Standard Effective Temperature (SET) Model. Figure 6.11 presents a graphical example of a comfort zone using the Elevated Air Speed Comfort Zone Method compared to the one using the Analytical Comfort Zone Method.

ASHRAE Standard 55 also includes a separate adaptive model for determining acceptable thermal conditions in occupant-controlled naturally conditioned spaces. Users are now allowed to use the adaptive

Table 6.5 Optimal operative temperature and acceptable range for light sedentary activity at 50% relative humidity and at mean velocity ≤0.15 m/s (30 ft/min).

Season	Typical clothing	I_{ct} [clo]	Optimum operative temperature	Acceptable range
Winter	Heavy slacks, long-sleeve shirt, and sweater	0.9	22 °C	20–23.5 °C
			71 °F	68–75 °F
Summer	Light slacks and short-sleeve shirt	0.5	24.5 °C	23–26 °C
			76 °F	73–79 °F
	Minimal	0.05	27 °C	26–29 °C
			81 °F	79–84 °F

Source: ASHRAE (2013).

Table 6.6 Applicability of methods for determining acceptable thermal environments in occupied spaces.

Average air speed, m/s (fpm)	Humidity ratio	met	clo	Comfort zone method
<0.20 (40)	All	1.0–2.0	0–1.5	Section 5.3.1, "Analytical Comfort Zone Method"
>0.20 (40)	All	1.0–2.0	0–1.5	Section 5.3.2, "Elevated Air Speed Comfort Zone Method"

Source: ASHRAE 55-2020.

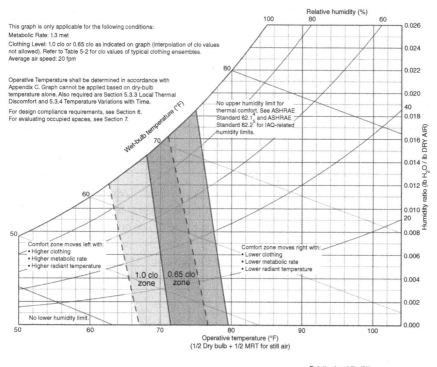

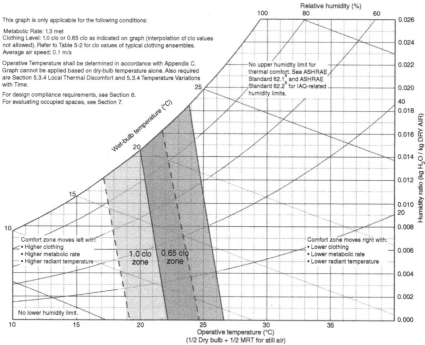

Figure 6.9 Examples of comfort zones using the Analytical Comfort Zone Method (in SI and IP units). *Source:* ASHRAE 55-2020.

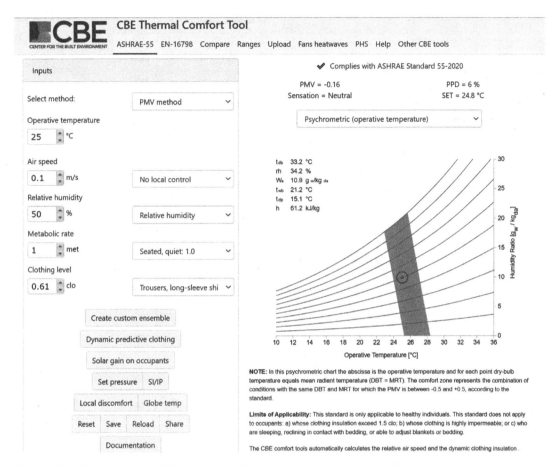

Figure 6.10 Snapshot of the CBE thermal comfort tool.

model when an air-conditioning unit is installed but not operating. The allowed indoor OT shall be determined from Figure 6.12 using the 80% acceptability limits (the 90% limits are included for information only), or using the following equations:

$$\text{Upper 80\% acceptability limit } (^{\circ}\text{C}) = 0.31\overline{T_{\text{pma(out)}}} + 21.3 \tag{6.20}$$

$$\text{Upper 80\% acceptability limit } (^{\circ}\text{F}) = 0.31\overline{T_{\text{pma(out)}}} + 60.5 \tag{6.21}$$

$$\text{Lower 80\% acceptability limit } (^{\circ}\text{C}) = 0.31\overline{T_{\text{pma(out)}}} + 14.3 \tag{6.22}$$

$$\text{Lower 80\% acceptability limit } (^{\circ}\text{F}) = 0.31\overline{T_{\text{pma(out)}}} + 47.9 \tag{6.23}$$

where $\overline{T_{\text{pma(out)}}}$ is the prevailing mean outdoor air temperature.

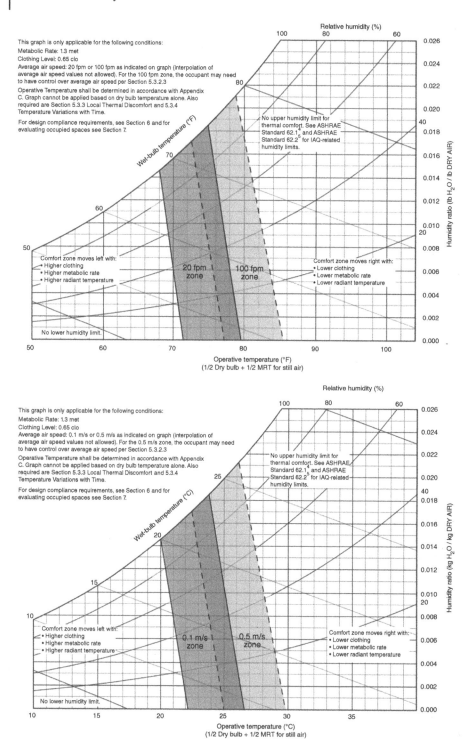

Figure 6.11 Examples of comfort zones using the Elevated Air Speed Comfort Zone Method (in SI and IP units). Source: ASHRAE 55-2020.

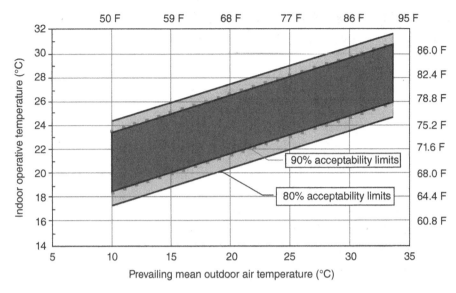

Figure 6.12 Acceptable indoor operating temperature ranges for naturally conditioned spaces.

6.4 Approaches to Improving Indoor Thermal Comfort

As indicated arlier, the main indoor comfort parameters include:

- **Air temperature**: T_{air} that affects convection
- **Relative humidity**: ϕ that affects evaporation
- **Air speed near a human body**: V that affects convection and evaporation
- **Mean radiant temperature**: T_{mrt} that affects radiation
- **Clothing level**: I_{cl}
- **Activity**: M

Variation of these parameters can play significant and integrated roles in regulating indoor comfort. Among these parameters, I_{cl} and M are subjective and up to the occupant activities and choices, while the others are objective and can be adjusted and controlled by building thermal systems.

Figure 6.13 illustrates the influence of increased air speed to the comfort zone, which moves or extends the comfort zone to the right (a higher temperature) but with a constant humidity ratio. Air speed increases both the mass and heat convective heat transfer coefficients and thus enhances the convective and evaporative heat transfer from the body to the environment. Obviously, this is favorable for the summer, which allows for a higher indoor air temperature, but unfavorable for the winter. Similarly, increasing the surface temperature (mean radiant temperature) increases the radiative heat transfer and requires a lower surrounding air temperature. Thus, the comfort zone is extended or moved to the left (a lower temperature) with a constant humidity ratio (as no water vapor is added to or removed from the air) (Figure 6.14).

Different clothing levels and human activities also affect the thermal comfort sensation, as anticipated. For different I_{cl} values and met values between 1.2 and 3, ASHRAE 55-2013 recommends that the comfort OT be determined from:

$$T_{o, corrected} = 27.2 - 5.9 \times I_{cl} - 3.0 \times (1.0 + I_{cl}) \times (M - 1.2) \, (^{\circ}C) \tag{6.24}$$

$$T_{o, corrected} = 81.0 - 10.6 \times I_{cl} - 5.4 \times (1.0 + I_{cl}) \times (M - 1.2) \, (^{\circ}F) \tag{6.25}$$

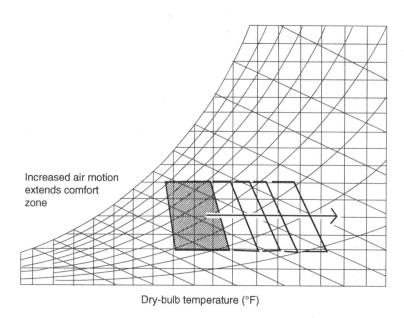

Increased air motion extends comfort zone

Dry-bulb temperature (°F)

Figure 6.13 Influence of increasing air motion to comfort zone.

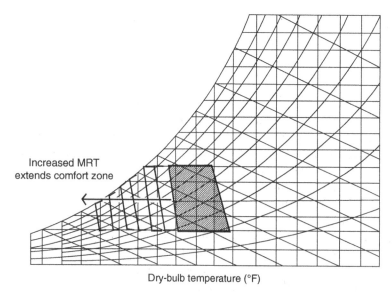

Increased MRT extends comfort zone

Dry-bulb temperature (°F)

Figure 6.14 Influence of increasing mean radiant temperature to comfort zone.

Example 6.3

Consider a gymnasium where the metabolic rate of the occupants is 3.0 and $I_{cl} = 0.3$. If the occupants are to experience the same level of comfort as when sedentary, what should be the operative temperature in this space? Given: $M = 3.0$, $I_{cl} = 0.3$; Find $T_{o,active}$

Solution

From Eq. (6.24): $T_{o,active} = 27.2 - 5.9 \times 0.3 - 3.0 \times (1.0 + 0.3) \times (3.0 - 1.2) = 18.4\,^{\circ}C$ (65.1 °F)

Note that the gymnasium need not be kept at this low temperature since typically people do not stay longer than about an hour and do not exercise continuously, while the ASHRAE comfort chart applies to a steady state (occupancy of 3 hours).

6.5 Other Thermal Comfort Factors

6.5.1 Draft

Draft is the unwanted, local cooling of a body caused by air movement. Draft can be a big concern when supplying a cold stream of air into occupied zones. The feeling of draft is determined by multiple air properties including air temperature, speed, and turbulence, as well as personal thermal sensation level. Sensitivity to draft is greatest on body parts not covered by clothing, especially the head, shoulder, and foot regions.

The percentage dissatisfied (PD) people due to draft (%) can be expressed as:

$$PD = 3.143 \times (34 - T_{air}) \times (V - 0.05)^{0.622} + 0.3696 \times (34 - T_{air}) \times (V - 0.05)^{0.622} \times V \times Tu \quad (6.26)$$

where T_{air} is the dry-bulb air temperature (°C), V is the air velocity (m/s), and Tu is the air turbulence intensity (%) that can be expressed as:

$$Tu = 100 \times V_{standard\ deviation}/V \quad (6.27)$$

When $V < 0.05$ m/s, use $V = 0.05$ m/s; when PD > 100%, assign PD = 100%.

Equation (6.26) indicates that draft is a concern when air temperature is below 34 °C and air velocity is above 0.05 m/s. Turbulence intensity can also increase the draft effect, which is directly related to the type of supply diffuser used. Avoiding the direct supply of cooling air toward an occupant's position, e.g. by adjusting the supply angle, is a major consideration when mitigating the impact of drafts. Keeping a proper range of supply air temperatures is another critical aspect in designing a thermally comfortable indoor environment.

6.5.2 Asymmetry of Radiation

The air environments of multiple spaces can be quickly mixed via convection. The environment is thus mostly uniform (especially in small spaces) for thermal sensation. However, unidirectional radiation may cause significant asymmetry that may lead to strong discomfort, such as the radiation from cold and hot windows. The following presents a severity ranking list due to asymmetry of radiation:

- Warming ceiling (****)
- Cool wall (***)

- Cool ceiling (∗∗)
- Warm wall (∗)

Avoiding a large surface temperature gradient is the primary remedy for addressing the issue, when possible. This includes installing shading devices for hot summer windows and providing movable insulations for cold winter windows. Adjusting the distance from heating/cooling sources to the receiver (i.e. the radiation view factor) is another effective method. Table 6.7 lists the allowable radiant temperature asymmetry from ASHRAE 55-2020. Figure 6.15 shows the local thermal discomfort levels caused by radiant asymmetry.

6.5.3 Thermal Stratification

Thermal stratification is commonly experienced in spaces with high ceilings and those with special heating and cooling systems (e.g. displacement ventilation system). Significant thermal stratification between the head

Table 6.7 Allowable radiant temperature asymmetry

Radiant temperature asymmetry °C (°F)			
Ceiling warmer than floor	**Ceiling cooler than floor**	**Wall warmer than air**	**Wall cooler than air**
<5 (9.0)	<14 (25.2)	<23 (41.4)	<10 (18.0)

Source: From ASHRAE 55-2020.

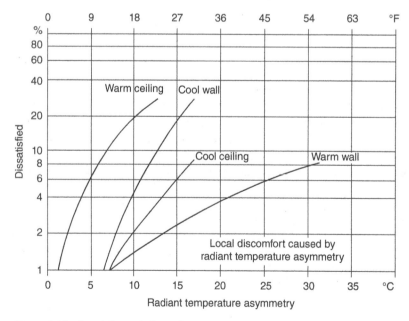

Figure 6.15 Local thermal discomfort caused by radiant asymmetry. *Source:* ASHRAE 55-2020.

and feet can cause notable discomfort. A warm head is even worse than warm feet. ASHRAE 55-2020 states that the air temperature difference between head level and ankle level (at 0.1 m or 4 inch above the floor) should not exceed 3 °C (5.4 °F) for seated occupants or 4 °C (7.2 °F) for standing occupants. Figure 6.16 shows the local thermal discomfort caused by vertical temperature differences (ASHRAE 55-2020).

6.5.4 Thermal Variations with Time

The indoor air temperature can be either passively or actively fluctuating. The temperature drift or ramp should meet some requirements to avoid noticeable discomfort. Cyclic variations in OT that have a period of less than 15 minutes shall have a peak-to-peak amplitude not greater than 1.1 °C (2.0 °F). Noncyclic changes in OT and cyclic variations with a period greater than 15 minutes shall not exceed the most restrictive requirements from Table 6.8.

6.5.5 Floor Surface Temperature

Considering seated occupants, whose feet contact the floor, the floor surface temperature within the occupied zone shall be within the range of 19–29 °C (66.2–84.2 °F) (based on 10% dissatisfied people). Figure 6.17

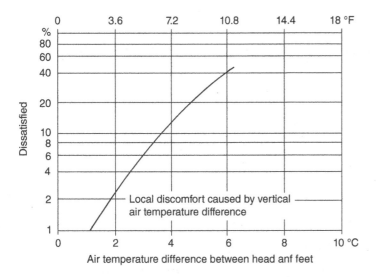

Figure 6.16 Local thermal discomfort caused by vertical temperature differences. *Source:* ASHRAE 55-2020.

Table 6.8 Limits on temperature drifts and ramps.

Time period, h	0.25	0.5	1	2	4
Maximum operative temperature t_o Change allowed, °C (°F)	1.1 (2.0)	1.7 (3.0)	2.2 (4.0)	2.8 (5.0)	3.3 (6.0)

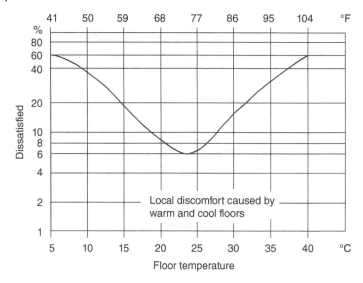

Figure 6.17 Local discomfort caused by floor temperature.

shows the percentage of people who are wearing lightweight indoor shoes and are dissatisfied due to the floor temperature. This figure does depict people with bare feet or those who are sitting on the floor.

Homework Problems

1   To save energy, an office building cooling system is controlled so that the minimum dry-bulb temperature in the space is 28 °C with a dewpoint temperature of 19 °C.
   (a) Comment on the comfort expectations for the space.
   (b) What general type of clothing would you recommend?
   (c) Assuming that the dry-bulb temperature is fixed, what could be done to improve comfort conditions?

2   Consider a room at sea level that measures 20 ft by 20 ft with 8 ft ceiling height. One wall, measuring 8 ft by 20 ft, is all glass and has an interior surface temperature of 50 °F. All other surfaces are at 65 °F. The room air is at 70 °F and 30% RH.
   (a) What are the mean radiant temperature and the operative temperature?
   (b) Will condensate occur on the window glass?
   (c) Discuss whether the conditions are within ASHRAE comfort limits.

3   Consider a relaxingly standing person of 75 kg weight and 1.80 m height with walking shorts and short-sleeve shirt. The room conditions are 72 °F DB and 70% RH. Calculate the PMV and PPD of the person. Compare the results against those obtained using the CBE Thermal Comfort Tool (under ASHRAE-55). Make assumptions as necessary.

4 Staged Project Assignment (Teamwork)

   (a) Check ASHRAE-55 to determine and justify your indoor design conditions for both summer and winter.

   (b) Design the building geometry and layout (basic drawings for deliverables with functions and circulations described and justified).

   (c) Determine building orientation with justifications (e.g., view, convenience, solar, wind, etc.).

References

ASHRAE (1981). *ASHRAE Standard 55: Thermal Environmental Conditions for Human Occupancy*. Atlanta: ASHRAE.

ASHRAE (1989). *ASHRAE Standard 55: Thermal Environmental Conditions for Human Occupancy*. Atlanta: ASHRAE.

ASHRAE (1997). *ASHRAE Standard 55: Thermal Environmental Conditions for Human Occupancy*. Atlanta: ASHRAE.

ASHRAE (2001). *ASHRAE Standard 55: Thermal Environmental Conditions for Human Occupancy*. Atlanta: ASHRAE.

ASHRAE (2004). *ASHRAE Standard 55: Thermal Environmental Conditions for Human Occupancy*. Atlanta: ASHRAE.

ASHRAE (2013). *ASHRAE Standard 55: Thermal Environmental Conditions for Human Occupancy*. Atlanta: ASHRAE.

ASHRAE (2020). *ASHRAE Standard 55: Thermal Environmental Conditions for Human Occupancy*. Atlanta: ASHRAE.

7

Indoor Air Quality, Ventilation, and Infiltration

7.1 Indoor Air Quality

The purpose of HVAC systems is to create an ideal indoor environment with appropriate air speed, temperature, humidity, and contaminant concentrations, since people usually spend 80–90% of their time indoors. However, the improper design and use of building HVAC systems not only waste energy but also cause thermal discomfort and indoor air quality (IAQ) problems. For example, an inappropriate installation location of a supply air diffuser may cause "old" air to be circulated in an occupied zone. Reports of sick building syndrome (SBS) and other health complaints related to indoor environments have been increasing recently. Evidence from the literature (https://www.cdc.gov/niosh/topics/indoor-env/default.html) shows that poor indoor environments significantly increase the rate of respiratory illness, allergy and asthma symptoms, and sick building symptoms; consequently, worker performance is adversely affected (Figure 7.1).

SBS is a condition in which people in a building suffer from symptoms of illness or become infected with chronic disease from the building in which they work or reside. SBS is often comparable to a cold or influenza in terms of symptoms which include headaches, drowsiness, eye irritation, and nose and throat infection. The major distinction between SBS and a cold is that a cold typically lasts for 1–2 weeks even without treatments, while SBS can last for months and seasons and won't change if the working or living environments are not altered.

The majority of studies indicate an average productivity loss of 10% due to poor indoor environments, although a conservative value of 6% is widely accepted (Dorgan et al. 1998). The overall economic losses due to poor indoor environments in US commercial buildings are estimated to be about $40–$160 billion per year (Fisk 2000) in lost wages and productivity, administrative expenses, and health care costs.

7.1.1 Causes of Sickness

The causes of SBS and IAQ issues can generally be categorized into three groups:

- **Toxic gases:** CO_x, NO_x, NO_x, radon, Ozone, etc. from various indoor and outdoor sources.
- **Volatile organic compounds (VOCs):** gases containing a variety of chemicals emitted from liquids or solids, such as formaldehyde, benzene, or ethylene glycol.
- **Particulate matters:** carbonaceous particles in association with adsorbed organic chemicals and reactive metals; depending on the particle size, are generally classified into (i) coarse particles, PM10 of

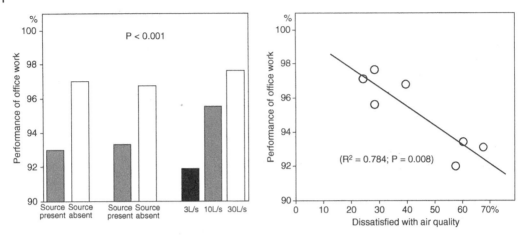

Figure 7.1 Performance influence of IAQ.

diameter <10 µm; (ii) fine particles, PM2.5 of diameter <2.5 µm; and (iii) ultrafine particles, PM0.1 of diameter <0.1 µm.

Other categorizations may classify indoor air pollutants into biologic and chemical categories, and some further into organic, inorganic, biological, or radioactive categories. Biologic pollutants may include pollen, mold, various aerosols (e.g. virus and bacteria), and biologic agents (e.g. VX). Table 7.1 shows the potential sources, permitted levels, and health effects of the three types of contaminants. The health effects of indoor contaminants are related to not only the contaminant concentration but also the exposure time. Table 7.2 presents the regulated concentration limits for major indoor air pollutants with variable exposure times. Different thresholds are set for coarse and fine particles due to different exposure risks and health impacts (Table 7.3).

It should be noted that some contaminants (e.g. VOCs) may last for a very long period of time in a space (Figure 7.2). This is especially true when the contaminant sources (e.g. furniture for VOCs) are not

Table 7.1 Sources, permitted levels, and health effects of IAQ contaminants.

Contaminants	Sources	Permitted level	Health effects
CO_2	Human, combustion	1000 ppm	Stuffing
CO	Combustion, ETS	15 ppm	Body chemistry
SO_x	Combustion		Irritation, asthma
NO_x	Combustion	100 mg/m³	Not very clear
Ra	Soil	4 picocuries/l	Lung cancer
VOCs (Formaldehyde)	Combustion, pesticides, building materials, etc.	0.1 ppm	Eyes and mucous membrane irritation
Particulate matters	Outdoor air, activities, ETS, furnishings, pets, etc.		Lung diseases Cancer (ETS)

Table 7.2 Indoor air quality regulations for major indoor air pollutants.

Pollutants	Concentration levels (mg/m^3)	Exposure time	Organization
CO	100	15 min	WHO
	60	30 min	
	30	1 h	
	10	8 h	
	29	1 h	USEPA
	10	8 h	
CO_2	1800	1 h	WHO
	0.4	1 h	
NO_2	0.15	24 h	WHO
	0.1	1 yr	USEPA
	0.15	24 h	
PM	0.05	1 yr	USEPA
	0.15–0.2	1 h	WHO
O_3	0.1–0.12	8 h	
	0.235	1 h	USEPA
SO_2	0.5	10 min	WHO
	0.35	1 h	
	0.365	24 h	USEPA
	0.08	1 yr	
Pb	0.0005–0.001	1 yr	WHO
	0.0015	3 mo	USEPA
Xylene	8	24 h	WHO
Formaldehyde	0.1	30 min	WHO
Radon	100 Bq/m^3	1 yr	WHO

Source: From European Commission DG XVII: https://www.europeansources.info/corporate-author/europeancommission-dg-xvii/.

Table 7.3 PM10 and 2.5 standards.

$\mu g/m^3$	Standard's issue	24 h	Annual
PM10	WHO 2006	50 Day average	20 Year average
PM10	EU 2008–2010	50 90.4th percentile over 1 yr	40 Year average
PM10	USA 2012	150 99.9th percentile over 3 yr	
PM2.5	WHO 2006	25	10
PM2.5	EU 2008–2010		25 Target to become a limit in 2015
PM2.5	USA 2012	35 98th percentile over 3 yr	12 primary 15 secondary over 3 yr

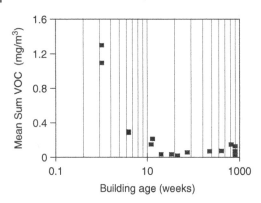

Figure 7.2 Sum VOC from EPA public buildings study (by age of the building).

removed. Cleaning building surfaces and indoor spaces with chemical cleaning products may significantly increase the VOC and other IAQ problems.

7.1.2 Control of Indoor Contaminants

Three general approaches are used to control indoor contaminants:

- **Source control:** this is the first important and most effective method to control indoor pollutants. Identifying contaminant sources (either indoor or outdoor) can help determine proper mitigation measures, e.g. to remove, isolate, dilute, and/or clean the sources. Selecting low-VOC emitting materials such as paint, furniture, hard-wood floors, etc. can effectively reduce indoor pollutant sources in the first place. Using non-VOC cleaning solvents and products can eliminate the risk of secondary contamination. Planning house layouts with kitchens and bathrooms located downwind can avoid the cross-flow of contaminants through the whole house and remove indoor-generated contaminants (e.g. from cooking) rapidly to the outdoors or to exhaust vents.
- **Ventilation:** diluting the contaminated air with fresh/clean air is the fastest way to reduce the contaminant concentration in confined spaces. This can be conducted for spaces with known or unknown sources of contaminants. Ventilation can be supplied passively (e.g. through open windows, called natural ventilation), actively (e.g. through mechanical systems), or in a hybrid way. When the outdoor air is clean, it can be used directly for ventilation; on the contrary, dirty outdoor air needs be cleaned before it can be supplied into indoor spaces or cannot be used at all due to the elevated contamination risk. This is true for most particulate matter controls where outdoor air is the primary source for indoor particles. Effective ventilation requires adequate airflow rates, appropriate supply air temperature and humidity, and proper layouts of supply inlets and exhaust outlets. The effectiveness of a ventilation system also relates to the source conditions (e.g. nature, location, time period, release pattern, strength and frequency, deposition and sorption, etc.).
- **Air purification:** purifying air is necessary to achieve clean indoor air. This is important for both cleaning the contaminated indoor air and supplying clean outdoor/ventilation air. Different techniques have been developed to clean different contaminants such as using HEPA (high efficiency particulate air) filters to remove particles and using photocatalytic oxidation devices to treat VOCs. A variety of techniques are available to handle one type of contaminant; for instance, electrostatic precipitation can also be applied to remove indoor particles. Purification devices can be installed in central air systems and can also be used alone as mobile/portable units. When choosing suitable air cleaning techniques and products, one has to consider multiple factors such as safety (e.g. for UV light), energy cost

(e.g. for better HEPA), and environmental impacts (e.g. byproducts of electrostatic precipitation), besides just the purification performance.

7.2 Ventilation

Ventilation purposely delivers the required amount of clean/fresh air to the spaces of interest. Ventilation air not only supplies the oxygen required for living but also dilutes and removes various contaminants including CO_2 exhaled from the human body and particles released from combustion. It also exhausts moisture from plants, cooking, and bathing. This is extremely important for energy efficient buildings that tend to be built air tightly in order to reduce the undesired heat gain and loss through infiltration and exfiltration.

Based on the supply mechanisms, ventilation can be divided into natural ventilation and mechanical ventilation. The former counts on natural-driven forces such as natural wind and thermal buoyancy due to air temperature gradient, while the latter uses mechanical fans to circulate air throughout a building. In practice, combined ventilation, called hybrid ventilation, is commonly designed, where natural ventilation is used when there is adequate wind and/or buoyancy and mechanical ventilation is turned on when natural forces are not sufficient. The mode switch is often automatically operated by monitoring indices such as indoor CO_2 concentration, indoor–outdoor temperature differences, outdoor air temperature and humidity, and outdoor wind speed and rain conditions, etc.

Standards have been developed in regulating minimum ventilation rates for different types of buildings and spaces. Table 7.4 summarizes the ventilation requirements for residential buildings in a few countries based on a floor area of 100 m^2 (1000 ft^2). The prevalent ventilation standards are ANSI/ASHRAE Standard 62.1 "Ventilation for Acceptable Indoor Air Quality" and 62.2 "Ventilation and Acceptable Indoor Air Quality in Residential Buildings."

First published in 1973 as Standard 62, Standard 62.1 guides the improvement of IAQ in new and existing buildings by specifying minimum ventilation rates and other measures for new and existing buildings. This intends to provide IAQ that is acceptable to human occupants and that minimizes adverse health effects. Standard 62.1 has been fully revised for the first time since 2004 to include three procedures for ventilation design: the IAQ Procedure, the Ventilation Rate Procedure, and the Natural Ventilation Procedure. Standard 62.2, first sent out for public review in 1999, defines the roles of and minimum requirements for mechanical and natural ventilation systems and the building envelope, intended to provide acceptable IAQ in residential buildings. This chapter is focused on mechanical ventilation while natural ventilation is included in Chapter 10 "Passive Building Systems." The contents and methods in Standard 62.1, related to mechanical ventilation, are cited in this chapter. Figure 7.3 illustrates the basic structure of the ASHRAE Standard 62.1-2007. Using a similar structure, the standards have undergone key changes over the years, reflecting the ever-expanding knowledge, experience, and research related to ventilation and air quality. Two procedures are introduced to specify minimum ventilation requirements: the ventilation rate procedure and the indoor air quality procedure (IAQP).

7.2.1 Ventilation Rate Procedure (VRP)

The ventilation rate procedure describes the minimum ventilation (outdoor) rate requirements for different spaces. Table 7.5 shows the evolution of the requirements from 2001 to 2007 to 2019 for office buildings. Example 7.1 demonstrates how to use these standards to specify the minimum ventilation rates. In

Table 7.4 Ventilation requirements for residential buildings on a floor area of 100 m² (1000 ft²).

Country (standard)	General (whole dwelling)	Bedroom (10 m²)	Kitchen (7 m²)	Bathroom (5 m²)
Canada (CSA Standard F326)	0.3 ACH	10 l/s (double)	30 l/s (continuous)	15 l/s (continuous) 25 l/s (intermittent)
France (CSTB)	0.5 ACH	N/A	12.5–25 l/s	8.3–16.6 l/s
Netherlands (NEN 1087)	N/A	10 l/s	21–28 l/s	14 l/s
Norway (BF)	N/A	N/A	22 l/s	16 l/s
Sweden (SBN 1980)	0.5 ACH	N/A	10 l/s	10 l/s
Sweden (National Code)	N/A	N/A	15 l/s	10 l/s
Switzerland (SIA 380)	N/A	N/A	22–33 l/s	17 l/s
UK (Building regulations)	N/A	6 l/s-single 12 l/s-double	7–30 l/s	10 l/s
U.S. (ASHRAE 62-1999)	0.35 ACH, not less than 15 cfm (8 l/s) per person	N/A	20 cfm (12 l/s) (continuous)	50 cfm (25 l/s) (intermittent) 20 cfm (10 l/s) (continuous)

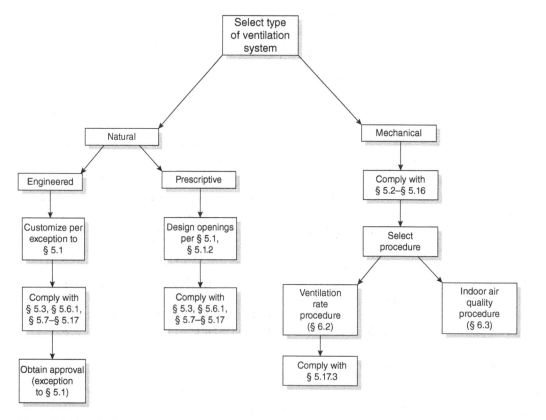

Figure 7.3 Ventilation requirements. *Source:* Based on ASHRAE Standard 62.1-2007.

Table 7.5 Minimum ventilation rate requirements for office buildings.

(a) ASHRAE 62.1-2001

Application	Estimated maximum occupancy P/1000 ft² or 100 m²	Outdoor air requirements				Comments
		cfm/person	l/s·person	cfm/ft²	l/s·m²	
Offices						
Office space	7	20	10			Some office equipment may require local exhaust
Reception areas	60	15	8			
Telecommunication centers and data entry areas	60	20	10			
Conference rooms	50	20	10			

(b) ASHRAE 62.1-2007

Occupancy category	People outdoor air rate R$_p$		Area outdoor air rate R$_a$			Default values			
						Occupant density (see Note 4)	Combined outdoor air rate (see Note 5)		Air class
	cfm/person	l/s·person	cfm/ft²	l/s·m²	Notes	#/1000 ft² or #/100 m²	cfm/person	l/s·person	
Office buildings									
Office space	5	2.5	0.06	0.3		5	17	8.5	1
Reception areas	5	2.5	0.06	0.3		30	7	3.5	1
Telephone/data entry	5	2.5	0.06	0.3		60	6	3.0	1
Main entry lobbies	5	2.5	0.06	0.3		10	11	5.5	1

(c) ASHRAE 62.1-2019

Occupancy category	People outdoor air rate R$_p$		Area outdoor air rate R$_a$		Default values occupant density	Air class	OS (6.2.6.1.4)
	cfm/person	l/s·person	cfm/ft²	l/s·m²	#/1000 ft² or #/100 m²		
Office buildings							
Breakrooms	5	2.5	0.12	0.6	50	1	
Main entry lobbies	5	2.5	0.06	0.3	10	1	✓
Occupiable storage rooms for dry materials	5	2.5	0.06	0.3	2	1	
Office space	5	2.5	0.06	0.3	5	1	✓
Reception areas	5	2.5	0.06	0.3	30	1	✓
Telephone/data entry	5	2.5	0.06	0.3	60	1	✓

fact, Standard 62.1-2019 calls the calculated V, in the example, the minimum breathing zone outdoor airflow rate V_{bz}. To obtain the minimum zone outdoor airflow rate, zone air distribution effectiveness E_z should be considered.

$$V_{oz} = V_{bz}/E_z \tag{7.1}$$

Table 7.6 lists the zone air distribution effectiveness E_z for different air distribution systems.

The ventilation rate procedure is the most widely used method in practice to determine the minimum space ventilation requirements, attributed to its perceived simplicity (thus reduced design time) and perceived risk reduction (thus avoided hassles such as lawsuits). It was developed for and is applicable to

Table 7.6 Zone air distribution effectiveness E_z for different air distribution systems.

Air distribution configuration	E_z
Well-mixed air distribution systems	
Ceiling supply of cool air	1.0
Ceiling supply of warm air and floor return	1.0
Ceiling supply of warm air 15 °F (8 °C) or more above space temperature and ceiling return	0.8
Ceiling supply of warm air less than 15 °F (8 °C) above average space temperature where the supply air-jet velocity is less than 150 fpm (0.8 m/s) within 4.5 ft (1.4 m) of the floor and ceiling return	0.8
Ceiling supply of warm air less than 15 °F (8 °C) above average space temperature where the supply air-jet velocity is equal to or greater than 150 fpm (0.8 m/s) within 4.5 ft (1.4 m) of the floor and ceiling return	1.0
Floor supply of warm air and floor return	1.0
Floor supply of warm air and ceiling return	0.7
Makeup supply outlet located more than half the length of the space from the exhaust, return, or both	0.8
Makeup supply outlet located less than half the length of the space from the exhaust, return, or both	0.5
Stratified air distribution systems	
Floor supply of cool air where the vertical throw is greater than or equal to 60 fpm (0.25 m/s) at a height of 4.5 ft (1.4 m) above the floor and ceiling return at a height less than or equal to 18 ft (5.5 m) above the floor	1.05
Floor supply of cool air where the vertical throw is less than or equal to 60 fpm (0.25 m/s) at a height of 4.5 ft (1.4 m) above the floor and ceiling return at a height less than or equal to 18 ft (5.5 m) above the floor	1.2
Floor supply of cool air where the vertical throw is less than or equal to 60 fpm (0.25 m/s) at a height of 4.5 ft (1.4 m) above the floor and ceiling return at a height greater than 18 ft (5.5 m) above the floor	1.5
Personalized ventilation systems	
Personalized air at a height of 4.5 ft (1.4 m) above the floor combined with ceiling supply of cool air and ceiling return	1.40
Personalized air at a height of 4.5 ft (1.4 m) above the floor combined with ceiling supply of warm air and ceiling return	1.40
Personalized air at a height of 4.5 ft (1.4 m) above the floor combined with a stratified air distribution system with nonaspirating floor supply devices and ceiling return	1.20
Personalized air at a height of 4.5 ft (1.4 m) above the floor combined with a stratified air distribution system with aspirating floor supply devices and ceiling return	1.50

typical indoor environments with common contaminant sources and strengths, based on a reasonable level of adapted occupant satisfaction (i.e. 80%). The procedure is most commonly used for code enforcement (e.g. for LEED certification).

Example 7.1

An office space with a floor area of 1000 ft^2 is designed for 15 staff members, conditioned with a well-mixed air distribution system. Using the ventilation rate procedure,

1) What is the minimum amount of outdoor air required by ASHRAE Standard 62.1-2001?
2) What is the minimum amount of outdoor air required by ASHRAE Standard 62.1-2019?

Solution

1) V = 15 person × 20 cfm/person = 300 cfm (or 15 person × 10 l/s·person = 150 l/s)

V = 15 person × 5 cfm/person + 1000 ft^2 × 0.06 cfm/ft^2 = 135 cfm (or 15 person × 2.5 l/s · person + 100 m^2 × 0.3 l/s · m^2 = 67.5 l/s)

7.2.2 Indoor Air Quality Procedure (IAQP)

Space ventilation rate requirements can also be determined by calculating the mass conservation of a particular species of contaminant in a confined space (Figure 7.4):

$$VC_I + N = V C_R \tag{7.2}$$

where V is the total ventilation supply air (ft$_a$^{3}/min, or m$_a$^{3}/s), C_I is the contaminant concentration of the supplied air (ft$_c$^{3}/ft$_a$^{3}, or m$_c$^{3}/m$_a$^{3}), N is the contaminant sources generated within the space (ft$_c$^{3}/min, or m$_c$^{3}/s), and C_R is the average contaminant concentration in the space (ft$_c$^{3}/ft$_a$^{3}, or m$_c$^{3}/m$_a$^{3}). Equation (7.2) assumes a steady state perfect-mixing indoor environment, where the room shares the same air properties including contaminant concentration. Therefore, the exhausted/returned contaminant concentration is

Figure 7.4 Mass conservation of a contaminant in a confined space.

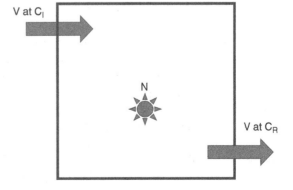

the same as the space contaminant concentration, both represented as C_R. The procedure for determining the ventilation rate V using Equation (7.2) is called the IAQP.

To use the IAQP requires prior knowledge of specific contaminant(s). This includes:

- The concentration of contaminant(s) in the ambient or supply air (e.g. 400 ppm for CO_2 in typical outdoor environments);
- The sources and generation rates of contaminant(s) in indoor environments (e.g. exhaled CO_2 at 5 ml/s·person);
- The allowed (maximum) concentration of contaminant(s) in specific indoor environments (e.g. 1000 ppm for CO_2 in typical indoor environments).

With such information, the required ventilation rate can be calculated:

$$V = N/(C_R - C_I) \tag{7.3}$$

Example 7.2 demonstrates the application of the IAQP.

Example 7.2

An office space with a floor area of 1000 ft^2 is designed for 15 staff members, conditioned with a well-mixed air distribution system. What is the minimum amount of outdoor air required if using the indoor air quality procedure and using CO_2 as the contaminant indicator? Assume each person is producing carbon dioxide at the average rate of 5 ml/s·person, the supply air has 400 ppm CO_2, and the maximum indoor CO_2 concentration is 1000 ppm.

Solution

Using Equation (7.3),

$$V = N/(C_R - C_I) = (15 \text{ person} \times 5 \text{ ml/s person})/(1000 \text{ ppm} - 400 \text{ ppm})$$
$$= (75 \times 10^{-6} \text{m}_c^3/\text{s})/(600 \times 10^{-6} \text{m}_c^3/\text{m}_a^3) = 0.125 \text{ m}_a^3/\text{s} \, (= 265 \text{ cfm})$$

For spaces with stricter indoor air quality requirements (e.g. less than 1000 ppm CO_2), such as for health care facilities, this will result in higher ventilation rates. The same is true for "dirty" outdoor environments (e.g. more than 400 ppm CO_2).

The IAQP is often used for unusual contaminant sources (or strength) such as SARS-CoV-2 (COVID-19) and/or for spaces with a different desired level of acceptable air quality (especially for specific target concentrations, e.g. for unadapted occupants). The procedure requires more assumptions (or prior knowledge) for calculations (e.g. contaminant type and criteria). Note that the designed IAQ thresholds may alter depending on the exposure/occupying time in the space. For instance, for CO, 40 mg/m^3 (35 ppm) is the threshold for one-hour exposure while 10 mg/m^3 (9 ppm) is the threshold for 8-hour exposure. The IAQP may take credits from LEED or similar rating systems for special source-control and removal measures (e.g. using low-emitting materials and air cleaning devices), and sometimes may lead to lower ventilation rate requirements (and thus energy savings).

7.3 Air Purification

Many technologies have been developed and commercialized to clean contaminants in the air, which may include:

- fibrous media filter
- adsorbent media filter
- ionizers
- ultraviolet germicidal irradiation (UVGI)
- photocatalytic oxidation (PCO)
- electrostatic precipitation (ESP)
- plasma
- intentional ozone generator

These technologies can handle either gaseous pollutants (e.g. carbon monoxide, nitrogen dioxide, sulfur dioxide, ozone, radon, VOCs) or particulate pollutants (e.g. PM1, PM2.5, PM10, dust, pollen, mold spores, smoke residue, and airborne viruses/bacteria). The cleaning mechanisms generally fall into two major categories: mechanical and electronic. Mechanical cleaning is safe and effective, while electronic cleaning is less efficient and may generate harmful byproducts (e.g. ozone). Table 7.7 compares the mechanisms, advantages, and disadvantages of various air cleaning technologies (https://www.epa.gov/sites/default/files/2018-07/documents/residential_air_cleaners_-_a_technical_summary_3rd_edition.pdf).

Among various cleaning technologies, mechanical-based air filtration is still one of the primary air purification technologies, widely used in both residential and commercial buildings and in both central and portable ventilation systems. Figure 7.5 illustrates the typical cleaning layers in an air purifier. Figure 7.6 presents the collection mechanisms and capture efficiency of particles of different sizes by air filters. It is noted that both small and large particles can be captured efficiently while particles at the size of 0.1–1 μm present major challenges.

An air filter's performance is rated by the Minimum Efficiency Reporting Value (MERV), which quantifies a filter's ability to capture larger particles between 0.3 and 10 microns (μm). The values are helpful in comparing the performance of different filters, which are derived from a test method developed by ASHRAE. The higher the MERV rating, the better the filter is at trapping specific types of particles. Table 7.8 summarizes the MERV ratings and associated efficiencies as well as filter types and the types of removed particles. Table 7.9 further provides the suggested applications for air filters with different MERV ratings.

Generally, using an air filter with a MERV rating of about 5–8 is appropriate for most residential homes. MERV 5-8 filters provide good filtration and can remove most pollen, mold spores, and dust mites. Using MERV 5-8 filters will keep HVAC systems more efficient than using higher rated filters, as long as filters are replaced regularly. If the occupants are susceptible to allergies, asthma, or other respiratory conditions, a filter with a MERV rating of about 10–12 may be desired. These types of filters are sufficient to trap particles as small as 1 micron such as automobile exhaust, pollen, and mold spores. Air filters with a MERV of 8–10 are commonly used for commercial buildings, while MERV 13-20 is typically found in hospital and pharmaceutical manufacturing settings. For special conditions such as SARS-CoV-2 (COVID-19), ASHRAE recommends using a filter with a MERV of 13, while a MERV 14 (or better) filter is preferred.

Table 7.7 Mechanisms, advantages, and disadvantages of various air cleaning technologies.

Air-cleaning technology	Targeted indoor air pollutant(s)	Mechanism(s) of action	Advantages	Disadvantages	Test standards (and rating metrics)
Fibrous filter media	Particles	Collection: filter fibers capture particles • Mechanical filtration media rely on mechanical forces alone • Electrostatically charged (i.e. "electret") media use mechanical fibers with an electrostatic charge applied to collect oppositely charged particles, enhancing removal efficiency	• If rated efficiency is high, they can have excellent removal capabilities for many particle sizes • Mechanical media filters see improved efficiency with loading	• Regular replacement is required • Used particle filters can be a source of sensory pollution/odors • High pressure drops on some fibrous media filters can negatively impact HVAC systems • Electret media filters see reduced efficiency with loading • Confusing number of test standards and rating metrics	Filters: • ANSI/ASHRAE Standard 52.2 (MERV) • ISO 16890 (ePM) • ISO 29463 (HEPA) • Proprietary test standards (FPR, MPR) Portable air cleaners: • AHAM AC-1 (CADR)
Electrostatic precipitation (ESP)	Particles	Collection: Corona discharge wire charges incoming particles, which collect on oppositely charged plates	• Can have high removal efficiency for a wide range of particle sizes • Low pressure drop and minimal impacts on HVAC systems • Low maintenance requirements	• Sometimes ESPs have high ozone and nitrogen oxide generation rates • Efficiency typically decreases with loading and plates require cleaning • High electric power draw requirements	ANSI/UL Standard 867 for electrical safety and ozone emissions (similar to IEC 60335-2-65) (pass/fail; no rating metric)
Ionizers (i.e. ion generators)	Particles	Collection: similar to ESP, ionizers use a high-voltage wire or carbon fiber brush to electrically charge air molecules, which produces negative ions that attach to airborne particles; the charged particles are then collected either on oppositely charged plates in the air cleaner or become attracted to other surfaces in the room and deposited elsewhere	• Typically low power draw requirements • Quiet • Low maintenance	• Generates ozone • Typically low effectiveness because of very low airflow rates and clean air delivery rates (CADRs)	None specific to ionizers, although AHAM AC-1 can be used to measure CADR

Ultraviolet germicidal irradiation (UVGI)	Microbes	Destruction: UV light kills/inactivates airborne microbes	• Can be effective at high intensity with sufficient contact time • Can be used to inactive microbes on cooling coils and other surfaces	• Uncoated lamps can generate ozone • Potential for eye injury • Effectiveness increases with lamp intensity, which is typically low in residential UVGI air cleaners • High electrical power draw requirements • Inactivates but does not remove microbes	Air irradiation: • ANSI/ASHRAE Standard 185.1 Surface irradiation: • ANSI/ASHRAE Standard 185.2
Adsorbent media	Gases	Collection: gases physically adsorb onto high-surface-area media (typically activated carbon)	• Potential for high removal efficiency for many gaseous pollutants in air cleaners with a sufficient amount of media for the application • No byproduct formation	• Regular replacement is required because its adsorption capacity is exhausted and physical adsorption is a reversible process, meaning pollutants may not be permanently captured • Effectiveness of many consumer-grade systems with small amounts of activated carbon is unknown • High pressure drops on some sorbent media filters can negatively impact HVAC systems • Different removal efficiency for different gases at different concentrations • Standard test methods are not widely used	Media: • ANSI/ASHRAE Standard 145.1 (no rating metric) In-duct air cleaners: • ANSI/ASHRAE Standard 145.2 (no rating metric) No effectiveness standards
Chemisorbent media	Gases	Collection: gases chemically adsorb onto media coated or impregnated with reactive compounds	• Potential for high removal efficiency for many gaseous pollutants • Chemisorption is an irreversible process, meaning pollutants are permanently captured	• Regular replacement is required because its chemisorption capacity is exhausted • Effectiveness of many consumer-grade systems is unknown • High pressure drops on some sorbent media filters can negatively impact HVAC systems • Different removal efficiency for different gases at different concentrations	Media: • ANSI/ASHRAE Standard 145.1 (no rating metric) In-duct air cleaners: • ANSI/ASHRAE Standard 145.2 (no rating metric) No effectiveness standards

(Continued)

Table 7.7 (Continued)

Air-cleaning technology	Targeted indoor air pollutant(s)	Mechanism(s) of action	Advantages	Disadvantages	Test standards (and rating metrics)
Catalytic oxidation	Gases	Conversion: most utilized photocatalytic oxidation (PCO) in which a high-surface-area medium is coated with titanium dioxide as a catalyst; incoming gases adsorb onto the media and UV lamps irradiate and activate the titanium dioxide, which reacts with the adsorbed gases to chemically transform them	• Can degrade a wide array of gaseous pollutants (e.g. aldehydes, aromatics, alkanes, olefins, halogenated hydrocarbons) • Can be combined with adsorbent media to improve effectiveness	• Can generate harmful byproduct such as formaldehyde, and acetaldehyde, and ozone • No standard test methods • Often relatively low removal efficiency for many indoor gases, but high variability in removal for different gases • Lack of field studies to validate performance • Catalyst often has a finite lifespan	None specific to PCO
Plasma	Gases	Conversion: electric current is applied to create an electric arc; incoming gases are ionized and bonds are broken to chemically transform the gaseous pollutants	• Can have high removal efficiency • Can be combined with other air-cleaning technologies (e.g. PCO) to improve performance and minimize byproduct formation	• Wide variety of plasma generation types yields confusion on how a product actually works • Byproducts are formed from many plasma technologies, including particles, ozone, formaldehyde, carbon monoxide, chloroform, nitrogen oxides, and a large number of other organic gases • Most studies have investigated gaseous removal while fewer have evaluated particle removal	None specific to plasma
Intentional ozone generation	Gases	Conversion: intentional generation of ozone using corona discharge, UV, or other method to oxidize odorous compounds and other gases	• Reacts with many indoor gases • Can be combined with other less-harmful technologies such as adsorbent media	• High ozone generation rates • High amounts of byproduct formation • Can cause degradation to indoor materials	None specific to ozone generators

Note that "Gases" are inorganic gases (e.g. carbon monoxide, nitrogen dioxide, ozone) and organic gases (e.g. VOCs, aldehydes).

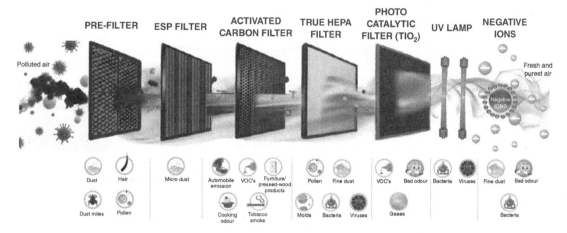

Figure 7.5 Typical cleaning layers in an air purifier (https://cdn.shopify.com/s/files/1/0508/0697/articles/Air_filter.png?v=1601411583).

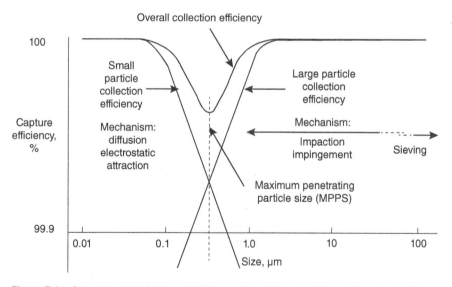

Figure 7.6 Capture mechanisms and efficiency of particles of different sizes by air filters.

Table 7.8 MERV rating and efficiency and filter type and removed particles.

MERV rating	Air filter will trap air particles size 0.03–1.0 microns	Air filter will trap air particles size 1.0–3.0 microns	Air filter will trap air particles size 3–10 microns	Filter type ~ Removes these particles
MERV 1	<20%	<20%	<20%	Fiberglass and aluminum Mesh
MERV 2	<20%	<20%	<20%	~
MERV 3	<20%	<20%	<20%	Pollen, dust mites, spray
MERV 4	<20%	<20%	<20%	Paint, carpet fibers
MERV 5	<20%	<20%	20–34%	Cheap disposable filters
MERV 6	<20%	<20%	35–49%	~
MERV 7	<20%	<20%	50–69%	Mold spores, cooking Dusts, hair spray, furniture polish
MERV 8	<20%	<20%	70–85%	
MERV 9	<20%	Less than 50%	85% or Better	Better home box filters
MERV10	<20%	50–64%	85% or Better	~
MERV 11	<20%	65–79%	85% or Better	Lead dust, milled flour, auto fumes, welding fumes
MERV 12	<20%	80–90%	90% or Better	
MERV 13	Less than 75%	90% or Better	90% or Better	Superior commercial Filters
MERV 14	75–84%	90% or Better	90% or Better	~
MERV 15	85–94%	90% or Better	90% or Better	Bacteria, smoke, many viruses
MERV 16	95% or Better	90% or Better	90% or Better	

Source: Illustration Provided by LakeAir/ www.lakeair.com.

Table 7.9 Applications for air filters with different MERV ratings.

	Suggested application table				
MERV Std 52.2	Dust spot efficiency	Ability to remove dust	Dust particle sizes	Common uses	Filter types
MERV 1-4	Less than 20% <20%	60–80%	10 micron +	Residential filters Light commercial Minimum equipment protection	Washable metal Loose fiberglass Disposable panels
MERV 5-8	20–60%	80–95%	3–10 micron	Industrial work spaces Commercial application Better residential application Paint booths	Pleated filters Extended filters Media panel
MERV 9-12	40–85%	90–98%	1–3 micron	Superior residential Better industrial Better commercial	Pocket filters Rigid box Rigid cell Cartridge filters
MERV 13-16	70–98%	95–99%	0.3–1	Smoke removal General surgery Hospitals & health care Superior Commercial	Ridge cell V-cells Pocket filters H10–H12 HEPA
MERV 17-20	More than 98%	99% or	0.3 micron	Clean rooms Radioactive materials Pharmaceutical Carcinogens	HEPA 13 HEPA 14 ULPA 15 ULPA 16

7.4 Infiltration

Infiltration is an important building characteristic, which is defined as the flow of outdoor air into a building through cracks and other unintentional openings and through the normal use of exterior doors for entrances and egress. This airflow is different from ventilation which is the intentional (and thus manageable) introduction of air from the outside into a building. Infiltration represents the airtightness level of a space. When air flows from the inside to outside through unintentional openings, this is called exfiltration.

Infiltration can bring in outdoor air and thus outdoor contaminants, while it may also dilute and purge indoor contaminants out of the building. Infiltration can bring in outdoor hot or cold air that increases a building's cooling and heating loads (and thus energy cost), but it may also introduce outdoor fresh air to complement the indoor ventilation requirements and thus reduce active ventilation energy need. In the 1970s, the energy crisis led to the development of supertight building envelopes that largely reduced building heating and cooling energy requirements but caused IAQ problems. As a remedy, mechanical ventilation through duct systems was implemented and operated. However, this increases the initial and operating costs of buildings.

Because of its uncontrollable nature, infiltration is considered to be a disadvantage most of the time. Reducing infiltration is always a top priority in developing energy efficient buildings. Many studies were conducted to understand the physics of infiltration to develop an infiltration prediction model. Figure 7.7 shows the systematic factors that affect building air filtration: indoor–outdoor pressure differences caused by outdoor wind, temperature gradients (buoyancy or stack effect), and mechanical systems. Figure 7.8 illustrates the airflow directions through a building envelope caused by stack, wind, and combined effects.

Statistics show that most US houses have an infiltration rate ranging from 0.25 to 1 ACH (air change rate per hour), while new constructions have a tighter envelope (0.3~0.7 ACH with the median of 0.5 ACH) and low-income houses have a relatively large infiltration (0.3~2 ACH with the median of 0.9 ACH; Figure 7.9). Apartment buildings usually have compact configurations and thus less infiltration. Recent tests in China show that most middle-rise and high-rise new apartment buildings have an average infiltration of 0.25~0.5 ACH. ASHRAE recommends a leakage of 0.35 ACH for the proper ventilation of a house (via infiltration) with minimum heat loss.

Generally, three sources of air leakage exist:

- **Component perforations:** which are relatively easy to identify (e.g. vents, stacks, chimneys);
- Openings, which are also easy to identify (e.g. windows, doors);
- Background or fabric leakage, which highly depends on construction. It is challenging or almost impossible to identify all cracks. Therefore, these are often assumed to be uniformly distributed over the surface area of a building and represented in terms of -"leakage per unit area of the envelope."

Many studies attempted to develop various models to quantify the infiltration rate, which is a critical input for building energy simulation and prediction. The LBL infiltration method is the most popular one. The LBL infiltration method, developed by the Lawrence Berkeley National Laboratory, originally was created for residential buildings. The LBL infiltration model uses several parameters including the leakage area, height of the building, inside–outside temperature difference, terrain class of the structure, and

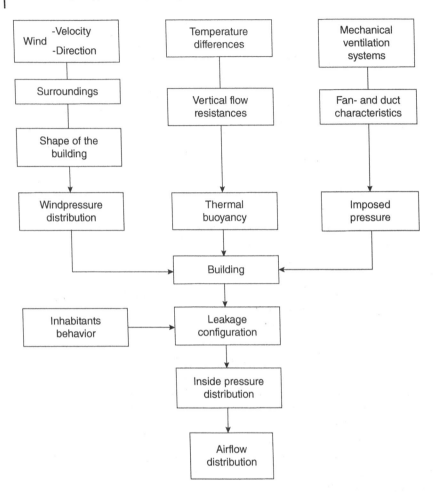

Figure 7.7 Factors affecting building air infiltration. *Source:* Feustel and Rayner-Hooson (1990)/U.S. Department of Energy/Public Domain.

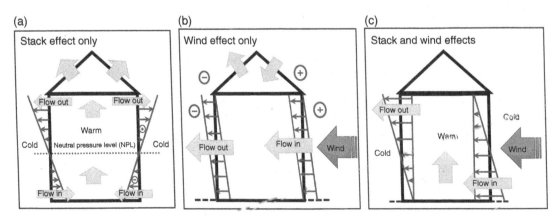

Figure 7.8 Airflow through the building envelope caused by stack, wind, and combined effects.

(a)

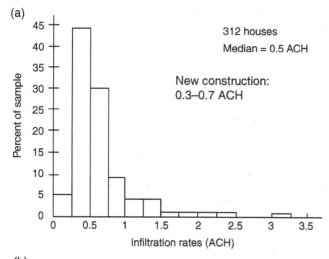

(b)

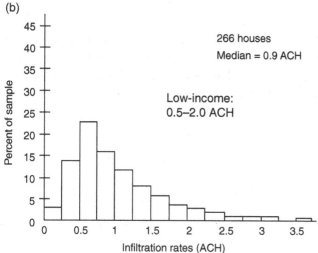

Figure 7.9 Histogram of infiltration values in US houses (a) new houses; (b) low-income houses.

wind speed (Sherman and Grimsrud 1980a). The infiltration airflow rate through a leakage can be expressed as:

$$Q = A\sqrt{\frac{2}{\rho}\Delta P}$$

(7.4)

Or more generally,

$$Q = C(\Delta P)^n$$

(7.5)

where Q is the infiltration rate (m^3/s); A is the effective leakage area (m^2); ρ is the air density (kg/m^3); ΔP is the indoor-outdoor pressure difference (Pa); C is the building resistance coefficient; and n is the exponent index. Equation (7.4) or (7.5) can be further divided into two parts according to the driving forces.

$$Q_{wind} = f_w^* A_o v' \tag{7.6}$$

$$Q_{stack} = f_s^* A_o \sqrt{\Delta T} \tag{7.7}$$

where Q_{wind} is the infiltration caused by the wind (m^3/s); Q_{stack} is the infiltration caused by the stack (m^3/s); A_o is the total leakage area of the structure (m^2); f_w^* is the reduced wind parameter; f_s^* is the reduced stack parameter; v' is the weather station wind speed (m/s); and ΔT is the inside–outside temperature difference (K).

The "reduced" term on the reduced wind parameter is used to distinguish it from the original wind parameter (f_w) that does not incorporate the shielding coefficient (Sherman and Grimsrud 1980b). The value of f_w^* can be obtained as follows:

$$f_w^* = C' \left((1-R)^{1/3} \right)' \left(\frac{\alpha(H/10)^{\gamma}}{\alpha'(H'/10)^{\gamma'}} \right) \tag{7.8}$$

where C' is the generalized shielding coefficient; R is the vertical leakage fraction; α' and γ' are terrain parameters of the wind data measurement site; α and γ are the terrain parameters of the building site; H is the height of the ceiling (m); and H' is the height of the wind measurement (m). The value of the vertical leakage fraction (R) can be calculated using the following equation:

$$R = \frac{(A_c + A_f)}{A_o} \tag{7.9}$$

where A_c is the area of ceiling leakage (m^2); A_f is the area of floor leakage (m^2); and A_o is the total effective leakage area (m^2).

The terrain parameters value (α, γ) and the generalized shielding coefficient (C') can be found in Tables 7.10 and 7.11, respectively.

The reduced stack parameter is similar to the reduced wind parameter; however, the term "reduced" is not because of the shielding coefficient, as the stack infiltration is caused by a temperature difference and has little to do with the terrain or geometry of surrounding environments. Rather, the reduced stack

Table 7.10 Terrain parameters for standard terrain classes.

Class	γ	α	Description
I	0.10	1.30	Ocean or other body of water with at least 5 km of unrestricted expanse
II	0.15	1.00	Flat terrain with some isolated obstacles (e.g. building or trees well separated from each other)
III	0.20	0.85	Rural areas with low buildings, trees, etc.
IV	0.25	0.67	Urban, industrial, or forest areas
V	0.35	0.47	Center of large city

Table 7.11 Generalized shielding coefficients.

Shielding Class	C′	Description
I	0.324	No obstructions or local shielding whatsoever
II	0.285	Light local shielding with few obstructions
III	0.240	Moderate local shielding, some obstructions within two house heights
IV	0.185	Heavy shielding, obstruction around most of perimeter
V	0.102	Very heavy shielding, large obstruction surrounding perimeter within two house heights

parameter includes a variable X which is the ceiling-floor leakage difference. The reduced stack parameter can be calculated as follows:

$$f_s^* = \frac{(1 + (R/2))}{3}\left(1 - \frac{X^2}{(2-R)^2}\right)^{3/2}\sqrt{\frac{gH}{T}} \tag{7.10}$$

where R is the vertical leakage fraction; X is the ceiling-floor leakage difference; g is the acceleration of gravity (9.8 m/s^2); H is the height of the structure; and T is the indoor air temperature. The ceiling-floor leakage parameter (X) can be calculated using the following equation:

$$X = \frac{A_c - A_f}{A_o} \tag{7.11}$$

where A_c is the area of ceiling leakage (m^2); A_f is the area of floor leakage (m^2); and A_o is the total effective leakage area (m^2).

The total infiltration rate can then be calculated by combining Q_{stack} and Q_{wind}:

$$Q = \sqrt{Q^2_{stack} + Q^2_{wind}} \tag{7.12}$$

7.5 Blower Door Test

The blower door test is conducted, mostly for small buildings (larger buildings may require multiple fans), to estimate the airtightness of the building and its aggregate envelope leakage and to locate and fix leaks. Figure 7.10 shows a photo of a blower door test rig, which consists of

- a calibrated, variable-speed fan, capable of inducing a range of airflows sufficient to pressurize and depressurize a variety of building sizes,
- a pressure measurement instrument, called a manometer, to simultaneously measure the pressure differential induced across the face of the fan and across the building envelope, as a result of fan airflow, and
- a mounting system, used to mount the fan in a building opening, such as a door or a window.

Figure 7.10 Photo of a blower door test rig.

The blower door test is typically performed by a trained energy auditor. It takes between 10 and 20 minutes to install the blower door inside an existing door frame, an additional 10–15 minutes to prepare the house for the test, and about 15 minutes to run the test. The entire procedure of a blower door test is summarized as follows:

1) Install and fit the blower door inside the main door frame;
2) Close all exterior doors (except the one where the blower door is to be installed) and close all windows;
3) Leave interior doors open to ensure that all rooms in the home are included in the test;
4) Shut down all pilot lights from stoves and water heaters and flues in wood-burning stoves and fireplaces;
5) Measure and calculate the total volume of airspace inside the house;
6) Turn on the fan in the blower door to depressurize the house, monitor the air pressure inside and outside the house to reach a certain ΔP, and record the ΔP and associated fan flow rate Q;
7) Vary the fan power/speed at a 5–10 Pa incremental step and repeat the measurement until fairly high pressures (about 50–60 Pa) are reached;
8) (Optional) Reverse the fan direction and pressurize the house and repeat Steps (6) and (7) to test the exfiltration characteristics of the house;
9) Calculate the effective leakage area using Equation (7.13) and the total infiltration rate using Equation (7.5).

Another auditor may walk through the house during the test to check for drafts at specific leakage locations. The fan is so powerful that as it sucks the air out of the house, it draws additional air in through

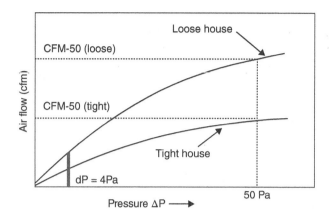

Figure 7.11 The power law curve obtained from the blower door test.

unsealed gaps in the house. These are called "penetrations." The auditor will feel around doors, windows, and vents for drafts and will also use a small fog generator to pinpoint drafts that are too weak to feel. When there is no draft, the fog will hover in the air, but if a draft is present, the fog will quickly blow away.

The effective leakage area (ELA) (in^2 or m^2) is calculated as:

$$ELA = C_r \times Q_r \times \sqrt{\rho/(2\Delta P)} \tag{7.13}$$

where C_r is the air leakage coefficient ($C_r = 0.186$ for IP units or 0.544 for SI units); ρ is the air density (lb_m/ft^3, or, kg/m^3); ΔP is the reference pressure, typically 4 Pa in the United States and 10 Pa in Canada; and Q_r is the reference volume flow rate (cfm, or, m^3/s) determined from the power law curve/equation obtained from the test (as shown in Figure 7.11) when using $\Delta P = 4$ Pa. Note that the power law curve is regressed using the experimental data in the range of 10–50 Pa and extrapolation is needed to obtain the lower values at $\Delta P = 4$ Pa. This extrapolation may result in an inaccurate estimation of Q_r due to a somewhat imprecise form of the assumed flow-pressure relation at lower levels. The calculated ELA can then be used in infiltration models such as the LBL model to predict the hourly infiltration rate under dynamic weather conditions (e.g. TMY3).

Equation (7.5) can be converted into a linearized form:

$$Log(Q) = Log(C) + n \times Log(\Delta P) \tag{7.14}$$

Using the collected data set (ΔP, Q) and regressing the obtained curve between Q and ΔP, as illustrated in Figure 7.12, can yield both C and n. With these values, the actual building infiltration rate can be estimated using Equation (7.5) with $\Delta P = 4$ Pa. Note that infiltration and exfiltration may have different performances (curves) due to different resistance characteristics at the cracks and openings between inflow and outflow.

ACH_{50}, which stands for the air change rate per hour (ACH) at $\Delta P = 50$ Pa, is an important index for building tightness, and can be calculated as:

$$ACH_{50} = CFM_{50} \times 60/V_{building} \tag{7.15}$$

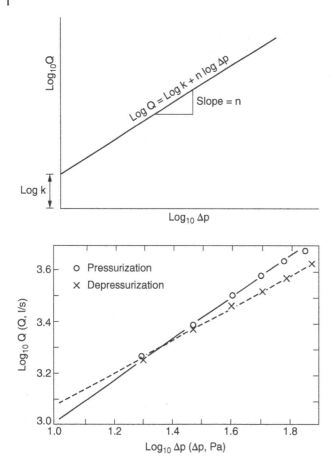

Figure 7.12 The power law curve obtained from the blower door test in log format (where k=C in Equations (7.14) and (7.5)).

where CFM_{50} is the fan supply air flow rate at $\Delta P = 50$ Pa and $V_{building}$ is the building volume.

The German passive house requires a value of 0.6 ACH_{50}. The US Department of Energy's Zero Energy Ready Home program requires that whole house leakage must be tested and meet the following infiltration limits:

- Climate zones 1–2: ≤ 3 ACH_{50};
- Climate zones 3–4: ≤ 2.5 ACH_{50};
- Climate zones 5–7: ≤ 2 ACH_{50};
- Climate zone 8: ≤ 1.5 ACH_{50};
- Attached dwellings: ≤ 3 ACH_{50}.

Homework Problems

1 A classroom at CU is occupied with 39 students and 1 teacher. Each person is producing carbon dioxide at the average rate of 5 ml/s/person. Determine the concentration level of carbon dioxide in the classroom if the outdoor air rate supplied is 8 l/s per person (ASHRAE Standard 62-1999) and the outdoor air level of carbon dioxide at CU is 300 ppm.

2 A school lecture hall with a floor area of 1000 ft^2 is designed for 55 students.
   (a) What is the minimum amount of clean outdoor air required by ASHRAE Standard 62.1-**2001** according to the information provided in the standard?
   (b) What is the minimum amount of clean outdoor air required by ASHRAE Standard 62.1-**2019** (https://www.ashrae.org/technical-resources/standards-and-guidelines/read-only-versions-of-ashrae-standards)?
   (c) What is the minimum amount of clean outdoor air if using the indoor air quality approach based on CO_2 concentration (assuming outdoor concentration at 400 ppm, indoor upper limit at 1000 ppm, and indoor generation rate at 5 ml/s/person)?
   (d) Comment on the differences among (a)–(c) results and which one you would suggest using for this space and why.

3 The following table assembles the test results of an actual house of 200 m^2 (floor area) and 3 m (ceiling height) where the blower door tests were performed both before and after the weather stripping.
   (a) Determine the two sets of coefficients C and n for the tests done before and after the house tightening.
   (b) Calculate the ELA values for both sets.
   (c) Calculate the ACH_{50} for both sets.

Before weather stripping		After weather stripping	
Δp (Pa)	$\dot{V}$ (m^3/h)	Δp (Pa)	$\dot{V}$ (m^3/h)
3.0	365.0	2.2	99.2
5.0	445.9	5.5	170.4
5.8	492.7	6.7	185.6
6.7	601.8	8.2	208.5
8.2	699.2	11.6	263.2
9.0	757.5	13.5	283.1
10.0	812.4	15.6	310.2
11.0	854.1	18.2	346.2

4 Staged Project Assignment (Teamwork)
   (a) Check ASHRAE-62.1 (https://www.ashrae.org/technical-resources/standards-and-guidelines/read-only-versions-of-ashrae-standards) to calculate and determine the minimum outdoor ventilation air for your building.
   (b) Check references (e.g., ASHRAE Handbook Fundamentals) to make a reasonable assumption about infiltration rate of your building.

References

Dorgan, C.B., Dorgan, C.E., Kanarek, M.S., and Willman, A.J. (1998). Health and productivity benefits of improved indoor air quality. *ASHRAE Transactions* 104 (1): 4161.

Fisk, W.J. (2000). Health and productivity gains from better indoor environments and their relationship with building energy efficiency. *Annual Review of Energy and the Environment* 25: 537–566.

Feustel, H. E. and Rayner-Hooson, A. (1990). COMIS Fundamentals. Report LBL-28560, US Department of Energy, Lawrence Berkeley Laboratory.

Sherman, M.H. and Grimsrud, D.T. (1980a). Measurement of Infiltration Using Fan Pressurization and Weather Data. *1st Symposium of the Air Infiltration Centre on Instrumentation and Measurement Techniques*, Windsor, England (6–8 October).

Sherman, M.H. and Grimsrud, D.T. (1980b). Infiltration-Pressurization Correlation: Simplified Physical Modeling.*1980 Annual ASHRAE Conference*, Denver, CO (22–26 June).

8

Heat Transfer through Building Envelope

Heat and mass transfer through building envelopes have crucial impacts on the quality of the indoor environment (comfort and air quality) and determine the heating, cooling, and ventilation energy requirements to maintain a comfortable and healthy indoor environment. Mass transfer includes air, moisture, and pollutant movement in and out of various spaces. The movement of air and pollutants has been discussed in Chapter 7 "Ventilation, Infiltration and Air Quality." Moisture is typically moved by airflow and is often associated with phase changes (e.g. evaporation and condensation). Moisture is the key for estimating the latent heating and cooling load (either to remove the extra indoor moisture – dehumidification, or to deliver additional moisture to the spaces – humidification).

8.1 Latent Heat Transfer

Two types of latent heat transfer can be found in a building envelope. Phase change materials (PCM; e.g. salt and wax) can be used as building envelope elements/layers to provide additional heat storage for extra heat absorbed during the day while releasing the stored heat for night use as needed. Since the phase change heat is much more significant than the sensible heat (caused by temperature changes), a small size of PCM can deliver a large amount of heat storage.

The second type of latent heat transfer is attributed to the evaporation and condensation of water vapor in the air. When the air temperature falls below the dewpoint, the water vapor in the air will be condensed and become water droplets, during which the air temperature will increase due to the heat released from the condensation process (where water vapor becomes water and release heat). Condensation will result in significant consequences if this occurs in the building envelope. The resulting water in the envelope not only deteriorates the insulation's performance but also causes mold issues inside the envelope.

Moisture and latent heat transfer through a building envelope, although important, do not contribute to the major heating and cooling load of buildings but do require involved knowledge and calculations, which therefore are not included in this book. Latent heating and cooling load, however, will be introduced in Chapter 11 "Building Load Calculation."

Energy Efficient Buildings: Fundamentals of Building Science and Thermal Systems, First Edition. Zhiqiang (John) Zhai.
© 2023 John Wiley & Sons, Inc. Published 2023 by John Wiley & Sons, Inc.

8.2 Sensible Heat Transfer

Sensible heat transfer is driven by a temperature gradient. The 2nd law of thermodynamics indicates that energy is always transferred from a high temperature to a low temperature.

8.2.1 Heat/Thermal Storage

Every object has the capability to store energy. Adding heat to a material increases its energy level – usually *increasing temperature (and is thus called sensible)*. For a given amount of heat transferred, temperature rise depends on the mass and specific heat of the materials. For instance,

- Water, C_p = 1 Btu/(lb–°F): thus 1 Btu heat will cause 1 lb of water to raise 1 °F.
- Brick, C_p = 0.2 Btu/(lb–°F): thus 1 Btu heat will cause 1 lb of brick to raise 5 °F.
- Steel, C_p = 0.12 Btu/(lb–°F): thus 1 Btu heat will cause 1 lb of steel to raise 8 °F.
- Air, C_p = 0.24 Btu/(lb–°F): thus 1 Btu heat will cause 1 lb of air to raise 4 °F

The ability of an object to store energy can be expressed in:

- Heat capacity (Btu/ft^3-°F) = density (ρ) × specific heat (C_p); and
- Thermal mass (Btu/°F) = density (ρ) × volume (V) × specific heat (C_p)

where density and specific heat are the properties of the selected/used materials while volume is based on the design. Figures 8.1 and 8.2 show the specific heat and density, respectively, of some common building materials. Water appears to have a larger specific heat than most of the building materials. Most metallic materials have a higher density. Figure 8.3 shows the heat capacity ($\rho \times C_p$) of some common building materials. Water and steel present a higher heat capacity than other materials.

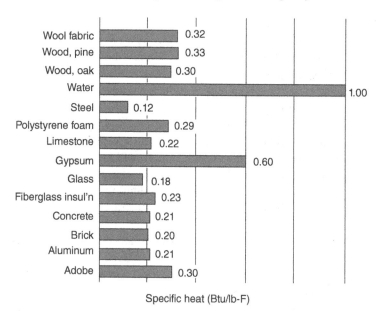

Figure 8.1 Specific heat of some common building materials.

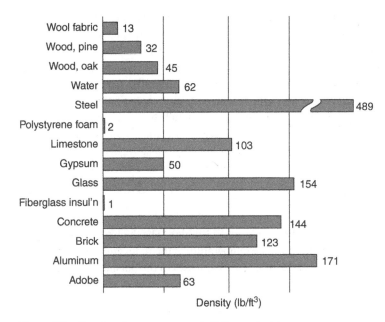

Figure 8.2 Density of some common building materials.

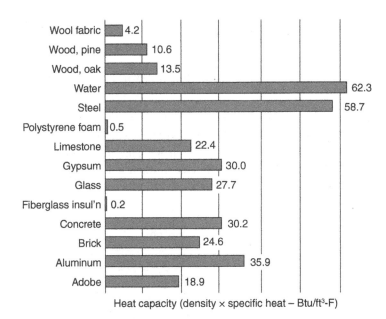

Figure 8.3 Heat capacity of some common building materials.

While the volume of a thermal mass is an important factor in a material's heat capacity, the form or distribution of a thermal mass in a space is more critical. A thermal mass is often best applied over a large area rather than in large volumes or thicknesses. This is because storage capacity is also related to how fast the heat received at the surface of a thermal mass object can penetrate into the object (which depends on the thermal conductivity of the material, to be introduced later). If the heat is accumulated at the surface with slow penetration into the center of the object, the surface will be thermally saturated and thus reflect/reject the external heat. In this situation, the center of the thermal mass, although being counted, may never receive any heat to serve the thermal storage purpose. A concrete slab foundation either left exposed or covered with conductive materials, e.g. tiles, is a favorable solution for thermal storage. Another method is to place the masonry facade of a timber-framed house on the inside ("reverse-brick veneer"). In both cases, 7.5–10 cm (3"–4") of the material thickness is often adequate while 15 cm (6") is usually the limit.

Example 8.1

The dimensions of a classroom at CU are 30-ft long, 18-ft wide, and 12-ft high. The thickness of walls, ceiling, and floor is 1.0 ft. Compare the energy needed to raise the temperature of the room envelope from 60 to 70 °F for a concrete structure and for an oak wood structure (the heat losses to the outside are neglected).

Solution

	Concrete	Oak wood
V (ft³)	$2\,(30 \times 18 + 18 \times 12 + 30 \times 12) \times 1 = 2232$	2232
ρ (lb/ft³)	144	45
C_p (Btu/lb°F)	0.21	0.3
ΔT (°F)	10	10
$Q = \rho V C_p\,\Delta T$ (Btu)	6.75×10^5 or 198 kWh	3.01×10^5 or 88 kWh

When heat flows from one side (e.g. outdoor) of a wall, through the wall, to the other side (e.g. indoor) of the wall, part of the heat will be absorbed by and stored in the wall (Figure 8.4) which will be released back to the spaces (through both sides). The fraction of the heat stored is a function of the thermal mass properties, the thermal saturation condition of the thermal mass, and the indoor–outdoor temperature difference. The calculation of this transient heat transfer process is introduced later in this chapter.

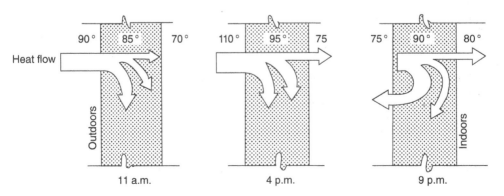

Figure 8.4 Thermal storage in a transient heat transfer through a wall.

8.2.2   Conduction: Conductive Heat Transfer

a) **Mechanism:** caused by material molecule collision and vibration; therefore it needs a still media to transfer heat (from a high temperature to a low temperature). The still media may include a solid, still liquid, and gas.

b) **Property:** thermal conductivity, K (W/(m · °C), or, Btu/(h · ft · °F)). This property describes how much heat can be transferred through a unit surface area of a unit thickness material for a unit temperature difference. Different materials have different thermal conductivity (Table 8.1). In general,

$$K_{gas} < K_{insulator} < K_{liquid} < K_{nonmetallic\ solid} < K_{metal\ alloy} < K_{pure\ metal} < K_{nonmetallic\ crystals}$$

In principle, thermal conductivity varies with material temperature. When the temperature increases, the thermal conductivity of a solid will decrease due to the escaping potential of connected molecules, while the thermal conductivity of a gas will increase due to the increasing collision of gas molecules. At typical room temperature, thermal conductivity can be treated as constant. Thermal conductivity can be expressed in other terms and units, such as,

- Conductance, U or C (W/m^2 · °C, or, Btu/(h · ft^2 · °F)): heat transfer per unit area, per unit temperature difference; for actual thickness, U = K/Δx;
- Resistance, R (m^2 · °C/W, or, (h · ft^2 · °F)/Btu: thermal resistance per unit area, per unit temperature difference; for actual thickness, R = 1/U = Δx/K.

c) Steady state heat conduction

Assume: (i) steady state: there is no thermal storage in the material – heat transfer into one side of an envelope is fully transferred out of the other side (Figure 8.5); (ii) one-dimensional heat transfer through the primary surface section – good for most parts of the building envelope but the corners; (iii) uniform surface temperatures at both sides – no temperature gradient in the other two directions perpendicular to the heat flow direction.

Table 8.1   Density and thermal conductivity of some common building materials.

No	Building material	Density (kg/m³)	K (W/m.K)
1	Concrete	2.400	1,448
2	Aerated concrete	960	0,303
3	Plastered clay brick	1.760	0,807
4	Exposed clay brick		1,154
5	Glass	2.512	1,053
6	Gypsum board	880	0,170
7	Steel	7.840	47,6
8	Granite	2.640	2,927
9	Marble/ceramic/terazzo	2.640	1,298

Source: SNI 03-6389-2000

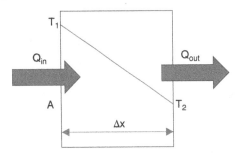

Figure 8.5 Steady state one-dimensional conductive heat transfer.

The steady state one-dimensional conductive heat transfer equation is:

$$Q_{in} = Q_{out} = Q_{cond} = \frac{K}{\Delta x} A(T_1 - T_2) \qquad (8.1)$$

where Q_{cond} is the conductive heat transfer (W or Btu/h); K is the thermal conductivity (W/m·°C, or, Btu/(h·ft·°F)); Δx is the thickness of the envelope (m or ft); A is the cross-section surface area of the envelope (m² or ft²); T_1 and T_2 are the surface temperature of two sides of the envelope (°C or °F). If the temperature difference is zero, there is no conductive (and any sensible) heat transfer. If the surface area doubles, the conductive heat transfer doubles; if the envelope thickness halves, the conductive heat transfer doubles.

Equation (8.1) can be rewritten as

$$Q_{cond} = \frac{K}{\Delta x} A(T_1 - T_2) = UA(T_1 - T_2) = \frac{A(T_1 - T_2)}{R} = \frac{(T_1 - T_2)}{R_t} \qquad (8.2)$$

where $U = K/\Delta x$ is the thermal conductance of the material, $R = \Delta x/K$ is the thermal resistance of the material; both are usually provided by manufacturers. $R_t = R/A$ is the total thermal resistance that includes the surface area A of the actual building envelope. If the R or R_t value doubles (e.g. more insulation), the conductive heat transfer halves.

Example 8.2

If the temperature difference between the inside and outside surface of a wall of 100 ft² is 10 °F, determine the heat flow for the R-13 wall. What if the wall is replaced with R-40?

Solution

R-13 wall:

$$Q = (\Delta T/R) \times A = (10/13) \times 100 = 77 \, Btu/h$$

R-40 wall:

$$Q = (\Delta T/R) \times A = (10/40) \times 100 = 25 \, Btu/h$$

d) Steady State Heat Conduction for Composite Envelope

Most building envelopes (e.g. walls, windows, roof) have more than one layer with multiple compositions, as illustrated in Figure 8.6. Calculating the overall conductive heat transfer through this kind of envelope requires some simplifications and assumptions. Typically, two approaches are used to compute the heat transfer in composite envelope: a series connection and a parallel connection.

• Series Connection

Considering a wall with multiple layers in series as shown in Figure 8.7, each layer has its own thermal conductivity and thickness, assume: (i) steady state heat flow (no storage at any layer);

Figure 8.6 Composite walls. *Source:* GDM photo and video/Adobe Stock.

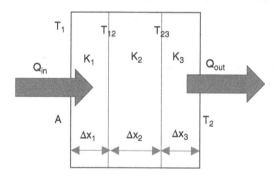

Figure 8.7 Multiple layered wall.

(ii) one-dimensional heat flow; (iii) uniform temperature at each surface (T_1 and T_2) and interface (T_{12} and T_{23}) (no heat flow in other two perpendicular directions).

Since

$$Q_{in} = Q_1 = Q_2 = Q_3 = Q_{out} \tag{8.3}$$

$$Q_1 = K_1/\Delta x_1 \times A \times (T_1 - T_{12}) = (T_1 - T_{12})/R_{t,1} \tag{8.4}$$

$$Q_2 = K_2/\Delta x_2 \times A \times (T_{12} - T_{23}) = (T_{12} - T_{23})/R_{t,2} \tag{8.5}$$

$$Q_3 = K_3/\Delta x_3 \times A \times (T_{23} - T_2) = (T_{23} - T_2)/R_{t,3} \tag{8.6}$$

where $R_{t,1} = \Delta x_1/(K_1 \times A)$, $R_{t,2} = \Delta x_2/(K_2 \times A)$, $R_{t,3} = \Delta x_3/(K_3 \times A)$. Rewriting Eqs. (8.4)–(8.6) as:

$$Q_1 R_{t,1} = T_1 - T_{12} \tag{8.7}$$

$$Q_2 R_{t,2} = T_{12} - T_{23} \tag{8.8}$$

$$Q_3 R_{t,3} = T_{23} - T_2 \tag{8.9}$$

Summarizing the left and right terms of Eqs. (8.7)–(8.9) provides

$$Q_1 R_{t,1} + Q_2 R_{t,2} + Q_3 R_{t,3} = Q(R_{t,1} + R_{t,2} + R_{t,3}) = Q R_{t,total} = T_1 - T_2 \tag{8.10}$$

Therefore, the overall heat flow rate (heat flux) can be calculated as

$$Q = (T_1 - T_2)/R_{t,total} \tag{8.11}$$

where $R_{t,\,total} = R_1 + R_2 + R_3$ is the sum of the individual thermal resistance (R_t) of all layers of the construction in series. Once the total heat flow rate is obtained, the interface temperatures can be calculated as:

$$T_{12} = T_1 - Q R_{t,1} \tag{8.12}$$
$$T_{23} = T_{12} - Q R_{t,2} \tag{8.13}$$

or

$$T_{23} = Q R_{t,3} + T_2 \tag{8.14}$$

The interface temperature could be important for evaluating whether the dewpoint temperature may be reached inside the envelope where condensation may occur and can further determine the proper location for vapor barriers.

This heat transfer process can be expressed in terms of a thermal network as shown below (Figure 8.8). This is analogous to the electrical network, where temperature (T) as the flow driving force corresponds to voltage (V) and heat flux (Q) is equivalent to current (I). Thermal resistance (R_t) is analogous to electrical resistance (R_e). The thermal network thus shares the same principles as the electrical network.

- Parallel Connection

When a wall has multiple parallel compositions, each with its own thermal conductivity and surface area that share the same thickness, as shown in Figure 8.9, the heat can flow along the individual paths, respectively, through the parallel compositions. The total heat flow rate is the sum of the individual heat fluxes along Path-1 and Path-2:

$$Q = Q_1 + Q_2 \tag{8.15}$$

Assuming that Path-1 and Path-2 are totally independent from each other, each path has its own surface temperature 1 and 2. This implies that the two parallel compositions (including the surfaces) are insulated/separated from each other, which is a reasonable approximation for paralleling compositions with distinct conductivities (e.g. metal and wood).

For the case in Figure 8.9,

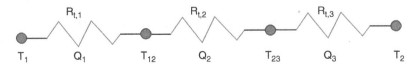

Figure 8.8 Thermal network in series connection.

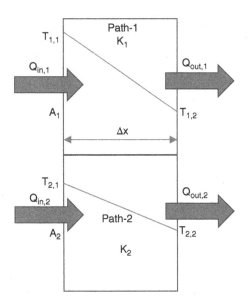

Figure 8.9 Parallel composite wall.

$$Q_1 = K_1/\Delta x \times A_1 \times (T_{1,1} - T_{1,2}) = (T_{1,1} - T_{1,2})/R_{t,1} \quad \text{(Path-1)} \tag{8.16}$$

$$Q_2 = K_2/\Delta x \times A_2 \times (T_{2,1} - T_{2,2}) = (T_{2,1} - T_{2,2})/R_{t,2} \quad \text{(Path-2)} \tag{8.17}$$

$$Q = Q_1 + Q_2 = (T_{1,1} - T_{1,2})/R_{t,1} + (T_{2,1} - T_{2,2})/R_{t,2} \quad \text{(Total)} \tag{8.18}$$

where $R_{t,1} = \Delta x/(K_1 \times A_1)$ and $R_{t,2} = \Delta x/(K_2 \times A_2)$. If using U values instead of R values, Eq. (8.18) can be rewritten as:

$$Q = Q_1 + Q_2 = U_{t,1} \times (T_{1,1} - T_{1,2}) + U_{t,2} \times (T_{2,1} - T_{2,2}) \tag{8.19}$$

where $U_{t,1} = 1/R_{t,1} = (K_1 \times A_1)/\Delta x$ and $U_{t,2} = 1/R_{t,2} = (K_2 \times A_2)/\Delta x$. If the paralleling compositions have similar conductivities (e.g. wood and insulation), it is acceptable to assume that the temperature at the external surfaces 1 and 2 is uniform (i.e. $T_{1,1} = T_{2,1}$ and $T_{1,2} = T_{2,2}$). This implies a full mixing of heat at each surface (e.g. by covering a surface in superconductive foil). The real heat transfer condition at the surfaces is between the full separation and full mixing approximations. Thus, the actual heat flux is somewhere in between these two results. Under the full mixing assumption, Eqs. (8.16)–(8.18) become:

$$Q_1 = K_1/\Delta x \times A_1 \times (T_1 - T_2) = (T_1 - T_2)/R_{t,1} \quad \text{(Path-1)} \tag{8.20}$$

$$Q_2 = K_2/\Delta x \times A_2 \times (T_1 - T_2) = (T_1 - T_2)/R_{t,2} \quad \text{(Path-2)} \tag{8.21}$$

$$Q = Q_1 + Q_2 = (T_1 - T_2)/R_{t,1} + (T_1 - T_2)/R_{t,2} = (T_1 - T_2)/R_{t,total} \quad \text{(Total)} \tag{8.22}$$

Therefore,

$$R_{t,total} = 1/(1/R_{t,1} + 1/R_{t,2}) \tag{8.23}$$

This is analogous to the calculation of the total electrical resistance with two parallel resistances. Figure 8.10 shows the corresponding thermal networks for both Eqs. (8.18) and (8.22).

(a)

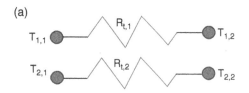

(b)

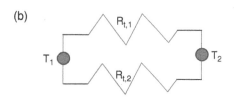

Figure 8.10 Thermal network in parallel connection (a) fully separate (b) fully mixing.

Example 8.3

Estimate the heat flow through a composite wall per unit area (m^2) as shown in Figure 8.11. $T_0 = 25\,°C$, $T_L = 5\,°C$, $K_1 = 2\,W/m \cdot °C$, $K_{2,1} = 10\,W/m \cdot °C$, $K_{2,2} = 5\,W/m \cdot °C$, $K_3 = 18\,W/m \cdot °C$, $A = 1\,m^2$, $A_{2,1} = 0.8\,m^2$, $A_{2,2} = 0.2\,m^2$, $\Delta X_1 = 0.025\,m$, $\Delta X_2 = 0.1\,m$, $\Delta X_3 = 0.18\,m$,

Solution

There are two approaches to calculate the heat flux:

1) Assuming the $K_{2,1}$ and $K_{2,2}$ material are fully separated, two individual heat flow paths are considered, split by the dash line. The resultant thermal network is shown in Figure 8.12a. You can calculate the individual path heat fluxes and then sum them up or you can calculate the total thermal resistance R between T_0 and T_L and then compute the total heat flux. The results are the same [*Approach-1*].
2) Assuming the surfaces of the $K_{2,1}$ and $K_{2,2}$ material share the same temperature at each side (the iso-surface approximation), the resultant thermal network is shown in Figure 8.12b [*Approach-2*].

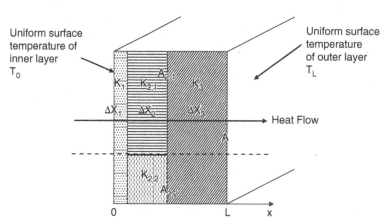

Uniform surface temperature of inner layer T_0

Uniform surface temperature of outer layer T_L

Heat Flow

Figure 8.11 Heat flow through a composite wall.

(a)

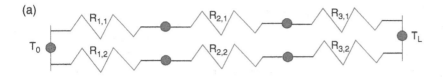

(b)

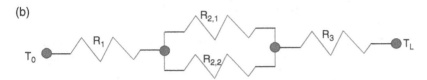

Figure 8.12 Thermal networks for heat flow through a composite wall. (a) Fully separate (b) fully mixing.

Approach-1:

$$R_{1,1} = \Delta X_1/(K_1 A_{2,1}) = 0.025/(2 \times 0.8) = 0.015625$$

$$R_{2,1} = \Delta X_2/(K_{2,1} A_{2,1}) = 0.1/(10 \times 0.8) = 0.0125$$

$$R_{3,1} = \Delta X_3/(K_3 A_{2,1}) = 0.18/(18 \times 0.8) = 0.0125$$

$$R_{path,1} = R_{1,1} + R_{2,1} + R_{3,1} = 0.040625$$

$$Q_{path,1} = (T_0 - T_L)/R_{path,1} = 20/0.040625 = 492 \, W/m^2$$

$$R_{1,2} = \Delta X_1/(K_1 A_{2,2}) = 0.025/(2 \times 0.2) = 0.0625$$

$$R_{2,2} = \Delta X_2/(K_{2,2} A_{2,2}) = 0.1/(5 \times 0.2) = 0.1$$

$$R_{3,2} = \Delta X_3/(K_3 A_{2,2}) = 0.18/(18 \times 0.2) = 0.05$$

$$R_{path,2} = R_{1,2} + R_{2,2} + R_{3,2} = 0.2125$$

$$Q_{path,2} = (T_0 - T_L)/R_{path,2} = 20/0.2125 = 94 \, W/m^2$$

$$\mathbf{Q_{total} = Q_{path,1} + Q_{path,2} = 492 + 94 = 586 \, W/m^2}$$

Or

$$\mathbf{R_{total} = 1/(1/R_{path,1} + 1/R_{path,2}) = 0.0341 \, (or \, U_{total} = 1/R_{total} = 29.3)}$$

$$\mathbf{Q_{total} = (T_0 - T_L)/R_{total} = 20/0.0341 = 586 \, W/m^2}$$

Approach-2:

$$R_{2,1} = \Delta X_2/(K_{2,1} A_{2,1}) = 0.1/(10 \times 0.8) = 0.0125$$

$$R_{2,2} = \Delta X_2/(K_{2,2} A_{2,2}) = 0.1/(5 \times 0.2) = 0.1$$

$$R_2 = 1/(1/R_{2,1} + 1/R_{2,2}) = 0.0111$$

$$R_1 = \Delta X_1/(K_1 A) = 0.025/(2 \times 1) = 0.0125$$

$$R_3 = \Delta X_3/(K_3 A) = 0.18/(18 \times 1) = 0.01$$

$$\mathbf{R_{total} = R_1 + R_2 + R_3 = 0.0336}$$

$$Q_{total} = (T_0 - T_L)/R_{total} = 20/0.0336 = 595\,W/m^2$$

Note that this R value is less than the one obtained in Approach-1, implying that the iso-surface approximation (in Approach-2) *underestimates* the thermal resistance and thus predicts a higher heat gain or heat loss than the parallel path approach, which is preferred for an energy load estimate for some safety rooms when sizing systems. ASHRAE suggests:

- Approach-1 is acceptable if all materials involved (including the bridge element and all building envelope elements in contact with it) are nonmetals (such as wood, drywall, concrete).
- Approach-2 is recommended if any portion of the envelope (e.g. the bridge) has high conductivity (e.g. structural steel or building skin materials or aluminum frames of windows) compared to the other elements.

The actual total R value is between these two R values.

e) Transient Heat Conduction

Steady state heat transfer is an over-simplified approximation, although it is still valuable in estimating a final (or long lasting) state heat transfer condition that can be used to size a system or a design. In real environments, especially with dynamic indoor and outdoor conditions, a building envelope will experience both thermal storage and heat transfer as illustrated in Figure 8.13. Predicting heat conduction with heat storage is thus of great interest and value. Figure 8.13 is similar to Figure 8.5, except that now $Q_{in} \neq Q_{out}$, rather, $Q_{in} = Q_{store} + Q_{out}$. Q_{in} is the total inflow heat flux, Q_{store} is the stored heat in the envelope, Q_{out} is the conductive heat transfer. As a result, the temperature change inside the envelope is not linear anymore. If splitting the envelop into two pieces (or multiple pieces if detail and accuracy of the envelope is of interest), the central temperature (T_m) is used to represent the envelope's overall temperature.

The heat transfer process can be presented as:

$$Q_{in} = Q_{cond,1} = (T_1 - T_m)/(0.5\,R) \qquad \text{[at the surface 1]} \qquad (8.24)$$

$$Q_{out} = Q_{cond,2} = (T_m - T_2)/(0.5\,R) \qquad \text{[at the surface 2]} \qquad (8.25)$$

$$Q_{cond,1} = \rho C_p V \times \Delta T_m/\Delta t + Q_{cond,2} \qquad \text{[in the envelope]} \qquad (8.26)$$

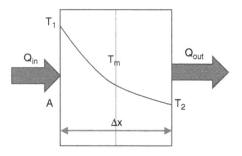

Figure 8.13 Transient one-dimensional conductive heat transfer.

where $R = \Delta x/KA$ is the total thermal resistance of the envelope, ρ is the density of the envelope's material, C_p is the specific heat of the envelope material, V is the volume of the envelope ($V = A \times \Delta x$), K is the thermal conductivity of the envelope material, T_m is the mean temperature of the envelope, and ΔT_m is the change of mean envelope temperature during a period of time Δt. Note that Q_{in}, Q_{out}, T_1, T_2, and T_m can all be a function of time. These are three independent energy conservation equations that can solve for three variables. Boundary conditions at both sides need be specified, which could be given surface temperatures (T_1 and T_2), surface heat flux (Q_{in} and Q_{out}), or a combination of these. Initial conditions for the envelope's temperature should also be provided to predict the transient behavior of the envelope.

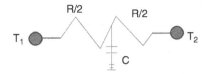

Figure 8.14 Thermal RC network.

$\alpha = K/\rho C_p$ is often used to find ρC_p based on given K and α values. α is a material property defined as thermal diffusivity, representing the rate of heat conducted to heat stored. $C = \rho C_p V$ is called the thermal capacitor, corresponding to the thermal resistance (resistor) R. Equation (8.26) can be rewritten as:

$$(T_1 - T_m)/(0.5\,R) = C \times \Delta T_m/\Delta t + (T_m - T_2)/(0.5\,R) \tag{8.27}$$

or

$$(T_1 - T_m) = 0.5\,RC \times \Delta T_m/\Delta t + (T_m - T_2) \tag{8.28}$$

$$T_1 + T_2 - 2T_m = 0.5\,RC \times \Delta T_m/\Delta t \tag{8.29}$$

This type of equation is often called the thermal RC network model, as illustrated in Figure 8.14. A similar thermal RC network model can be built for an envelope with multiple layers, which will be demonstrated later when combining convection and radiation.

Example 8.4

Predict the outdoor surface temperature T_1 and the wall temperature T_m change during the first hour if $Q_{in} = 500\,W$ (e.g. from solar), and $T_2 = 20\,°C$ is the constant indoor surface temperature during this hour. Assume R = 0.05 °C/W, C = 100 000 J/°C, and $T_{m,0} = 20\,°C$ (initial wall temperature). What is the conductive heat transfer Q_{out} through the wall into the indoor space during this hour?

Solution

From Eqs. (8.24) and (8.29):

$$Q_{in} = Q_{cond,1} = (T_1 - T_m)/(0.5R) = > 500 = (T_1 - T_m)/0.025 = > 12.5 = T_1 - T_m = > T_1 = 12.5 + T_m$$

$$T_1 + T_2 - 2T_m = 0.5RC \times \Delta T_m/\Delta t = > T_1 + 20 - 2T_m = 2500 \times \Delta T_m/\Delta t$$

$$= > 32.5 - T_m = 2500 \times \Delta T_m/\Delta t$$

This can be solved analytically or numerically (e.g. in Excel or Engineering Equation Solver (EES)).

Analytical solution:

$$32.5 - T_m = 2500 \times dT_m/dt = > 32.5 - T_m = -2500 \times d(32.5 - T_m)/dt$$

$$T_m = 32.5 - \exp(2.526 - t/2500)$$

$$T_{m,0} = 20°C; T_{m,3600} = 29.5°C$$

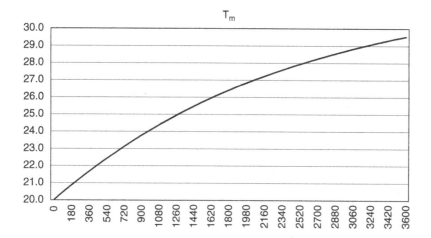

$$T_1 = 12.5 + T_m = 45 - \exp(2.526 - t/2500)$$
$$T_{1,0} = 32.5°C; T_{1,3600} = 42°C$$

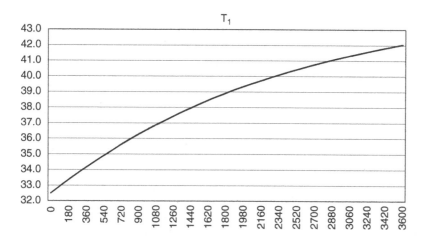

From Eq. (8.25),

$$Q_{out} = (T_m - T_2)/(0.5\,R) = [12.5 - \exp(2.526 - t/2500)]/0.025 = 500 - 40 \times \exp(2.526 - t/2500)$$
$$Q_{out,0} = 0W; Q_{out,3600} = 382\,W$$

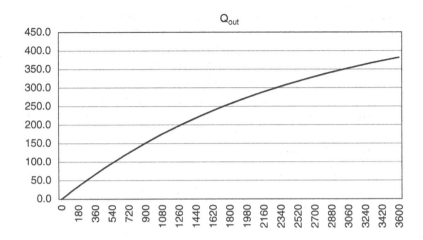

8.2.3 Convection: Convective Heat Transfer

a) **Mechanism:** caused by both a material's molecular collisions/vibration and movement; therefore, convection also needs a media to transfer heat (from high temperature to low temperature). Convective heat transfer typically occurs between a solid (e.g. wall surface) and a fluid (liquid and gas, e.g. water and air) (Figure 8.15). Convection includes the combined effect of conduction and fluid motion. When convection is considered, the conduction of the media does not need to be separately counted. If a fluid has no motion, convection becomes pure conduction.

b) **Classification:** according to the driving force of fluid motion, convection can be divided into:
   - forced convection: driven by external forces such as wind, a fan, or a pump;
   - natural convection: driven by a temperature gradient (or buoyancy).

   According to the flow domain of the fluid, convection can be categorized into:

   - internal flow (e.g. inside a pipe, a duct, or a room);
   - external flow (e.g. around a plate, a cylinder, a building, or a hill).

   Real scenarios may have a combined driving force (forced and natural) with a combined flow domain (internal and external).

c) **Convective Heat Transfer Equation:**

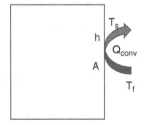

$$Q_{conv} = hA(T_s - T_f) = \frac{(T_s - T_f)}{R_{conv}} \tag{8.30}$$

where Q_{conv} is the convective heat transfer (W or Btu/h); h is the convective heat transfer coefficient (W/m$^2 \cdot$°C, or, Btu/(h $\cdot$ ft$^2 \cdot$°F));

Figure 8.15 Convective heat transfer.

A is the contacting surface area between the solid and the fluid (m^2 or ft^2); T_s and T_f are, respectively, the temperature of the solid surface and the fluid (°C or °F); and $R_{conv} = 1/(hA)$ is the convective thermal resistance (or film resistance) (°C/W or °F/Btu/h). If the temperature difference is zero, there is no convective heat transfer. If the surface area doubles, the convective heat transfer doubles. Note that, different from conductivity K, the convective heat transfer coefficient h is not a property of a material; rather, h is influenced by flow conditions (e.g. flow speed, temperature) and flow geometries (e.g. pipe size) as well as fluid properties (e.g. viscosity). In general, the higher the fluid speed, the larger the convective heat transfer coefficient (or the smaller the film resistance). Hence, forced convection usually has a higher h than natural convection.

d) **Determining Convective Heat Transfer Coefficient h**

The key in computing convective heat transfer is to determine or estimate proper convective heat transfer coefficient h. There are two approaches to determining the h value:

o *Theoretical approach:* which is often used for generic heat transfer scenarios by using classic regressions with dimensionless groups where the h values depend on flow, geometry, and fluid properties. The h value is represented by the dimensionless convective heat transfer coefficient, the Nusselt number (Nu).

$$Nu = (h \times L_c)/K \tag{8.31}$$

where L_c is the characteristic length of the surface and K is the conductivity of the fluid. The Nu number expresses the ratio of convection to conduction of the fluid at the surface where ΔT is the temperature difference between the surface and the fluid.

$$Nu = (h \times L_c)/K = (h\Delta T)/(K\Delta T/L_c) = \dot{q}_{conv}/\dot{q}_{cond} \tag{8.32}$$

When Nu = 1, it represents pure conduction.

For different convection scenarios, e.g. forced vs natural, internal vs external, Nu can be found as a function of the Reynolds (Re) number (for forced convection) or the Grashof (Gr) number, where,

$$Re = U \times L_c/\upsilon \tag{8.33}$$

$$Gr = (g \times \Delta\rho \times V)/(\rho \times \upsilon^2) = (g \times \beta \times \Delta T \times V)/\upsilon^2 \tag{8.34}$$

where U is the forced driving velocity; L_c is the characteristic length of the geometry; υ is the kinematic viscosity of the fluid that equals the absolute viscosity of the fluid divided by the fluid mass density; g is the gravity; ρ is the fluid density; $\Delta\rho$ is the density change due to the temperature ΔT; $V = L_c^3$ is the characteristic volume; and β is the volume expansion coefficient due to the temperature (for ideal gases, $\beta = 1/T$, where T is the fluid temperature in Kelvin). Re represents the ratio of the inertial force to the viscous force, and Gr represents the ratio of the buoyancy force to the viscous force. Tables 8.2 and 8.3 present some of the regressions for common external and internal forced convections. In both tables, Pr is the Prandtl number and $Pr = \upsilon/\alpha = \mu C_p/K$ = Momentum diffusivity/Thermal diffusivity. $Pr \ll 1$ for liquid metals, $Pr \approx 1$ for gases, and $Pr \gg 1$ for nonmetallic liquids. For air, $Pr \approx 0.7$. Tables 8.4 and 8.5 provide the regressions for some common external and internal natural convections.

Table 8.2 Empirical correlations for the average Nusselt number for forced convection over circular and noncircular cylinder in cross-flow.

Cross-section at the cylinder	Fluid	Range of Re	Nusselt number
Circle	Gas or liquid	0.4–4	$Nu = 0.989Re^{0.330}\ Pr^{1/3}$
		4–40	$Nu = 0.911Re^{0.385}\ Pr^{1/3}$
		40–4000	$Nu = 0.683Re^{0.466}\ Pr^{1/3}$
		4000–40 000	$Nu = 0.193Re^{0.618}\ Pr^{1/3}$
		40 000–400 000	$Nu = 0.027Re^{0.805}\ Pr^{1/3}$
Square	Gas	5000–100 000	$Nu = 0.102Re^{0.675}\ Pr^{1/3}$
Square (tilted 45°)	Gas	5000–100 000	$Nu = 0.246Re^{0.588}\ Pr^{1/3}$
Hexagon	Gas	5000–100 000	$Nu = 0.153Re^{0.638}\ Pr^{1/3}$
Hexagon (tilted 45°)	Gas	5000–19 500	$Nu = 0.160Re^{0.638}\ Pr^{1/3}$
		19 500–100 000	$Nu = 0.0385Re^{0.782}\ Pr^{1/3}$
Vertical plate	Gas	4000–15 000	$Nu = 0.228Re^{0.731}\ Pr^{1/3}$
Ellipse	Gas	2500–15 000	$Nu = 0.248Re^{0.612}\ Pr^{1/3}$

Table 8.3 Empirical correlations for the average Nusselt number for forced convection in pipes/ducts (Laminar: Re < 2300; Turbulent: Re > 4000; for noncircular pipe: D = 4A/P where A is the cross-section area and P is the perimeter; f is the flow friction factor).

Pipe/Duct $Nu_D = hD/K$	Given Ts	Given Qs
Laminar	$Nu_D = 3.66$	$Nu_D = 4.36$
Turbulent	$Nu_D = 0.023Re^{0.8}Pr^n$ (n = 0.4 heating; n = 0.3 cooling)	
Turbulent	$Nu_D = [(f/8)(Re-1000)Pr]/[1 + 12.7(f/8)^{0.5}(Pr^{2/3}-1)]$	

Table 8.4 Empirical correlations for the average Nusselt number for natural convection over surfaces.

Geometry	Characteristic length L_c	Range of Ra	Nu
Vertical plate	L	10^4-10^9 $10^{20}-10^{13}$ Entire range	$Nu = 0.59Ra_L^{1/4}$ $Nu = 0.59Ra_L^{1/3}$ $Nu = \left\{0.825 + \dfrac{0.387Ra_L^{1/6}}{\left[1 + (0.492/Pr)^{9/16}\right]^{8/27}}\right\}^2$ (complex but more accurate)
Inclined plate	L		Use vertical plate equations for the upper surface of a cold plate and the lower surface of a hot plate Replace g by g cosθ for Ra < 10^9
Horizontal plate (Surface area A and perimeter p) (a) Upper surface of a hot plate (or lower surface of a cold plate) (b) Lower surface of a hot plate (or upper surface of a cold plate)	A_s/p	10^4-10^7 10^7-10^{11} 10^5-10^{11}	$Nu = 0.54Ra_L^{1/4}$ $Nu = 0.15Ra_L^{1/3}$ $Nu = 0.27Ra_L^{1/4}$

Table 8.4 (Continued)

Geometry	Characteristic length L_c	Range of Ra	Nu
Vertical cylinder 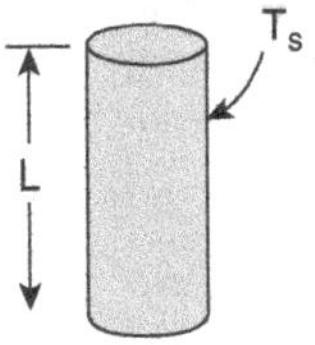	L		A vertical cylinder can be treated as a vertical plate when $$D \geq \frac{35L}{Gr_L^{1/4}}$$
Horizontal cylinder 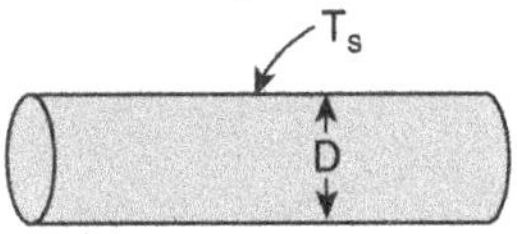	D	$Ra_D \leq 10^{12}$	$$Nu = \left\{ 0.6 + \frac{0.387 Ra_D^{1/6}}{\left[1 + (0.559/Pr)^{9/16} \right]^{8/27}} \right\}^2$$
Sphere 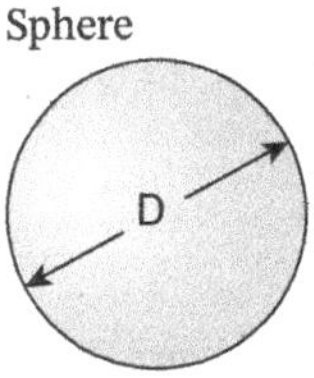	D	$Ra_D \leq 10^{11}$ $(Pr \geq 0.7)$	$$Nu = 2 + \frac{0.589 Ra_D^{1/4}}{\left[1 + (0.469/Pr)^{9/16} \right]^{4/9}}$$

Table 8.5 Empirical correlations for the average Nusselt number for natural convection in enclosures (δ in the diagram is the characteristic length L_c, Ra = Gr × Pr is called Rayleigh number).

Geometry	Fluid	H/δ	Range of Pr	Range of Ra	Nusselt number
Vertical rectangular enclosure (or vertical cylindrical enclosure)	Gas or liquid	—	—	Ra < 2000	Nu = 1
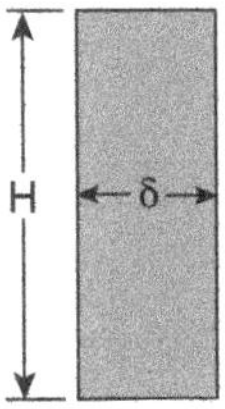	Gas	11–42	0.5–2	2×10^3–2×10^5	$Nu = 0.197 \, Ra^{1/4} \left(\frac{H}{d} \right)^{-1/9}$
		11–42	0.5–2	2×10^5–10^7	$Nu = 0.073 \, Ra^{1/3} \left(\frac{H}{d} \right)^{-1/9}$
	Liquid	10–40	1–20 000	10^4–10^7	$Nu = 0.042 Pr^{0.012} Ra^{1/4} \left(\frac{H}{d} \right)^{-0.3}$
		1–40	1–20	10^6–10^9	$Nu = 0.046 \, Ra^{1/3}$

(Continued)

Table 8.5 (Continued)

Geometry	Fluid	H/δ	Range of Pr	Range of Ra	Nusselt number
Inclined rectangular enclosure					Use the correlations for vertical enclosures as a first-degree approximation for $\theta \leq 20°$ by replacing g in the Ra relation by g $\cos \theta$
Horizontal rectangular enclosure (hot surface at the top)	Gas or liquid	—	—	—	$Nu = 1$
Horizontal rectangular enclosure (hot surface at the bottom)	Gas or liquid	—	—	$Ra < 1700$	$Nu = 1$
	Gas	—	0.5–2	1.7×10^3–7×10^3	$Nu = 0.059\,Ra^{0.4}$
		—	0.5–2	7×10^3–3.2×10^5	$Nu = 0.212\,Ra^{1/4}$
		—	0.5–2	$Ra > 3.2 \times 10^5$	$Nu = 0.061\,Ra^{1/3}$
	Liquid	—	1–5000	1.7×10^3–6×10^3	$Nu = 0.012\,Ra^{0.6}$
		—	1–5000	6×10^3–3.7×10^4	$Nu = 0.375\,Ra^{0.2}$
		—	1–20	3.7×10^4–10^8	$Nu = 0.13\,Ra^{0.3}$
		—	1–20	$Ra > 10^8$	$Nu = 0.057\,Ra^{1/3}$
Concentric rectangular cylinders	Gas or liquid	—	1–5000	6.3×10^3–10^6	$Nu = 0.11\,Ra^{0.29}$
		—	1–5000	10^6–10^8	$Nu = 0.40\,Ra^{0.20}$
Concentric spheres	Gas or liquid	—	0.7–4000	10^2–10^9	$Nu = 0.228\,Ra^{0.226}$

Table 8.6 Range of practical convection coefficients.

Arrangement	W/(m² · K)	Btu/(h · ft² · °F)
Air, free convection	6–30	1–5
Superheated steam or air, forced convection	30–300	5–50
Oil, forced convection	60–1800	10–300
Water, forced convection	300–6000	50–1000
Water, boiling	3000–60 000	500–10 000
Steam, condensing	6000–120 000	1000–20 000

The conversion between SI and USCS units is 5.678 W/(m² · K) = 1 Btu/(h · ft² · °F)

- o **Practical approach**: for most realistic heat transfers, the geometries and flow conditions are much more complicated than the simplified theoretic cases. The h value highly depends on geometry and flow, and thus seeking a specially developed empirical formula or values for such cases (or similar cases) will be the most suitable and convenient method for engineering calculations. Most of these calculations use typical fluid properties, e.g. $Pr_{air} = 0.7$ and assume turbulent flow etc. For instance, Eq. (8.35) shows the empirical h formula for the turbulent flow of air in ducts where v is the flow speed in the duct and D_h is the hydraulic diameter of the duct:

$$h = 0.50 \left(\frac{v^4}{D_h} \right)^{1/5} \quad \text{US}$$

$$h = 8.80 \left(\frac{v^4}{D_h} \right)^{1/5} \quad \text{SI}$$

(8.35)

These kinds of empirical h formulas, for the specific applications that the formulas were created for, are often more accurate than the general ones. For rough/quick engineering estimates, the typical ranges of h value as shown in Table 8.6 can also be useful.

Example 8.5

Can air be used as an insulation material?

Solution

- **Air spaces**
  Air conduction resistance

  $$R = \delta/\lambda$$

where

R = air conduction resistance
δ = air layer thickness
λ = thermal conductivity (0.023 W/m · K for air)

"Ideal case" for 0.02 m thick air layer

$$R = \delta/\lambda = 0.02/0.023 = 0.87 \text{m}^2 \text{ K/W}$$

Because of convection and radiation effect, the actual R is between 0.1 and 0.7 m² K/W.

- **Windows**

Single glazing   U = 5.5 W/m² · K   R-1 (ft² · °F · h/Btu)
Double glazing   U = 2.7 W/m² · K   R-2
Triple glazing   U = 1.8 W m² · K   R-3

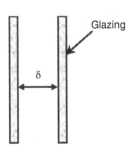

Considering both inside and outside surface convections, the steady state one-dimensional heat transfer through a composite wall (e.g. a stud wall consisting of wooden studs with insulation in between that is sandwiched between two layers) can be represented in two possible heat transfer network configurations as shown in Figure 8.16. Similar to what was presented earlier in conduction (Figures 8.11 and 8.12), the thermal networks based on the isothermal plane (IP) assumption and fully parallel path (PP) assumption can be found in Figure 8.16.

Heat flow is two dimensional in reality since the heat flow lines would not be perfectly parallel in the X-direction but rather would get distorted and bend so as to preferentially favor flow through the studs which have a higher material conductivity (the thermal bridge effect). As a result, less heat will flow through the insulation (lower conductivity material) and more through the wooden stud section. Hence, the one-dimensional network model IP would underestimate the total resistance and predict a higher heat loss (through the insulation and thus the wall) than the network PP, as confirmed in Example 8.3.

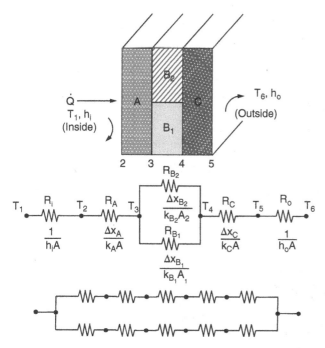

Figure 8.16 Heat transfer through a composite wall including surface convections.

8.2.4 Radiation: Radiative Heat Transfer

a) **Mechanism:** caused by the electromagnetic waves (or photons) between surfaces of different temperatures. Any object with a temperature (over 0 K) emits electromagnetic waves (radiation) and also absorbs radiation from other objects. Radiation does not require an intervening medium (i.e. radiation can occur in vacuum environments, such as the exosphere for solar radiation) and is not affected by the surrounding fluid temperature but rather the surface temperature difference. Radiation follows all electromagnetic wave theories, such as,

$$\lambda = C/f \tag{8.36}$$

where λ is the wavelength, C is the light speed (3×10^8 m/s), and f is the frequency. Therefore, radiation heat transfer is at the light speed.

b) **Electromagnetic wave spectrum:** the radiation emitted from a body exhibits as a wide electromagnetic wave spectrum as shown in Figure 8.17, in which only a small fraction is the thermal radiation (radiative heat transfer) while the majority of the radiation cannot be sensed directly by the human body. Radiation energy varies with object temperature and radiation wavelength. Figure 8.18 shows the maximum (black body) radiation energy distribution $E_{b\lambda}$ by wavelength at different temperatures. Note that a room temperature of 300 K does not provide radiation that reaches the visible wavelength range, and thus objects at room temperature (e.g. the human body temperature) are not visible in a

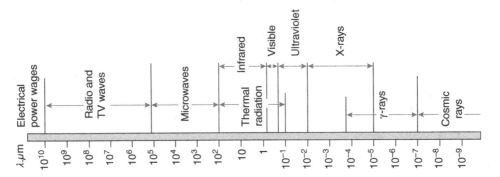

Figure 8.17 Electromagnetic wave spectrum of radiation.

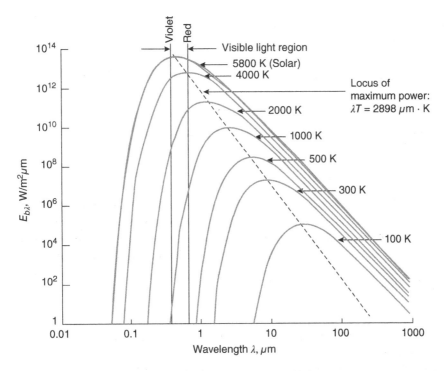

Figure 8.18 Electromagnetic wave energy spectrum of radiation.

fully dark environment (i.e. cannot glow). On the contrary, solar of 5800 K provides light and heat (and UV and others). Therefore, solar radiation and thermal radiation present different thermal and lighting performances (to be discussed later).

c) **Total radiation energy:** integrating the radiation energy distribution over the entire wavelength provides the total radiation energy (W/m^2) of a (black) body at temperature T.

$$E_b(T) = \int_0^\infty E_{b\lambda}(\lambda, T)d\lambda = \sigma T^4 \qquad (8.37)$$

where $\sigma = 5.67 \times 10^{-8}$ W/(m$^2 \cdot$ K^4) $= 0.1714 \times 10^{-8}$ Btu/ (h $\cdot$ ft$^2 \cdot$ R^4) is called Stefan-Boltzmann constant and T is the body temperature in the unit of K or R. Equation (8.37) shows the maximum radiation energy per unit surface area that can be emitted by an object at temperature T, which is defined as a black body. A black body is a perfect emitter and absorber of radiation. The total radiation energy (W) released from a body with a surface area A is then:

$$\dot{Q} = AE_b(T) = \sigma AT^4 \qquad (8.38)$$

d) **Radiation properties:** when a radiation hits on a surface (Figure 8.19), a fraction of the radiation will be reflected by the surface, a fraction will be transmitted

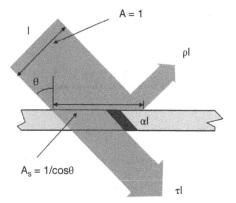

Figure 8.19 Surface radiation properties: reflectivity, transmissivity, and absorptivity (I is the total incident radiation in W/m^2).

(if the surface is transparent or semi-transparent), and the rest will be absorbed. These fractions in percentiles are expressed by the surface radiation properties: reflectivity (ρ), transmissivity (τ), and absorptivity (α).

$$\alpha + \rho + \tau = 1 \qquad (8.39)$$

If a material or surface is opaque, $\tau = 0$ and $\alpha + \rho = 1$. Note that the total incident area A is increased to $A_s = A/\cos\theta$ when it hits the surface. This implies that the incoming radiation strength (in W/m^2) will be spread over a larger area and will thus lead to a smaller strength (in W/m^2). The main reason why summer solar is much stronger than winter is due to a higher solar angle (to be introduced in Chapter 9).

Another important radiation property is emissivity (ε), which expresses the capability (between 0 and 1) of a body or a surface to emit radiation energy. $\varepsilon = 1$ means the body can emit full radiation with no loss (i.e. a perfect emitter), which is called a black body. The actual radiation E(T) (W/m^2) emitted from a real body at temperature T with $0 \le \varepsilon \le 1$ is:

$$E(T) = \varepsilon E_b(T) = \varepsilon\sigma T^4 \qquad (8.40)$$

Emissivity is a function of temperature T and electromagnetic wavelength λ: $\varepsilon(T, \lambda)$. As shown in Figure 8.20, although $\varepsilon(T, \lambda)$ varies with λ, it only fluctuates around a mean value for a given T. This mean value is solely a function of T: $\varepsilon(T)$. Using $\varepsilon(T)$ instead of $\varepsilon(T, \lambda)$ for most engineering calculations is a common approximation. A body or surface under this assumption is called a grey body. Table 8.7 lists the emissivities of some common building materials at specified temperatures. According to Kirchhoff's Law, at a steady state,

$$E_{absorb} = \alpha\sigma T^4 = E_{emit} = \varepsilon\sigma T^4 \qquad (8.41)$$

Therefore,

$$\alpha(T) = \varepsilon(T) \qquad (8.42)$$

Most material manufacturers provide these radiation properties: ε, ρ, and τ. Figure 8.21 demonstrates the different solar radiation properties of three types of glasses, as examples, which will surely

(a)

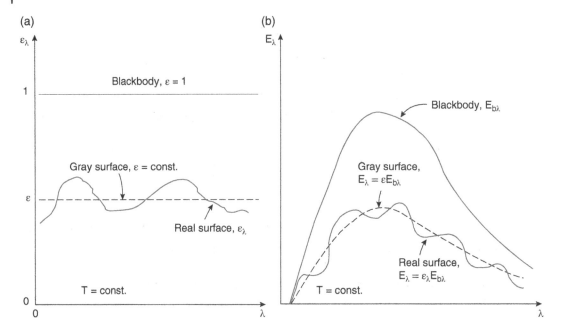

(b)

Figure 8.20 Emissivity for black body, gray body, and real body. (a) Emissivity for black, gray, and real bodies at a constant temperature over the wavelength and (b) radiation energy distribution for black, gray, and real bodies at a constant temperature over the wavelength.

affect the energy use of the space behind the glass. Note that all of these properties vary with temperature and thus their values are different for solar (5800 K) and room (300 K) temperatures, which are respectively called solar radiation (short-wavelength) and thermal radiation (long-wavelength). Figure 8.22 illustrates the variance of glass transmissivity (or transmittance) with different wavelengths. Due to the high transmissivity at short-wavelengths and the low transmissivity at long-wavelengths, solar radiation (short-wavelength) can easily penetrate through the glass while thermal radiation at room temperature (long-wavelength) cannot escape through the glass. This is the well-known greenhouse effect (Figure 8.23).

e) **Radiative heat transfer between two surfaces:** the net radiative heat transfer Q_{12} (W or Btu/h) between surfaces 1 and 2 can be calculated as:

$$Q_{12} = Q_{1 \to 2} - Q_{2 \to 1} = \frac{\sigma\left(T_1^4 - T_2^4\right)}{\left((1 - \varepsilon_1)/(A_1\varepsilon_1)\right) + \left(1/(A_1 F_{12})\right) + \left((1 - \varepsilon_2)/(A_2\varepsilon_2)\right)} \tag{8.43}$$

where $Q_{1 \to 2}$ is emitted by surface 1 and strikes surface 2 and $Q_{2 \to 1}$ is emitted by surface 2 and is absorbed by surface 1; T_1 and T_2 are the temperatures of surfaces 1 and 2 in K or R; A_1 and A_2 are the areas of surfaces 1 and 2 in m^2 or ft^2; and ε_1 and ε_2 are the emissivities of surfaces 1 and 2.

$(1 - \varepsilon_1)/(A_1\varepsilon_1)$ and $(1 - \varepsilon_2)/(A_2\varepsilon_2)$ are called the surface resistances of surfaces 1 and 2 and $1/(A_1 F_{12})$ is called the space resistance. When the surface is black, surface resistance becomes zero (no loss). If the surface is fully insulated (adiabatic), this is called the reradiating surface (assume negligible convection at the surface), which also has no surface heat loss.

Table 8.7 Emissivity of some common building materials at specified temperatures.

Surface	Temperature, °C	Temperature, °F	ε
Aluminum foil			
Bright	40	100	0.05
Brick			
Red, rough	40	100	0.93
Concrete			
Rough	40	100	0.94
Glass			
Smooth	40	100	0.94
Ice			
Smooth	0	32	0.97
Marble			
White	40	100	0.95
Paints			
Black gloss	40	100	0.90
White	40	100	0.89–0.97
Various oil paints	40	100	0.92–0.96
Paper			
White	40	100	0.95
Sandstone	40–250	100–500	0.83–0.90
Snow	−12 to −6	10–20	0.82
Water			
0.1 mm or more thick	40	100	0.96
Wood			
Oak. planed	40	100	0.90
Walnut, sanded	40	100	0.83
Spruce, sanded	40	100	0.82
Beech	40	100	0.94

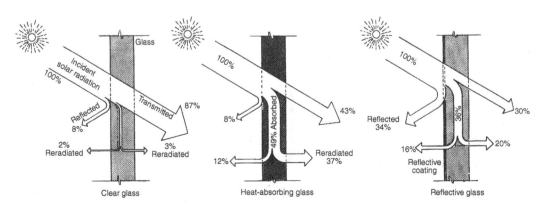

Figure 8.21 Solar radiation properties of three types of glasses.

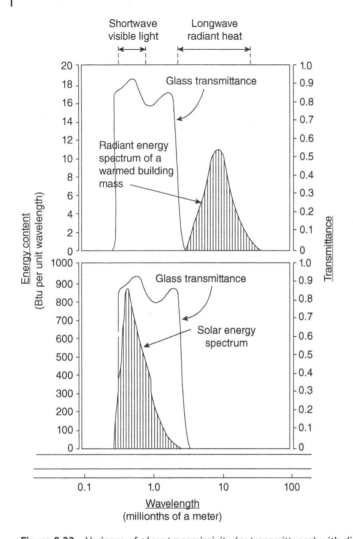

Figure 8.22 Variance of glass transmissivity (or transmittance) with different wavelengths.

Figure 8.23 Greenhouse effect.

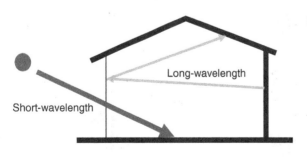

F_{12} is the view factor between surface 1 and 2 ($0 \leq F_{12} \leq 1$). The view factor represents the fraction of the radiation leaving surface 1 that strikes surface 2 directly. $F_{ii} = 0$ (the fraction of radiation leaving surface i that strikes surface i) if the surface i is plane or convex, while $F_{ii} > 0$ if the surface is concave. Other useful view factors include:

i) Reciprocity rule:

$$A_i F_{ij} = A_j F_{ji} \tag{8.44}$$

If $A_i = A_j$, $F_{ij} = F_i$.

ii) Summation rule:

$$\sum_{j=1}^{N} F_{ij} = 1 \tag{8.45}$$

The total view factor from surface 1 to all other surrounding surfaces is 1.

iii) Superposition rule:

$$F_{i,j} = F_{i,j1} + F_{i,j2} \tag{8.46}$$

Surfaces j1 and j2 are two sub-surfaces of surface j.

iv) Symmetry rule:

$$F_{i,j} = F_{i,k} \tag{8.47}$$

when surfaces j and k are symmetric about surface i.

Example 8.6

Calculate all of the view factors between the two spheric surfaces A_1 and A_2 shown below.

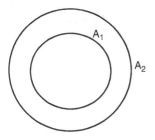

Solution

$$F_{11} = 0$$
$$F_{12} = 1 - F_{11} = 1$$
$$F_{21} = A_1 F_{12}/A_2 = A_1/A_2$$
$$F_{22} = 1 - F_{21} = 1 - A_1/A_2$$

Equation (8.43) can be simplified as follows if $F_{12} = 1$, and $A_1 = A_2 = A$ (which represents two surfaces of identical area, facing closely toward each other),

$$Q_{12} = \frac{\sigma(T_1^4 - T_2^4)A}{(1/\varepsilon_1) + (1/\varepsilon_2) - 1} \tag{8.48}$$

If one surface (say surface 2) is a black body ($\varepsilon_2 = 1$) and $F_{12} = 1$, Eq. (8.43) can be simplified as:

$$Q_{12} = \varepsilon_1 \sigma(T_1^4 - T_2^4)A_1 \tag{8.49}$$

This equation is commonly used when a small body (e.g. a person) is surrounded by a large isothermal enclosure (e.g. a room). The surrounding isothermal surfaces can be treated as effectively black as the surroundings will absorb all photons leaving the small body (A_1) and the chance of the reflected radiation from the surrounding surfaces back to A_1 is almost zero. Equation (8.49) can be expressed in a similar format as conduction and convection,

$$Q_{12} = \varepsilon_1 \sigma(T_1^4 - T_2^4)A_1 = \varepsilon_1 \sigma(T_1^2 - T_2^2)(T_1^2 + T_2^2)A_1 = \varepsilon_1 \sigma(T_1 + T_2)(T_1^2 + T_2^2)(T_1 - T_2)A_1$$
$$= h_{rad}A_1(T_1 - T_2) \tag{8.49}$$

where h_{rad} is the equivalent radiative heat transfer coefficient,

$$h_{rad} = \varepsilon_1 \sigma(T_1 + T_2)(T_1^2 + T_2^2) \tag{8.50}$$

If T_1 and T_2 are within the range of normal living conditions (around 300 K), Eq. (8.50) can be approximated as,

$$h_{rad} \approx 6\varepsilon_1 \tag{8.51}$$

Equation (8.51) is a very useful equation for most building engineering calculations. The radiative heat transfer coefficient is often combined with the convective heat transfer coefficient at building envelope surfaces to simplify the calculation.

Example 8.7

A person is standing at the center of a large classroom. The room air and surface temperatures are 20 °C, the body temperature is 29 °C, and the body area is 1.6 m². Assume the convective heat transfer coefficient is 6 W/m² · °C (ignore the conductive heat transfer via direct body contacts and assume the body surface emissivity is 0.95). What is the total heat loss from the person?

Solution

$$Q_{total} = Q_{conv} + Q_{rad} = h_{conv}A(T_{body} - T_{air}) + h_{rad}A(T_{body} - T_{surface})$$

$$Q_{total} = (h_{conv} + h_{rad})A(T_{body} - T_{air}) = (6 + 6 \times 0.95) \times 1.6 \times 9 = 168.48\ \text{W}$$

Since the typical heat generation from a human body is about 100 W, a person in this room will feel cold due to more heat loss than heat generation.

Example 8.8

A room of 5 m (W) × 5 m (L) × 3 m (H) has a large external window 5 m (W) × 3 m (H) at a temperature of 4 °C. The emissivity of the window is 0.9. The surface temperature of other interior walls is 25 °C. Assuming these walls are adiabatic (and thus nominally black), what is the heat loss through the window via radiation? If the room air temperature is 16 °C, and assuming the interior convective heat transfer coefficient is 2 W/m$^2 \cdot$°C, what is the total heat loss through the window?

Solution

Radiation:

$$Q_{rad} = 6\varepsilon \times (T_{window} - T_{wall}) \times A_{window} = 6 \times 0.9 \times 21 \times 15 = 1701 \text{ W}$$

Convection:

$$Q_{conv} = h \times A_{window} \times (T_{window} - T_{room}) = 2 \times 15 \times 9 = 270 \text{ W}$$

Total heat loss:

$$Q_{total} = 1701 + 270 = 1971 \text{ W}$$

8.3 Practical Heat Transfer through Building Envelope

The actual sensible heat transfer through a building envelope is the result of the combined conduction in the envelope with convection and radiation at both indoor and outdoor surfaces (as shown in Figure 8.24). The envelope can have multiple layers, and each layer may have multiple sections. A general approach to calculating the total heat transfer through the envelope includes these steps:

- Identify driving temperature potential (ΔT)
- Identify parallel heat transfer paths
- Calculate resistance for each parallel path, including radiation
- Calculate $U_i A_i$ for each parallel path

Figure 8.24 Heat transfer through building envelope with combined conduction, convection, and radiation.

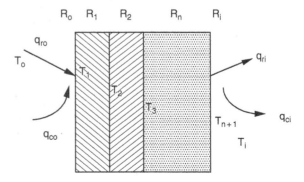

- Calculate overall UA
- Calculate total heat transfer

Again, this parallel path approach overestimates the thermal resistance and thus predicts a lower heat gain or heat loss than the isothermal plane approach, which is acceptable if all materials involved (including the bridge element and all building envelope elements in contact with it) are nonmetals (such as wood, drywall, concrete). Example 8.9 was used to demonstrate the engineering process to calculate the overall U and R values of an envelope and the total heat transfer through a building envelope.

Example 8.9

Determine the thermal conductance of the solid masonry walls shown in the following figure. Calculate the temperature distribution in the wall. Indoor air temperature is 70 °F and outdoor design air temperature is 5 °F. Solar radiation is neglected in this case.

Solution

The calculation considers both convection and radiation at both indoor and outdoor surfaces and conduction through the masonry wall.

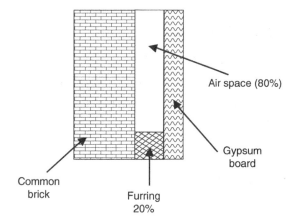

Two parallel paths were split by the boundary of furring and air space. Associated coefficients and properties were identified and filled in the calculation form below.

	Construction	Resistance (R)	
1.	Outside surface (15 mph wind)	0.17	0.17
2.	Common brick	1.60	1.60
3.	Nominal 1-in. × 3 in. vertical furring	–	0.94
4	Nonreflective air space 0.75 in. (50 °F mean, 10 °F difference)	1.01	–
5.	Gypsum wall board, 0.5 in.	0.45	0.45
6.	Inside surface (still air)	0.68	0.68
	Total path thermal resistance (R$_i$)	**3.91**	**3.84**

Table 8.8 Surface conductances and resistances for air.

		Surface emittance, ε					
		Nonreflective		Reflective			
		$\varepsilon = 0.90$		$\varepsilon = 0.20$		$\varepsilon = 0.05$	
Position of surface	Direction of heat flow	h_i	R	h_i	R	h_i	R
Still air							
Horizontal	Upward	9.26	0.11	5.17	0.19	4.32	0.23
Sloping—45°	Upward	9.09	0.11	5.00	0.20	4.15	0.24
Vertical	Horizontal	8.29	0.12	4.20	0.24	3.35	0.30
Sloping—45°	Downward	7.50	0.13	3.41	0.29	2.56	0.39
Horizontal	Downward	6.13	0.16	2.10	0.48	1.25	0.80
Moving air (any position)		h_o	R				
Wind (for winter) 6.7 m/s (24 km/h)	Any	34.0	0.030	—	—	—	—
Wind (for summer) 3.4 m/s (12 km/h)	Any	22.7	0.044	—	—	—	—

Notes:
1) Surface conductance h_i and h_o measured in W/(m$^2 \cdot$ K); resistance R in (m$^2 \cdot$ K)/W.
2) No surface has both an air space resistance value and a surface resistance value.
3) Conductances are for surfaces of the stated emittance facing virtual black body surroundings at the same temperature as the ambient air. Values are based on a surface-air temperature difference of 5.5 °C and for surface temperatures of 21 °C.

The surface resistances (or conductance $h = 1/R = h_c + h_r$) at inside and outside surfaces (Layers 1 and 6 in Example 8.9) are determined based on the ASHRAE standard shown in Table 8.8. The ASHRAE fundamental handbook also provides typical thermal properties (design values) of common building and insulation materials, as partly illustrated in Table 8.9 (used for Layers 2, 3, and 5 in Example 8.9). For an air cavity/space inside of the envelope (Layer 4), both convection and radiation need be considered, and thus the thermal conductance or resistance value is determined by the temperatures of two surfaces forming the air space, air space depth, and surface emissivity (emittance). Table 8.10 shows the thermal resistance of plane air spaces. To determine the resistance, the effective emittance ε_{eff} needs be calculated according to Table 8.11, where

$$\frac{1}{\varepsilon_{eff}} = \frac{1}{\varepsilon_1} + \frac{1}{\varepsilon_2} - 1 \tag{8.52}$$

The path R is respectively: 3.91 and 3.84
The total R: $1/R = A_1/R_1 + A_2/R_2 = 0.8/R_1 + 0.2/R_2 = 3.896$ °F $\cdot$ ft$^2 \cdot$ h/Btu

or

The total U : $U = 1/R = A_1 \times U_1 + A_2 \times U_2$

Table 8.9 Typical thermal properties (design values) of common building and insulation materials.

Description		Density, kg/m^3	Conductivity (k), W/(m·K)	Conductance (C), W/(m^2·K)	Resistance (R)		Specific heat, kJ/(kg·K)
					1/k, (m·K)/ W	For thickness listed (1/C), (m^2·K)/W	
Building board							
Asbestos-cement board		1900	0.58	—	1.73	—	1.00
Asbestos-cement board	3.2 mm	1900	—	187.4	—	0.005	—
Asbestos-cement board	6.4 mm	1900	—	93.7	—	0.011	—
Gypsum or plaster board	9.5 mm	800	—	17.6	—	0.056	1.09
Gypsum or plaster board	12.7 mm	800	—	12.6	—	0.079	—
Gypsum or plaster board	15.9 mm	800	—	10.1	—	0.099	—
Plywood (Douglas Fir)		540	0.12	—	8.66	—	1.21
Plywood (Douglas Fir)	6.4 mm	540	—	18.2	—	0.055	—
Plywood (Douglas Fir)	9.5 mm	540	—	12.1	—	0.083	—
Plywood (Douglas Fir)	12.7 mm	540	—	9.1	—	0.11	—
Plywood (Douglas Fir)	15.9 mm	540	—	7.3	—	0.14	—
Plywood or wood panels	19.0 mm	540	—	6.1	—	0.16	1.21
Vegetable fiber board							
Sheathing, regular density	12.7 mm	290	—	4.3	—	0.23	1.30
	9.8 mm	290	—	2.8	—	0.36	—
Sheathing intermediate density	12.7 mm	350	—	5.2	—	0.19	1.30
Nail-base sheathing	12.7 mm	400	—	5.3	—	0.19	1.30
Shingle backer	9.5 mm	290	—	6.0	—	0.17	1.30
Shingle backer	7.9 mm	290	—	7.3	—	0.14	—
Sound deadening board	12.7 mm	240	—	4.2	—	0.24	1.26
Tile and lay-in panels, plain or acoustic		290	0.058	—	17	—	0.59

Table 8.10 Thermal resistance of plane air spaces ($m^2 \cdot K/W$).

Position of air space	Direction of heat flow	Air space Mean temp, °C	Temp. diff., °C	13 mm air space Effective emittance ε_{eff}					20 mm air space Effective emittance ε_{eff}				
				0.03	0.05	0.2	0.5	0.82	0.03	0.05	0.2	0.5	0.82
Horiz.	Up	32.2	5.6	0.37	0.36	0.27	0.17	0.13	0.41	0.39	0.28	0.18	0.13
		10.0	16.7	0.29	0.28	0.23	0.17	0.13	0.30	0.29	0.24	0.17	0.14
		10.0	5.6	0.37	0.36	0.28	0.20	0.15	0.40	0.39	0.30	0.20	0.15
		−17.8	11.1	0.30	0.30	0.26	0.20	0.16	0.32	0,32	0.27	0.20	0.16
		−17.8	5.6	0.37	0.36	0.30	0.22	0.18	0.39	0.38	0.31	0.23	0.18
		−45.6	11.1	0.30	0.29	0.26	0.22	0.18	0.31	0.31	0.27	0.22	0.19
		−45.6	5.6	0.36	0.35	0.31	0.25	0.20	0.38	0.37	0.32	0.26	0.21
45° Slope	Up	32.2	5.6	0.43	0.41	0.29	0.19	0.13	0.52	0.49	0.33	0.20	0.14
		10.0	16.7	0.36	0.35	0.27	0.19	0.15	0.35	0.34	0.27	0.19	0.14
		10.0	5.6	0.45	0.43	0.32	0.21	0.16	0.51	0.48	0.35	0.23	0.17
		−17.8	11.1	0.39	0.38	0.31	0.23	0.18	0.37	0.36	0.30	0 23	0.18
		−17.8	5.6	0.46	0.45	0.36	0.25	0.19	0.48	0.46	0.37	0.26	0.20
		−45.6	11.1	0.37	0.36	0.31	0.25	0.21	0.36	0.35	0.31	0.25	0.20
		−45.6	5.6	0.46	0.45	0.38	0.29	0.23	0.45	0.43	0.37	0.29	0.23
Vertical	Horiz.	32.2	5.6	0.43	0.41	0.29	0.19	0.14	0.62	0.57	0.37	0.21	0.15
		10.0	16.7	0.45	0.43	0.32	0.22	0.16	0.51	0.49	0.35	0.23	0.17
		10.0	5.6	0.47	0.45	0.33	0.22	0.16	0.65	0.61	0.41	0.25	0.18
		−17.8	11.1	0.50	0.48	0.38	0.26	0.20	0.55	0.53	0.41	0.28	0.21
		−17.8	5.6	0.52	0.50	0.39	0.27	0.20	0.66	0.63	0.46	0.30	0.22
		−45.6	11.1	0.51	0.50	0.41	0.31	0.24	0.51	0.50	0.42	0.31	0.24
		−45.6	5.6	0.56	0.55	0.45	0.33	0.26	0.65	0.63	0.51	0.36	0.27
45° Slope	Down	32.2	5.6	0.44	0.41	0.29	0.19	0.14	0.62	0.58	0.37	0.21	0.15
		10.0	16.7	0.46	0.44	0.33	0.22	0.16	0.60	0.57	0.39	0.24	0.17
		10.0	5.6	0.47	0.45	0.33	0.22	0.16	0.67	0.63	0.42	0.26	0.18
		−17.8	11.1	0 51	0.49	0.39	0.27	0.20	0.66	0.63	0.46	0.30	0.22
		−17.8	5.6	0.52	0.50	0.39	0.27	0.20	0.73	0.69	0.49	0.32	0.23
		−45.6	11.1	0.56	0.54	0.44	0.33	0.25	0.67	0.64	0.51	0.36	0.28
		−45.6	5.6	0.57	0.56	0.45	0.33	0.26	0.77	0.74	0.57	0.39	0.29

(Continued)

Table 8.10 (Continued)

Position of air space	Direction of heat flow	Air space Mean temp, °C	Temp. diff., °C	13 mm air space Effective emittance ε_{eff}					20 mm air space Effective emittance ε_{eff}				
				0.03	0.05	0.2	0.5	0.82	0.03	0.05	0.2	0.5	0.82
Horiz.	Down	32.2	5.6	0.44	0.41	0.29	0.19	0.14	0.62	0.58	0.37	0.21	0.15
		10.0	16.7	0.47	0.45	0.33	0.22	0.16	0.66	0.62	0.42	0.25	0.18
		10.0	5.6	0.47	0.45	0.33	0.22	0.16	0.68	0.63	0.42	0.26	0.18
		−17.8	11.1	0.52	0.50	0.39	0.27	0.20	0.74	0.70	0.50	0.32	0.23
		−17.8	5.6	0.52	0.50	0.39	0.27	0.20	0.75	0.71	0.51	0.32	0.23
		−45.6	11.1	0.57	0.55	0.45	0.33	0.26	0.81	0.78	0.59	0.40	0.30
		−45.6	5.6	0.58	0.56	0.46	0.33	0.26	0.83	0.79	0.60	0.40	0.30

Position of air space	Direction of heat flow	Air space Mean temp, °C	Temp. diff., °C	40 mm air space					90 mm air space				
Horiz.	Up	32.2	5.6	0.45	0.42	0.30	0.19	0.14	0.50	0.47	0.32	0.20	0.14
		10.0	16.7	0.33	0.32	0.26	0.18	0.14	0.27	0.35	0.28	0.19	0.15
		10.0	5.6	0.44	0.42	0.32	0.21	0.16	0.49	0.47	0.34	0.23	0.16
		−17.8	11.1	0.35	0.34	0.29	0.22	0.17	0.40	0.38	0.32	0.23	0.18
		−17.8	5.6	0.43	0.41	0.33	0.24	0.19	0.48	0.46	0.36	0.26	0.20
		−45.6	11.1	0.34	0.34	0.30	0.24	0.20	0.39	0.38	0.33	0.26	0.21
		−45.6	5.6	0.42	0.41	0.35	0.27	0.22	0.47	0.45	0.38	0.29	0.23
45° Slope	Up	32.2	5.6	0.51	0.48	0.33	0.20	0.14	0.56	0.52	0.35	0.21	0.14
		10.0	16.7	0.38	0.36	0.28	0.20	0.15	0.40	0.38	0.29	0.20	0.15
		10.0	5.6	0.51	0.48	0.35	0.23	0.17	0.55	0.52	0.37	0.24	0.17
		−17.8	11.1	0.40	0.39	0.32	0.24	0.18	0.43	0.41	0.33	0.24	0.19
		−17.8	5.6	0.49	0.47	0.37	0.26	0.20	0.52	0.51	0.39	0.27	0.20
		−45.6	11.1	0.39	0.38	0.33	0.26	0.21	0.41	0.40	0.35	0.27	0.22
		−45.6	5.6	0.48	0.46	0.39	0.30	0.24	0.51	0.49	0.41	0.31	0.24
Vertical	Horiz.	32.2	5.6	0.70	0.64	0.40	0.22	0.15	0.65	0.60	0.38	0.22	0.15
		10.0	16.7	0.45	0.43	0.32	0.22	0.16	0.47	0.45	0.33	0.22	0.16
		10.0	5.6	0.67	0.62	0.42	0.26	0.18	0.64	0.60	0.41	0.25	0.18
		−17.8	11.1	0.49	0.47	0.37	0.26	0.20	0.51	0.49	0.38	0.27	0.20
		−17.8	5.6	0.62	0.59	0.44	0.29	0.22	0.61	0.59	0.44	0.29	0.22
		−45.6	11.1	0.46	0.45	0.38	0.29	0.23	0.50	0.48	0.40	0.30	0.24
		−45.6	5.6	0.58	0.56	0.46	0.34	0.26	0.60	0.58	0.47	0.34	0.26

Table 8.10 (Continued)

Position of air space	Direction of heat flow	Air space Mean temp, °C	Temp. diff., °C	13 mm air space Effective emittance ε_{eff}					20 mm air space Effective emittance ε_{eff}				
				0.03	0.05	0.2	0.5	0.82	0.03	0.05	0.2	0.5	0.82
45° Slope	Down	32.2	5.6	0.89	0.80	0.45	0.24	0.16	0.85	0.76	0.44	0.24	0.16
		10.0	16.7	0.63	0.59	0.41	0.25	0.18	0.62	0.58	0.40	0.25	0.18
		10.0	5.6	0.90	0.82	0.50	0.28	0.19	0.83	0.77	0.48	0.28	0.19
		−17.8	11.1	0.68	0.64	0.47	0.31	0.22	0.67	0.64	0.47	0.31	0.22
		−17.8	5.6	0.87	0.81	0.56	0.34	0.24	0.81	0.76	0.53	0.33	0.24
		−45.6	11.1	0.64	0.62	0.49	0.35	0.27	0.66	0.64	0.51	0.36	0.28
		−45.6	5.6	0.82	0.79	0.60	0.40	0.30	0.79	0.76	0.58	0.40	0.30
Horiz.	Down	32.2	5.6	1.07	0.94	0.49	0.25	0.17	1.77	1.44	0.60	0.28	0.18
		10.0	16.7	1.10	0.99	0.56	0.30	0.20	1.69	1.44	0.68	0.33	0.21
		10.0	5.6	1.16	1.04	0.58	0.30	0.20	1.96	1.63	0.72	0.34	0.22
		−17.8	11.1	1.24	1.13	0.69	0.39	0.26	1.92	1.68	0.86	0.43	0.29
		−17.8	5.6	1.29	1.17	0.70	0.39	0.27	2.11	1.82	0.89	0.44	0.20
		−45.6	11.1	1.36	1.27	0.84	0.50	0.35	2.05	1.85	1.06	0.57	0.38
		−45.6	5.6	1.42	1.32	0.86	0.51	0.35	2.28	2.03	1.12	0.59	0.39

Table 8.11 Emittance values of various surfaces and effective emittances of air spaces.

Surface	Average emittance ε	Effective emittance ε_{eff} of air space One surface emittance ε; other, 0.9	Both surfaces emittance ε
Aluminum foil, bright	0.05	0.05	0.03
Aluminum foil, with condensate just visible ($>0.5\,g/m^2$)	0.30	0.29	—
Aluminum foil, with condensate clearly visible ($>2.0\,g/m^2$)	0.70	0.65	—
Aluminum sheet	0.12	0.12	0.06
Aluminum-coated paper, polished	0.20	0.20	0.11
Steel, galvanized, bright	0.25	0.24	0.15
Aluminum paint	0.50	0.47	0.25
Building materials: wood, paper, masonry, nonmetallic paints	0.90	0.82	0.82
Regular glass	0.84	0.77	0.72

$$U_1 = 1/R_1 = 1/3.91 = 0.2558; U_2 = 1/R_2 = 1/3.84 = 0.2604$$

$$U = 0.8 \times U_1 + 0.2 \times U_2 = 0.2567 \, \text{Btu}/\left(\text{h} \cdot^\circ \text{F} \cdot \text{ft}^2\right)$$

If the isothermal plane approach is used,

$$R_{\text{isothermal}} = 0.17 + 1.60 + 1/(0.8/1.01 + 0.2/0.94) + 0.45 + 0.68 = 3.895^\circ \text{F} \cdot \text{ft}^2 \cdot \text{h/Btu}$$

$$U = 1/R = 1/3.895 = 0.2567 \, \text{Btu}/^\circ \text{F} \cdot \text{ft}^2 \cdot \text{h}$$

This case has very close R values of air space and furring. Therefore, both approaches provide almost identical solutions, where the isothermal R value is slightly smaller.

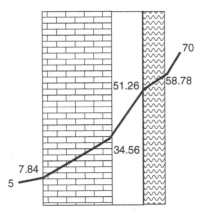

Under the steady state condition:

The total heat flux: $Q = [(T_i - T_o)/R_t] \times A = (T_i - T_o) \times U \times A = (70 - 5)/3.9 \times 1 = 16.7 \, (\text{Btu/h} \cdot \text{ft}^2)$

The temperature at different locations can be determined as:

$$T_1 = T_o + Q \times R_1 = 5 + 16.7 \times 0.17 = 7.84^\circ \text{F}$$

$$T_2 = T_1 + Q \times R_2 = 7.84 + 16.7 \times 1.6 = 34.56^\circ \text{F}$$

$$T_3 = T_2 + Q \times R_3 = 34.56 + 16.7 \times 1.0 = 51.26^\circ \text{F}$$

$$T_4 = T_3 + Q \times R_4 = 51.26 + 16.7 \times 0.45 = 58.78^\circ \text{F}$$

The demonstrated approaches can be used for all building envelopes such as a roof, wall, and door. Heat transfer through windows is somewhat complicated due to the transmitted heat and solar impact, which will be discussed in Chapter 9. Another important heat transfer route is through the building foundation.

8.4 Ground Heat Transfer

8.4.1 Slab-on-Grade

A slab-on-grade foundation is a type of shallow foundation in which a concrete slab rests directly on the ground below it. A slab-on-grade foundation usually consists of a thin layer of concrete across the entire area of the foundation with thickened footings at the edges or below load bearing walls in the middle of the building. Figure 8.25 illustrates the four typical slab-on-grade constructions.

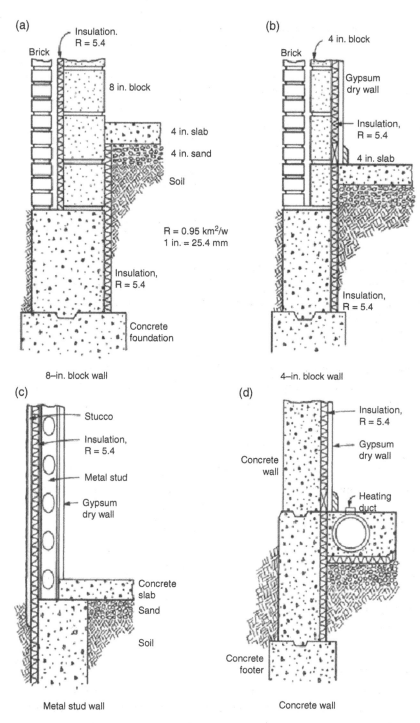

(a)

Brick

Insulation. R = 5.4

8 in. block

4 in. slab

4 in. sand

Soil

R = 0.95 km²/w
1 in. = 25.4 mm

Insulation, R = 5.4

Concrete foundation

8–in. block wall

(b)

4 in. block

Brick

Gypsum dry wall

Insulation, R = 5.4

4 in. slab

Insulation, R = 5.4

4–in. block wall

(c)

Stucco

Insulation, R = 5.4

Metal stud

Gypsum dry wall

Concrete slab

Sand

Soil

Metal stud wall

(d)

Insulation, R = 5.4

Gypsum dry wall

Concrete wall

Heating duct

Concrete footer

Concrete wall

Figure 8.25 Four typical slab-on-grade constructions. (a) 8-in. block wall, (b) 4-in. block wall, (c) Metal stud wall, (d) Concrete wall.

Table 8.12 Heat loss coefficient F_p of slab floor construction.

Construction	Insulation	F_p W/(m·K)
200 mm block wall, brick facing	Uninsulated	1.17
	R-0.95 (m²·K)/W from edge to footer	0.86
4 in. block wall, brick facing	Uninsulated	1.45
	R-0.95 (m²·K)/W from edge to footer	0.85
Metal stud wall, stucco	Uninsulated	2.07
	R-0.95 (m²·K)/W from edge to footer	0.92
Poured concrete wall with duct near perimeter	Uninsulated	3.67
	R-0.95 (m²·K)/W from edge to footer	1.24

Heat loss from an unheated concrete slab floor is mostly through the slab perimeter to the outdoor air rather than through the floor thickness into the ground. Therefore, a more accurate solution must consider the area-to-perimeter ratio. Equation (8.53) is often used to estimate the heat loss through slab-on-grade foundation $Q_{slab,loss}$(W or Btu/h):

$$Q_{slab,loss} = F_p \times P \times (T_i - T_o) \tag{8.53}$$

where T_i and T_o are the indoor and outdoor air temperature (°C or °F), F_p is the perimeter coefficient (W/m·°C or Btu/h·ft·°F), and P is the perimeter of the slab (m or ft). The values of F_p can be obtained from the ASHARE fundamental handbook as shown in Table 8.12 based on slab floor constructions.

8.4.2 Below-Grade Heat Transfer: Basement Wall and Floor

Heat transfer through an underground wall and floor is very complicated and involves two-dimensional and three-dimensional heat transfer mechanisms as demonstrated in Figure 8.26. It often requires detailed computer modeling of the heat transfer processes (e.g. via the finite element analysis). A quick estimate of below-grade heat transfer can be carried out using Eq. (8.54):

$$Q = U_{avg} A(T_{in} - T_{ground}) \tag{8.54}$$

where Q is the heat transfer through the basement wall or floor (W, Btu/h); U_{ave} is the empirical average U-factor for basement walls and floors (W/m²·°C or Btu/h·ft²·°F); A is the basement wall or floor area (m², ft²); T_{in} is the indoor air temperature in the basement; and T_{ground} is the ground temperature. The U-factors can be estimated based on ASHRAE standards (Tables 8.13 and 8.14). The ground temperature varies with building location and can be predicted as:

$$T_{ground} = T_{mean\ air\ in\ coldest} - A_{fluctuation\ of\ Tground} \tag{8.55}$$

where $T_{mean\ air\ in\ coldest}$ is the ambient mean air temperature in the coldest months of interest, and $A_{fluctuation\ of\ Tground}$ is the fluctuation amplitude of soil temperature at different locations as estimated in Figure 8.27. More detailed models to predict the soil and ground temperature were developed recently and some were implemented in building energy simulation models such as EnergyPlus.

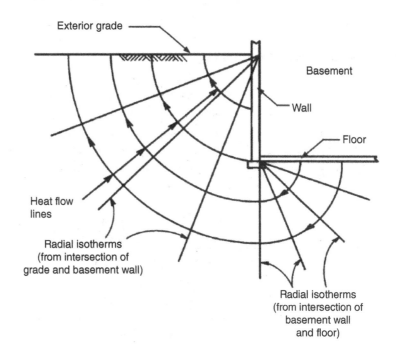

Figure 8.26 Two-dimensional heat transfer processes through basement wall and floor.

Table 8.13 Average U-factor for basement walls with uniform insulation.

Depth, m	$U_{avg,bw}$ from grade to depth, W/(m²·K)			
	Uninsulated	R-0.88	R-1.76	R-2.64
0.3	2.468	0.769	0.458	0.326
0.6	1.898	0.689	0.427	0.310
0.9	1.571	0.628	0.401	0.296
1.2	1.353	0.579	0.379	0.283
1.5	1.195	0.539	0.360	0.272
1.8	1.075	0.505	0.343	0.262
2.1	0.980	0.476	0.328	0.252
2.4	0.902	0.450	0.315	0.244

Table 8.14 Average U-factor for basement floor.

Z_f (depth of floor below grade), m	$U_{avg,bf}$, W/(m²·K)			
	w_b (shortest width of basement), m			
	6	7	8	9
0.3	0.370	0.335	0.307	0.283
0.6	0.310	0.283	0.261	0.242
0.9	0.271	0.249	0.230	0.215
1.2	0.242	0.224	0.208	0.195
1.5	0.220	0.204	0.190	0.179
1.8	0.202	0.188	0.176	0.166
2.1	0.187	0.175	0.164	0.155

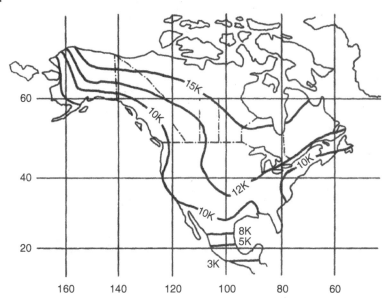

Figure 8.27 Ground temperature amplitude.

Example 8.10

A house is located in Denver. It has a rectangular floor plan, $60' \times 28'$, with $8'$ ceilings. The roof has a single ridge with a 6/12 pitch. The walls, ceiling, and roof have the construction detailed in the following table, with layers listed inside to outside.

The windows are operable double pane, clear, low-e ($\varepsilon = 0.1$ on one surface) with a 1/2" air space and wood frames in a rough-opening area of 240 ft^2. Both front and back doors are 1¾ in. solid core flush wood doors. Each door has an area of 20 ft^2. The floor is slab on grade with R-5 perimeter insulation (covering 3-ft). Infiltration is estimated to be 0.35 ACH.

Walls	Ceiling	Roof	Attic end walls
1/2" gypsum	1/2" gypsum	2 × 6 rafters	2 × 4 stud wall
2 × 4 stud wall with fiber glass insulation	2 × 6 rafters	3/4" plywood sheathing	1/2" plywood sheathing
1/2" plywood sheathing	9" blown cellulose (k = 0.286 Btu-in/fr.°F. ft²)	Felt paper	Aluminum siding
Aluminum siding		Asphalt shingles	

Calculate the design heating load under 99% winter design conditions if the indoor temperature is 70 °F.

Solution

Dimensional calculations

Length = 60 ft
Width = 28 ft
Height = 8 ft
Volume = 13 440 ft^3
DegPitch = ARCTAN(6/12)

Width Roof = Width/COS (DegPitch)

Perim = 2 × (Length + Width) = 176 ft

A_{wall} = Perim × Height-A_Window-A_Door = 1128 ft²

A_{Ceil} = Length × Width = 1680 ft²

A_{Roof} = Length × WidthRoof = 1878 ft²

A_{End} = Width × Width × TAN(DegPitch)/2 = 196 ft²

Design temperature

For Denver, ASHRAE states that the 99% design winter temperature is $T_{H,des}$ = 3 °F.

Calculation details

Walls

Layer	R-value between framing	R-value at framing
Inside convection, still air, nonreflective vertical wall	0.68	0.68
Gypsum, 1/2″	0.45	0.45
Stud, 2 × 4 nominal		4.38
Insulation, 3.5″	11	
Plywood, 1/2″	0.62	0.62
Siding	0.61	0.61
Outside convection, 15 mph	0.17	0.17
Path R-Value	13.53	6.91
Path U = 1/(R-value)	0.07391	0.1447
Path area ratio	0.85	0.15

Effective U: 0.08453 Btu/h · ft² · °F

Effective R-value: 11.83

Ceiling

Layer	R-value between framing	R-value at framing
Inside convection, still air, nonreflective, horizontal, up	0.61	0.61
Gypsum, 1/2″	0.45	0.45
Stud. 2 × 6 nominal		6.88
Insulation, 9″	31.47	
Insulation, 3.5″		12.24
Inside convection, still air, nonreflective, horizontal, up	0.61	0.61
Path R-value	31.14	20.78
Path U = 1 /(R-value)	0.0321	0.0481
Path area ratio	0.906	0.094

Effective U: 0.03362 Btu/h · ft² · °F

Effective R-value: 29.75

Attic end walls

Layer	R-value between framing	R-value at framing
Inside convection, still air, nonreflective vertical wall	0.68	0.68
Stud, 2 × 4 nominal		4.38
Plywood, 12″	0.62	0.62
Siding	0.61	0.61
Outside convection, 15 mph	0.17	0.17
Path R-value	2.08	6.46
Path U = 1/(R-value)	0.4808	0.1548
Path area ratio	0.85	0.15

Effective U: 0.4318 Btu/h · ft^2 · °F
Effective R-value: 2.31

Roof

Layer	R-value between framing	R-value at framing
Inside convection, still air, nonreflective, horizontal, up	0.61	0.61
Stud, 2 × 6 nominal		6.88
Plywood, 3/4″	0.94	0.94
Felt (vapor-permeable)	0.06	0.06
Asphalt shingles	0.44	0.44
Outside convection, 15 mph	0.17	0.17
Path R-value	2.22	9.01
Path U = 1 /(R-value)	0.4505	0.1099
Path area ratio	0.906	0.094

Effective U: 0.4184 Btu/h · ft^2 · °F
Effective R-value: 2.39

Ceiling + Roof

$$R_{top} = \frac{1}{(UA)_{top}} = \frac{R_{v,ceil}}{A_{ceil}} + \frac{1}{U_{roof}A_{roof} + U_{end}A_{end}}. \text{Thus, } (UA)_{top} = 53.03 \text{ Btu/h} \cdot °F$$

Windows and Door

$U_{window} = 0.51$ Btu/h · ft^2 · °F (from ASHRAE fundamental handbook)
$U_{door} = 0.40$ Btu/h · ft^2 · °F (from ASHRAE fundamental handbook)

Floor

$F_p = 0.77$ Btu/h · ft · °F (from ASHRAE fundamental handbook)

Infiltration

$$\dot{V}_{inf} = ACH \times V/60 = 78.4\,cfm$$

$$\rho = 0.061\,lb_m/ft^3\ (\text{at altitude})$$

The total UA value (called Building Load Coefficient, BLC) of the house

$$BLC = (UA)_{top} + (UA)_{walls} + (UA)_{window} + (UA)_{door} + F_pP + 60 \times \rho \times C_p \times \dot{V}_{inf}$$

$$BLC = 53.03 + 95.35 + 122.4 + 16.0 + 135.52 + 68.87 = 491.17\,Btu/h \cdot {}^\circ F$$

The design heating load

$$Q_{heat} = BLC \times (T_{in} - T_{H,des}) = 491.17 \times (70 - 3) = 32\,908(Btu/h)$$

Homework Problems

1 The total heat transfer per unit area $(1\,m^2)$ through the composite wall between the outdoor surface $T_{out,surface}$ and the room air T_{room} (shown in the figure below) is measured to be 50 W. $A_{x-1} = A_{x-2} = 0.5\,m^2$. Only convective heat transfer at the internal wall surface is considered, which has $h = 5\,W/m^2 \cdot {}^\circ C$. Assume that the thermal conductivity of the first layer is $K_y = 1\,W/m \cdot {}^\circ C$, and the surface temperatures $(T_x, T_{in,surface}\text{ and }T_{out,surface})$ are uniform.

(a) Draw the thermal resistance network between $T_{out,surface}$ and T_{room} (and provide equations or values for each resistance).

(b) Calculate the thermal conductivity K_{x-2} for the 5-cm-thick wall layer if knowing that
  (a) $K_{x-1} = 0.5 \times K_{x-2}$.

(c) Determine T_x, as shown in the figure below.

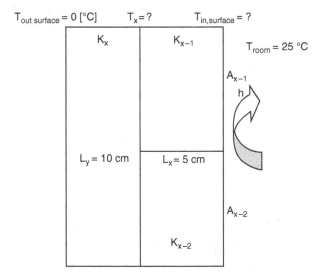

2 Figure below is a schematic diagram of a long, unventilated attic above a residence. What is the rate of heat transfer through the ceiling? What is the attic air temperature? The roof and ceiling properties are given in the table below.

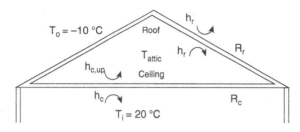

	Ceiling	Roof
Area, m^2	20	23
R value,$(K \cdot m^2)/W$	5.0	0.5
Convection coefficient, upper, $W/(m^2 \cdot K)$	10	25
Convection coefficient, lower, $W/(m^2 \cdot K)$	10	8

3 There is no wind blowing, so heat is removed only by free convection and radiation from the roof (a negligible amount passes through the roof to the interior). What is the roof temperature under these conditions if the air temperature is 90 °F (32 °C)? Assume that radiation takes place to the atmosphere, also at 90 °F (32 °C). Make assumptions as needed.

4 The following figure shows the side view of a ceiling of 5-m-wide suspended below a 6/12 pitch roof. The bottom surface of the roof is 7°C and the top of the ceiling is 37°C. The roof surface $\varepsilon = 0.9$ and the ceiling surface $\varepsilon = 0.7$. Find the radiation heat flux between the two surfaces per meter in depth (perpendicular to the paper).

$$\left(\rho_{air} = 1.2 \, kg/m^3, C_{p, air} = 1000 \, J/kg.K, \upsilon_{air} = 1.5 \times 10^{-5} \, m^2/s, K_{air} = 0.024 \, W/m.K, Pr_{air} = 0.7\right)$$

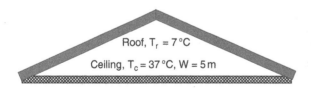

5 The exterior wall of an old building is constructed of face brick and common brick as shown in the figure below. The outdoor air temperature is 10 °F and the indoor air temperature is 65 °F.

(a) What is the resistance of the air gap?

(b) What is the R-value of the overall wall?

(c) What is the heat flux through the wall?

(d) If the inside surfaces of the air gap are covered with a reflective foil surface having a reflectivity of 80%, what is the percentage reduction in heat loss?

(e) If the air gap is filled with 2.0-in. polyisocyanurate board, what is the percentage reduction in heat loss compared to part c above?

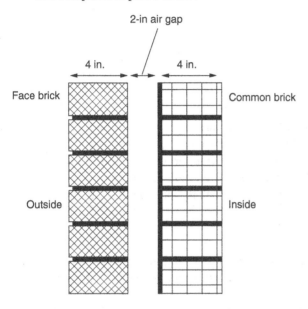

6 Consider a single building in which each room has its own heater. In one of the rooms, the heat has failed and the room is unheated. The physical arrangement of floor plan is shown in the attached figure. The heated indoor space surrounding the room is maintained at 72 °F. The outdoor temperature is −5 °F. The U-value of the exterior wall is 0.20 Btu/h · ft² °F, the U-value of the interior walls is 0.50 Btu/h · ft² °F, and the U-value of the roof is 0.07 Btu/h · ft² · °F. The wall height is 8 ft.

(a) Ignore floor slab and infiltration losses. What is the indoor temperature of the unheated room? What is the heat transfer through the exterior wall of the unheated room?

(b) If the floor slab losses are included, assume the floor is a slab-on-grade design, but that you know nothing of the construction details or the building location, please give the range of indoor temperatures based on the potential slab heat loss.

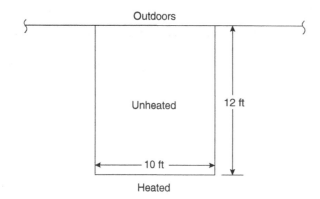

7 Upon **Example 8.4** in Chapter 8 of the textbook, if the indoor air temperature $T_{in} = 20\,°C$ and the outdoor air temperature $T_{out} = 35\,°C$ and there is no radiation Q_{in}, the initial wall temperature is $T_{m,0} = 20\,°C$. Assume the indoor and outdoor surface heat transfer coefficients are, respectively, $h_{in} = 5\,W/m^2 \cdot °C$ and $h_{out} = 15\,W/m^2 \cdot °C$ (including both convection and radiation). The wall cross-section area is $1\,m^2$. Calculate T_m and the heat transfer rate through the wall into the indoor space at $t = 3600\,s$.

8 **Staged Project Assignment (Teamwork)**
   (a) Propose envelope materials and properties (floor, walls, ceiling/roof): including type of materials, 3D sizes, construction layers, location for insulation (internal/external), thermal properties, installation consideration, etc.
   (b) Demonstrate the envelope compliance with the energy code identified in HW 3 (Chapter 3).
   (c) Develop an Excel tool that can calculate the heat gain/loss through the entire building envelope (including ceiling/roof, walls, floor, windows, doors) under indoor and outdoor design temperatures (refer to Example 8.10); present the snapshots of the developed tool with inputs and outputs highlighted.
   (d) Use the tool to evaluate and compare the heat transfer performance of alternative envelope solutions (at least three alternatives, e.g., for window, door, wall, and/or roof) for your design.

9

Sun and Solar Radiation

9.1 Sun and Solar

The sun is critical for all of the Earth's resources and human life, and it also heavily impacts thermal comfort, health, and the energy consumption of buildings. The amount of solar radiation released from the sun is tremendous. Above Earth's atmosphere, solar radiation has an intensity of approximately 1380 W/m^2. This value is known as the Solar Constant. At different latitudes on Earth, the value of the Solar Constant is reduced to approximately 1000 W/m^2 on a clear day at solar noon in the summer months. The reduction is attributed to various factors such as atmospheric absorption and scatterinsg (~25%), cloud reflection (~20%), and ground reflection (~5%). The solar spectrum at Earth's surface consists of about 47% visible radiation, 48% short-wave infrared radiation, and about 5% ultraviolet radiation.

The Earth revolves around the sun in an elliptical orbit and is therefore closer to the sun during part of the year. When the sun is near the Earth, the Earth's surface receives additional solar energy. Interestingly, the Earth is closer to the sun (within 3 million miles) when it is winter in the northern hemisphere and summer in the southern hemisphere, which is opposite to the seasonal temperatures that humans experience. In fact, the 23.5° tilt in the Earth's axis of rotation is a more significant factor in determining the amount of sunlight striking the Earth at a particular location. Tilting results in longer days in the northern hemisphere from the spring (vernal) equinox to the fall (autumnal) equinox and longer days in the southern hemisphere during the other six months out of the year. Days and nights are both exactly 12 hours long on the equinoxes, which occur each year on or around 21 March and 21 September (Figure 9.1). Tilting also results in a higher solar intensity at the Earth's surface in the summer (when solar irradiance is more perpendicular to the surface) than in the winter (with a lower solar incidence angle) (Figure 9.2).

The rotation of the Earth is also responsible for hourly variations in sunlight and solar radiation. In the early morning and late afternoon, the sun has a lower solar incidence angle and sun rays travel further through the atmosphere and thus experience more absorption and scattering loss than at noon, when the sun is at its highest point. As sunlight passes through the atmosphere, the absorbed, scattered, and reflected sun rays also impact the Earth, which is referred to as diffuse solar radiation. The solar radiation that reaches the Earth's surface without being diffused is called direct beam solar radiation, which has the most significant impacts to conditions on Earth. The sum of the diffuse and direct solar radiation is called the global solar radiation. Atmospheric conditions can reduce direct beam radiation by 10% on clear, dry days and by 100% during thick, cloudy days.

Energy Efficient Buildings: Fundamentals of Building Science and Thermal Systems, First Edition. Zhiqiang (John) Zhai.
© 2023 John Wiley & Sons, Inc. Published 2023 by John Wiley & Sons, Inc.

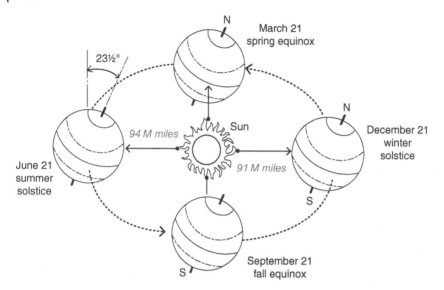

Figure 9.1 Rotation of the Earth around the sun with tilted axis that causes four seasons.

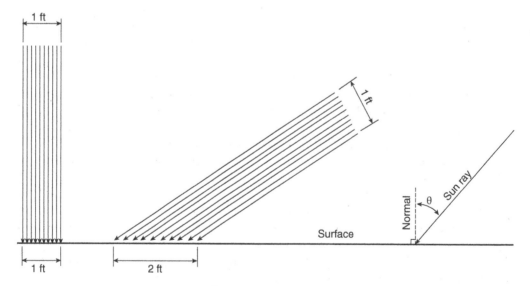

Figure 9.2 Higher solar intensity (Btu/h·ft²) at the Earth's surface in summer (when solar is more perpendicular to the surface) than in winter (with a lower solar incident angle).

All solar calculations should use solar time rather than clock or standard time. Clock time is the artificial time used in everyday life to standardize people's time measurements. It allows people in different locations to use the same time or to easily convert time from one location to another. Standard time is related to the time zone of the specified location defined by the reference value of the longitude, where

Greenwich, London in the United Kingdom is adopted as the reference meridian of zero longitude. Each region on Earth has its distance measured east or west of the prime meridian in Greenwich. The meridian also serves as the reference mark for the Coordinated Universal Time (UTC) with 1 hour for every 15 degrees longitude. In the United States, the reference meridians for the time zones are 75 °W for Eastern (UTC-5), 90 °W for Central (UTC-6), 105 °W for Mountain (UTC-7), and 120 °W (UTC-8) for Pacific standard time. In most regions, the clock time is set to the same standard time within a time zone.

A particular longitude in degrees has to be divided by 15 to determine the precise time zone in hours. Local civil time represents the exact time at a specific location that accounts for the difference in longitude between the reference meridian and the local meridian. Since one full cycle of a day corresponds to 360° longitude, each degree of longitude corresponds to (24 h × 60 min/h)/360° = 4 min.

Local solar time (or simply solar time) is the time according to the position of the sun in the sky relative to one specific location on the ground. In solar time, the sun is always due south (or north) at exactly noon. This means that people apart by a few miles east or west will experience a slightly different solar time even though their clock time is probably the same.

For the purpose of calculating local solar time, clock time must be modified to compensate for three adjustments: (i) the relationship between the local time zone and the local longitude, (ii) daylight savings time, if any, and (iii) the Earth's slightly irregular motion around the sun (corrected for using the equation of time). Local solar time (t_{sol}) (in hours and minutes) is calculated as follows:

$$t_{sol} = t_{std} \pm 4' \times (L_{std} - L_{loc}) + E_t \tag{9.1}$$

where t_{std} is the local standard/clock time in hours and minutes, which may be adjusted for daylight savings time if necessary; L_{loc} is the local longitude (in degrees); L_{std} is the local standard time meridian, measured in degrees, which runs through the center of each time zone. It can be calculated by multiplying the differences in hours from Greenwich Mean Time by 15 degrees per hour. E_t is the equation of time adjustment in minutes:

$$E_t = 9.87 \times \sin(2B) - 7.53 \times \cos(B) - 1.5 \times \sin(B) \tag{9.2}$$

where, $B = 360° \times (n - 81)/364$ for the nth day of year.

Note that the plus (+) sign in Equation (9.1) is to be used for locations west of Greenwich and the negative (−) sign for locations east of Greenwich. The "4'" in Equation (9.1) is the quotient of 60 minutes of time and the 15 degrees of longitude that the earth rotates in that time (i.e. the earth rotates one degree every 4 minutes). Converting from t_{sol} to t_{std} (i.e. clock time) requires the reverse calculation of Equation (9.1).

9.2 Solar Angles

To project the solar impacts on buildings (e.g. for heating and shading), it is important to determine the sun's position with respect to a local observer (or building) on the ground at any hour of any day during the year. The sun's location can be identified with two solar angles: altitude and azimuth (shown in Figure 9.3):

- The solar altitude angle α is the angle between the sun's rays and a horizontal plane. It is related to the solar zenith angle φ, which is the angle between the sun's rays and the vertical plane. $α + φ = 90°$.

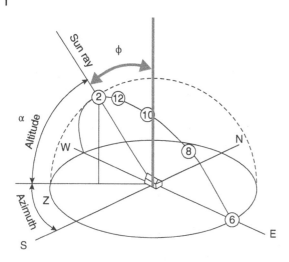

Figure 9.3 Solar angles.

- The solar azimuth angle z is the angle of the sun's rays measured in the horizontal plane from due south (true south) for the northern hemisphere or due north for the southern hemisphere; westward is designated as positive.

The solar altitude angle (α) is determined from the latitude (λ) of the site (values north of the equator are positive and those south are negative) and the solar declination (δ). The solar altitude angle (α) at solar noon (t_{sol} = 12 p.m.) is

$$\alpha = 90° - \lambda + \delta \tag{9.3}$$

$$\sin(\alpha) = \sin(90° - \lambda + \delta) \tag{9.4}$$

$$\sin(\alpha) = \cos(\phi) = \cos(\lambda - \delta) = \cos(\lambda)\cos(\delta) + \sin(\lambda)\sin(\delta) \tag{9.5}$$

where the solar declination (δ) ranges from $-23.45°$ (the winter solstice) to $0°$ (the spring and fall equinox) to $+23.45°$ (the summer solstice). The solar declination (δ) for the nth day of a year can be calculated by:

$$\delta = 23.45° \times \sin\left(\frac{360° \times (284 + n)}{365°}\right) \tag{9.6}$$

The solar altitude angle (α) at any time other than solar noon (t_{sol} = 12 p.m.) can be computed as:

$$\sin(\alpha) = \cos(\phi) = \cos(\lambda - \delta) = \cos(\lambda)\cos(\delta)\cos(\omega) + \sin(\lambda)\sin(\delta) \tag{9.7}$$

where ω is called the hour angle that expresses the time of day with respect to solar noon and is defined as the angle through which the Earth would turn to bring the meridian of the site directly under the sun. One hour of time is represented by $360°/24 = 15°$ of hour angle. Therefore,

$$\omega = (t_{sol} - 12) \times 15° \tag{9.8}$$

where 12 stands for the 12 p.m. of solar noon. Conventionally, the hour angles in the afternoon are positive and in the morning are negative. Therefore, t_{sol} is expressed in a 24-hour format.

The solar azimuth angle (z) can also be calculated as:

$$\sin(z) = \frac{\cos(\delta)\sin(\omega)}{\cos(\alpha)} \tag{9.9}$$

The azimuth angle for the morning hours is negative and for the afternoon hours is positive.

Using Equation (9.7), one can determine the times of the sunrise and sunset for a particular day when $\alpha = 0$:

$$\sin(0) = 0 = \cos(\lambda)\cos(\delta)\cos(\omega) + \sin(\lambda)\sin(\delta) \tag{9.10}$$

$$\cos(\omega) = -\frac{\sin(\lambda)\sin(\delta)}{\cos(\lambda)\cos(\delta)} = -\tan(\lambda)\tan(\delta) \tag{9.11}$$

where ω is taken as positive at sunset. Since each hour angle represents 15° of longitude,

$$T_{sunset} = -T_{sunrise} = \cos^{-1}[-\tan(\lambda)\tan(\delta)]/15 \tag{9.12}$$

The daylength is thus:

$$T_{day} = T_{sunset} - (-T_{sunrise}) = 2 \times \cos^{-1}[-\tan(\lambda)\tan(\delta)]/15 \tag{9.13}$$

Example 9.1

Find the azimuth and altitude angles at 9:30 a.m. on 31 May, and the sunrise and sunset time of the day for Boulder, CO (all in solar time).

Solution

Boulder, CO, is located at 40 °N Latitude ($\lambda = 40°$), 105 °W Longitude.

The local standard time meridian (Mountain Time Zone) is 105 °W Longitude.

From Equation (9.6) and n = 151,

$$\delta = 23.45° \times \sin\left(\frac{360° \times (284 + n)}{365°}\right) = 21.9°$$

From Equation (9.8),

$$\omega = (t_{sol} - 12) \times 15° = (9:30 - 12) \times 15° = -37.5°$$

From Equation (9.7),

$$\sin(\alpha) = \cos(\lambda)\cos(\delta)\cos(\omega) + \sin(\lambda)\sin(\delta)$$
$$= \cos(40)\cos(21.9)\cos(-37.5) + \sin(40)\sin(21.9) = 0.8$$

$$\alpha = 53.5°$$

From Equation (9.9),

$$\sin(z) = \frac{\cos(\delta)\sin(\omega)}{\cos(\alpha)} = \frac{\cos(21.9)\sin(-37.5)}{\cos(53.5)} = -0.95$$

$z = -72°$ (from south toward east)

From Equation (9.12),

$$T_{sunset} = \frac{\cos^{-1}[-\tan(\lambda)\tan(\delta)]}{15} = 7.314 = 7:19 \text{ p.m.}$$

$$T_{sunrise} = 12 - 7.314 = 4.686 = 4:41 \text{ a.m.}$$

9.3 Sky Dome and Sun-Path Diagrams

The sky dome refers to the sum of the components for the entire sky from horizon to zenith and in all azimuthal directions (Figure 9.4). Linking all of the solar positions at their altitude and azimuth of every hour for a given date yields a sun-path from sunrise to sunset. The lowest sun-path is on the winter solstice (21 December) while the highest is on the summer solstice (21 June). Typically, seven sun-paths are presented as 21 December, 21 January, 21 February, 21 March, 21 April, 21 May, and 21 June. Because of the symmetry of the solar movement related to the Earth, 21 July, 21 August, 21 September, 21 October, and 21 November, are, respectively, overlapped with 21 May, 21 April, 21 March, 21 February, and 21 January. Similarly, the hours in the morning are symmetric, with respect to the solar noon, to the hours in the afternoon (the numbers in the circles in Figure 9.4 represent the hours in a day).

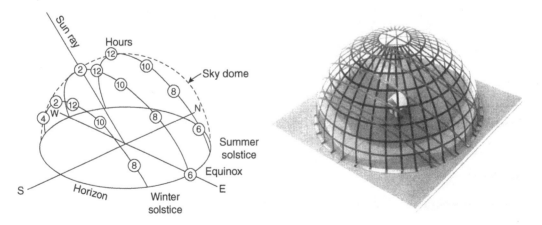

Figure 9.4 Sky dome.

The sky dome is useful to demonstrate the solar positions at different times and dates and their potential impacts (e.g. solar heating, daylighting and shading, etc.) to specific building sites. Note that because the solar altitude is dependent on the site latitude, the sun-paths on the sky dome vary with the latitude of the site in question. A 3D sky dome and 3D sun-paths are visually meaningful but less practical for engineering applications. Therefore, projecting the 3D models of the sky dome and sun-paths to a 2D plane (either a horizonal plane or a vertical plane) produces 2D models of horizonal and vertical sky domes and sun-paths, as shown in Figures 9.5–9.7. To use a proper 2D sun-path diagram, the first important thing is to check is if the altitude on the diagram is correct (e.g. 32 °N Latitude for Figure 9.6 and 40 °N Latitude for Figure 9.7). For a given date and time, the sun-path diagram can help identify the solar altitude and azimuth, as well as the sunrise and sunset times (solar time). Interpolation should be used for dates and

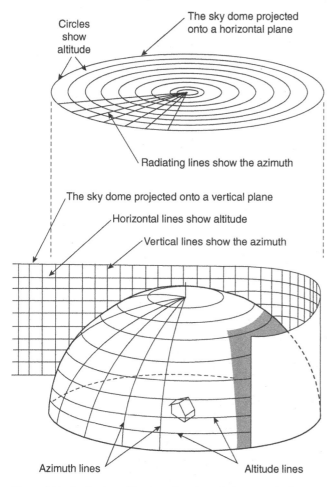

Figure 9.5 Derivation of horizonal and vertical sun-path diagrams.

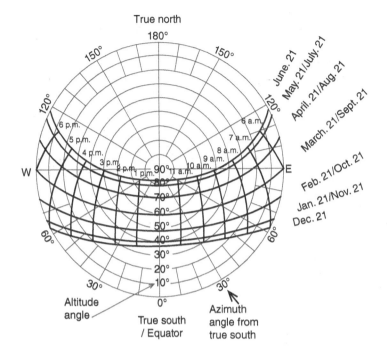

Sun path diagram, 32° N Latitude

Figure 9.6 Horizonal sun-path diagram for 32 °N Latitude.

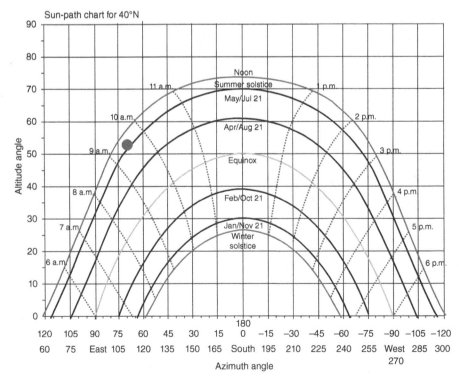

Figure 9.7 Vertical sun-path diagram for 40 °N Latitude.

times between the lines. Note that the times on the diagrams are in solar time. Equation (9.1) should be employed to convert local clock time to local solar time.

Example 9.2

Find the azimuth and altitude angles at 9:30 a.m. on 31 May, and the local sunrise and sunset times that day for Boulder, CO (without daylight saving time).

Solution

Boulder, CO, is located at 40 °N Latitude, 105 °W Longitude.

The local standard time meridian (Mountain Time Zone) is 105 °W Longitude.

The specified time is shown in Figure 9.7 as a solid dot.

Reading the chart provides,

$$\alpha = 52°; z = 70°\,(\text{east}); T_{sunrise} = 4{:}50\ \text{a.m.}; T_{sunset} = 7{:}10\ \text{p.m.}$$

Using Equation (9.1),

$$t_{std} = t_{sol} - 4' \times (L_{std} - L_{loc}) - E_t = t_{sol} - E_t$$

From Equation (9.2),

$$E_t = 9.87 \times \sin(2B) - 7.53 \times \cos(B) - 1.5 \times \sin(B) = 6.55 - 2.67 - 1.40 = 2.48\ (\text{mins}) \approx 2\ \text{mins}$$

where, $B = 360° \times (n - 81)/364 = 360° \times (151 - 81)/364 = 69.2°$

So $T_{sunrise,\ std} = 4{:}48$ am; $T_{sunset,\ std} = 7{:}08$ p.m.

9.4 Solar Shading

The solar positions (i.e. solar angles) at different dates and times can be used for various purposes, including to predict the solar heating impacts on a building's heating and cooling loads (to be introduced in upcoming sessions) and to predict the daylighting and shading consequences.

Figure 9.8 shows a wall with a window under direct sun rays. The wall's shadow can be projected using the obtained solar angles at specific times. The vertex of the shadow can be determined as:

$$dx = h \times \sin(z)/\tan(\alpha) \tag{9.14}$$

$$dy = h \times \cos(z)/\tan(\alpha) \tag{9.15}$$

Other vertices of the wall and the window can be similarly determined.

Shadows or shading from neighboring or surrounding objects (e.g. terrain, buildings, trees, etc.) on the site can also be calculated and projected onto the sun-path diagram, which can quantify the impact scope of the shading (e.g. the season and time periods) (Figure 9.9). As a result, appropriate design strategies can be considered. Example 9.3 demonstrates the calculation process using a simple case. Figure 9.9 also highlights a red/bold curved box that is called the solar window, representing the most effective solar positions and time periods (from 9 a.m. to 3 p.m.) for passive solar heating and photovoltaic (PV) applications.

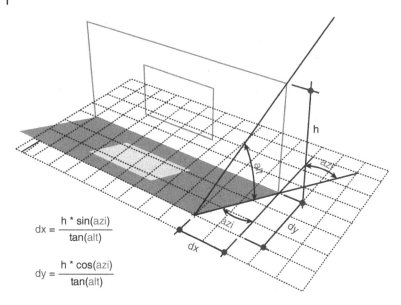

$$dx = \frac{h * \sin(\text{azi})}{\tan(\text{alt})}$$

$$dy = \frac{h * \cos(\text{azi})}{\tan(\text{alt})}$$

Figure 9.8 Shadow of a building.

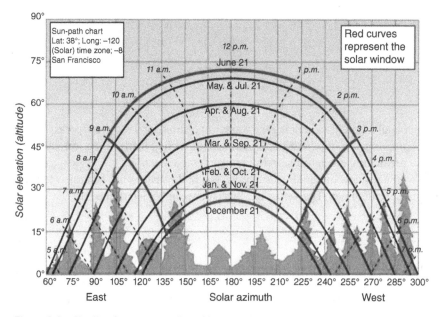

Figure 9.9 Shading from surrounding objects.

Example 9.3

Project the neighboring building of $40'(L) \times 15'(W) \times 20'(H)$ as shown below (Figure 9.10) onto the horizonal sun-path diagram for 36 °N Latitude. (The angle of O = 30° and the distance AB = 36')

Solution

Find altitude α and azimuth z for all eight vertices: namely C1–C4 (for top face) and B1–B4 (for bottom face).

$\alpha(B1) = \alpha(B2) = \alpha(B3) = \alpha(B4) = 0$ because B1–B4 are on the ground (i.e. the same height as the site/building).

$$z(B1) = z(C1) = \tan^{-1}([36 \times \sin[30] + 40]/[36 \times \cos[30]]) = 61.7°$$

$$z(B2) = z(C2) = 30°$$

$$z(B3) = z(C3) = \tan^{-1}([36 \times \sin[30]]/[36 \times \cos[30] + 15]) = 21.3°$$

$$z(B4) = z(C4) = \tan^{-1}([36 \times \sin[30] + 40]/[36 \times \cos[30] + 15]) = 51.5°$$

$$\alpha(C1) = \tan^{-1}(20/[36 \times \cos[30]/\cos[61.7]]) = 17°$$

$$\alpha(C2) = \tan^{-1}(20/36) = 29°$$

$$\alpha(C3) = \tan^{-1}(20/[[36 \times \cos[30] + 15]/\cos[21.3]]) = 22°$$

$$\alpha(C4) = \tan^{-1}(20/[[36 \times \cos[30] + 15]/\cos[51.5]]) = 15°$$

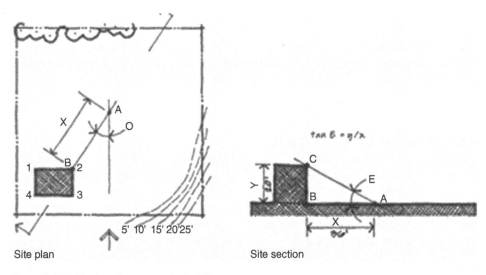

Site plan   Site section

Figure 9.10 Shadow from a nearby building.

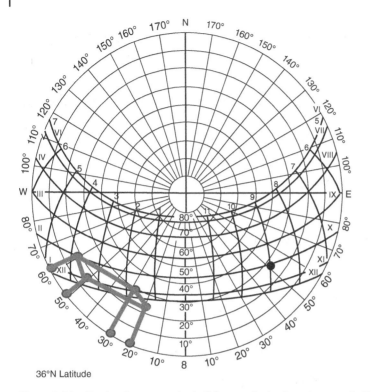

36°N Latitude

Figure 9.11 Shadow from a nearby building on the horizonal sun-path diagram.

Plot these vertices on the horizonal sun-path diagram for 36 °N Latitude and link the vertices as seen in Figure 9.11. The affected time period by the shadow will be between 1:30 and 5 p.m. for the winter seasons from 21 November to 21 January.

9.5 Solar Radiation on External Walls

Solar radiation has tremendous impacts on building heating, cooling, and lighting loads, as well as the electricity generation rate by PV systems. The total solar radiation released from the sun outside of the Earth's atmosphere at the mean sun–Earth distance on a surface normal to the solar rays is about 1367 W/m^2 (or 433.3 Btu/h·ft^2). This is called the Solar Constant, I_{sc}. This value varies slightly (about ±3.3%) with the date due to the eccentricity of the orbit, with a correction coefficient E_0:

$$I_{sc,c} = E_0 \times I_{sc} \tag{9.16}$$

where,

$$E_0 = 1 + 0.033 \times \cos\left(\frac{360 \times n}{365}\right) \tag{9.17}$$

When a building's surface is not perpendicular to the solar rays, the total direct (beam) radiation received by the surface, without considering the atmospheric loss, is:

$$I_0 = E_0 \times I_{sc} \times \cos(\theta) \tag{9.18}$$

where θ is the incidence angle of the sun on the surface at any tilt and orientation (which is the angle between the normal of the surface and the line to the sun) (Figure 9.12). $\cos(\theta)$ can be computed as follows:

$$\cos(\theta) = \sin(\phi) \sin(\beta) \cos(z - z_s) + \cos(\phi) \cos(\beta) \tag{9.19}$$

Or

$$\cos(\theta) = \cos(\alpha) \sin(\beta) \cos(z - z_s) + \sin(\alpha) \cos(\beta) \tag{9.20}$$

where β is the surface tilt angle and z_s is the azimuth of the surface normal (positive for orientations west of south) (Figure 9.12).

For a vertical plane ($\beta = 90°$), Equations (9.19) and (9.20) become:

$$\cos(\theta) = \sin(\phi) \cos(z - z_s) \tag{9.21}$$

Or

$$\cos(\theta) = \cos(\alpha) \cos(z - z_s) \tag{9.22}$$

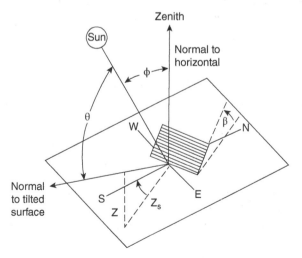

Figure 9.12 Angles on a tilted surface.

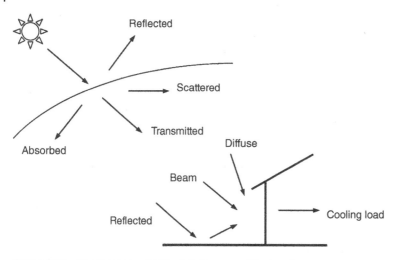

Figure 9.13 Components of solar radiation on a tilted surface.

The solar radiation reaching a tilted surface on Earth is comprised of three components: beam radiation, diffuse radiation (from the sky and surroundings), and reflected radiation (from the ground and surroundings) (Figure 9.13). The total radiation incident on the surface ($I_{t,s}$) is, thus,

$$I_{t,s} = I_{beam} \times \cos(\theta) + I_{diff} \times F_{sky} + I_{horizonal} \times \rho_g \times F_{grd} \tag{9.23}$$

where, I_{beam}, I_{diff}, and $I_{horizonal}$ are, respectively, the direct, diffuse, and horizonal/ground solar radiation insolation (W/m^2 or Btu/h·ft^2), which can be found from weather data (e.g. the TMY3 file). $I_{horizonal}$ can also be estimated as:

$$I_{horizonal} = I_{beam} \times \sin(\alpha) + I_{diff} \tag{9.24}$$

ρ_g is the reflectivity of the ground. F_{sky} and F_{grd} are the view factors of the sky and ground to the surface.

$$F_{sky} = [1 + \cos(\beta)]/2 \tag{9.25}$$

$$F_{grd} = [1 - \cos(\beta)]/2 \tag{9.26}$$

For a vertical surface ($\beta = 90°$), $F_{sky} = F_{grd} = 1/2$. For a horizontal surface ($\beta = 0°$), $F_{sky} = 1$ and $F_{grd} = 0$. The impacts of solar radiation on a building's external envelope can also be integrated into the total heat transfer through a building envelope by using the concept of solar-air temperature.

$$q = h_o \times (T_{os} - T_s) = h_o \times (T_o - T_s) + \alpha \times I_{solar} + \Delta q_{IR} \tag{9.27}$$

where q is the total heat transfer to the surface (W/m^2 or $Btu/h\cdot ft^2$), T_o is the outdoor air temperature (°C or °F), T_s is the surface temperature (°C or °F), T_{os} is the solar-air temperature (°C or °F), α is the solar absorptivity of the surface, I_{solar} is the incident solar radiation on the surface (W/m^2 or $Btu/h\cdot ft^2$), and Δq_{IR} is the difference in thermal (infrared) radiation exchange between the surface and the surroundings (including other surfaces, sky, ground, etc.). The solar-air temperature T_{os} is then defined as:

$$T_{os} = T_s + \alpha \times I_{solar}/h_o + \Delta q_{IR}/h_o = T_s + \alpha \times I_{solar}/h_o + \Delta T_{IR} \tag{9.28}$$

where, generally, $\Delta T_{IR} = 0$ for vertical walls (assuming the surroundings share the same temperature as the vertical walls); $\Delta T_{IR} = -7\,°F$ ($-4\,°C$) for horizonal roofs attributed to the cool sky temperature.

9.6 Solar Radiation on Windows

Windows are critical elements in buildings which not only provide views and a connection to the outdoor but also bring in natural light, sound, and ventilation. However, improperly designed windows are the primary weakness of a building envelope, creating additional heating, cooling, lighting, and ventilation costs. Heat transfer through windows is more complicated than through walls and roofs due to the diverse optical and thermal properties of window glazing. Fundamentally, a glazing should have the following characteristics:

- Transmit visible light – perhaps not the full amount since radiation becomes extra heat;
- Reject all ultraviolet – fades fabrics, hurts eyes;
- Reject all infrared wavelengths in summer – just heat;
- Admit infrared wavelengths in winter – provides passive heating.

Figure 9.14 illustrates the basic configuration of a double-glazed window and the heat transfer mechanisms at the window. Two glass panes form a cavity that can be a vacuum or filled with air or a special gas. Three glass panes may also be found in practice and provide better thermal performance but are also much more expensive. The glass panes are separated and supported by metal or insulated spacers. Desiccant is used to absorb water vapor in the window cavity.

The heat transfer mechanisms at a window include the conventional heat conduction through glazing panes and window frame/dividers, the convection at both interior and exterior glass surfaces and in the cavity, the infiltration through leaks around the frame and gaps around moveable window parts, and the radiation transmitted, reflected, and absorbed at each glazing pane. The heat transfer process becomes more sophisticated when multiple glazing panes are used and when the surface coatings or films are applied. Figure 9.15 demonstrates the variation of solar radiative heat transfer through three different glazing panes. With different values of reflectivity (ρ) at the exterior surface and absorptivity (α) and transmissivity (τ) of the glazing, the total heat transfer through the glazing is different. Note that the absorbed heat will eventually flow indoors or back outdoors, and the amount of heat that goes in each direction depends on the temperature difference of the indoor–outdoor environments.

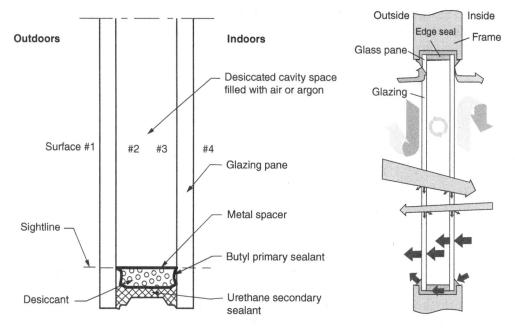

Figure 9.14 Basic configuration and heat transfer mechanisms of a double-glazing window.

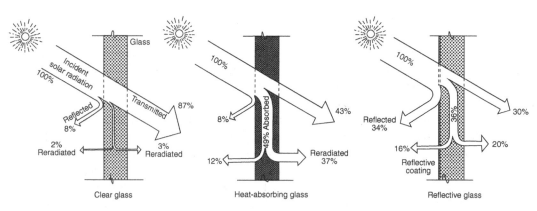

Figure 9.15 Variation of solar radiative heat transfer through three different glazing panes.

The total heat transfer through a window can then be calculated as:

$$Q_{win} = Q_{cond + conv} + Q_{rad} = UA_{win}(T_{out} - T_{in}) + (SHGC)\,A_{win}I_t \tag{9.29}$$

where Q_{win} is the total heat transfer through the window assembly (W or Btu/h); T_{out} and T_{in} are the outdoor and indoor air temperatures (°C or °F), respectively; A_{win} is the projected window area (m² or ft²); U is the overall heat transfer coefficient of the window (W/m²·C or Btu/h·ft²·°F); and I_t is the incident solar radiation on the window surface (W/m² or Btu/h·ft²), including both beam and diffuse solar radiation. SHGC stands for the solar heat gain coefficient which includes two components: (i) the transmitted solar radiation and (ii) the solar radiation absorbed by the glass that has finally passed into the building.

The overall heat transfer coefficient of the window U can be computed by weighting the individual U values for the window parts as shown below:

$$U = \left(U_{cg}A_{cg} + U_{eg}A_{eg} + U_f A_f + U_d A_d\right)/A_{win} \tag{9.30}$$

where U is the U value and A is the area of each element of a window. The subscript cg, eg, f, and d, respectively, stand for center of glass, edge of glass, frame, and divider. Note that the glass may have different properties at its edge and center (mostly due to the spacer material) (Figure 9.16). The edge area is defined as 65 mm (2.5″) from each edge of a piece of glass. The ASHRAE fundamental handbook provides detailed properties for various typical fenestration products, as demonstrated in Tables 9.1 and 9.2 as examples. Figure 9.17 illustrates the representative fenestration frame types as used in Table 9.2. The

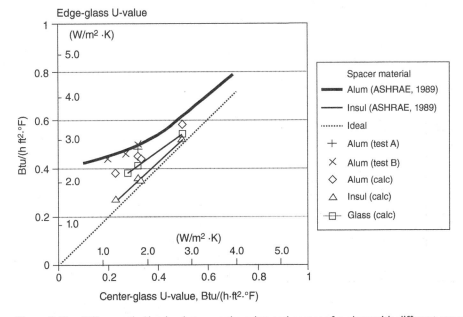

Figure 9.16 Difference in U value between the edge and center of a glass with different spacers.

Table 9.1 U values for several fenestration products (W/m²·K).

| | | Glass only | | Vertical installation | | | | | | | | | |
| | | | | Operable (including sliding and swinging glass doors) | | | | | Fixed | | | | |
ID	Frame type / glazing type	Center of glass	Edge of glass	Aluminum without thermal break	Aluminum with thermal break	Reinforced vinyl/ aluminum clad wood	Wood/ vinyl	Insulated fiberglass/ vinyl	Aluminum without thermal break	Aluminum with thermal break	Reinforced vinyl/ aluminum clad wood	Wood/ vinyl	Insulated fiberglass/ vinyl
	Single glazing												
1	3.2 mm glass	5.91	5.91	7.24	6.12	5.14	5.05	4.61	6.42	6.07	5.55	5.55	5.35
2	6.4 mm acrylic/ polycarb	5.00	5.00	6.49	5.43	4.51	4.42	4.01	5.60	5.25	4.75	4.75	4.58
3	3.2 mm acrylic/ polycarb	5.45	5.45	6.87	5.77	4.82	4.73	4.31	6.01	5.66	5.15	5.15	4.97
	Double glazing												
4	6.4 mm airspace	3.12	3.63	4.93	3.70	3.25	3.13	2.77	3.94	3.56	3.19	3.17	3.04
5	12.7 mm airspace	2.73	3.36	4.62	3.42	3.00	2.87	2.53	3.61	3.22	2.86	2.84	2.72
6	5.4 mm argon space	2.90	3.48	4.75	3.54	3.11	2.98	2.63	3.75	3.37	3.00	2.98	2.85
7	12.7 mm argon space	2.56	3.24	4.49	3.30	2.89	2.76	2.42	3.47	3.08	2.73	2.70	2.58
	Double glazing, e = 0.60 on surface 2 or 3												
8	5.4 mm airspace	2.95	3.52	4.80	3.58	3.14	3.02	2.67	3.80	3.41	3.05	3.03	2.90
9	12.7 mm airspace	2.50	3.20	4.45	3.26	2.85	2.73	2.39	3.42	3.03	2.68	2.66	2.54

#	Description												
10	6.4 mm argon space	2.67	3.32	4.58	3.38	2.96	2.84	2.49	3.56	3.17	2.82	2.80	2.67
11	12.7 mm argon space	2.33	3.08	4.31	3.13	2.74	2.62	2.28	3.28	2.89	2.54	2.52	2.40
	Double glazing, e = 0.40 on surface 2 or 3												
12	64 mm airspace	2.78	3.40	4.66	3.46	3.03	2.91	2.56	3.66	3.27	2.91	2.89	2.76
13	12.7 mm airspace	2.27	3.04	4.27	3.09	2.70	2.58	2.25	3.23	2.84	2.49	2.47	2.35
14	6.4 mm argon space	2.44	3.16	4.40	3.21	2.81	2.69	2.35	3.37	2.98	2.63	2.61	2.49
15	12.7 mm argon space	2.04	2.88	4.09	2.93	2.55	2.43	2.10	3.04	2.65	2.31	2.29	2.17

Source: From *ASHRAE Handbook: Fundamentals 2017.*

Table 9.2 U values of representative fenestration frames in vertical orientation (W/m²·K).

Frame material	Type of spacer	Operable			Fixed			Garden window		Plant-assembled skylight			Curtain wall[e]			Sloped/Overhead glazing[e]		
		Single[b]	Double[c]	Triple[d]	Single[b]	Double[c]	Triple[d]	Single[b]	Double[c]	Single[b]	Double[c]	Triple[d]	Single[f]	Double[g]	Triple[h]	Single[f]	Double[g]	Triple[h]
Aluminum without thermal break	All	13.51	12.89	12.49	10.90	10.22	9.88	10.67	10.39	44.57	39.86	39.01	17.09	16.81	16.07	17.32	17.03	16.30
Aluminum with thermal break[a]	Metal	6.81	5.22	4.71	7.49	6.42	6.30			39.46	28.67	26.01	10.22	9.94	9.37	10.33	9.99	9.43
	Insulated	n/a	5.00	4.37	n/a	5.91	5.79			n/a	26.97	23.39	n/a	9.26	8.57	n/a	9.31	8.63
Aluminum-clad wood/reinforced vinyl	Metal	3.41	3.29	2.90	3.12	2.90	2.73			27.60	22.31	20.78						
	Insulated	n/a	3.12	2.73	n/a	2.73	2.50			n/a	21.29	19.48						
Wood/vinyl	Metal	3.12	2.90	2.73	3.12	2.73	2.38	5.11	4.83	14.20	11.81	10.11						
	Insulated	n/a	2.78	2.27	n/a	2.38	1.99	n/a	4.71	n/a	11.47	9.71						
Insulated fiberglass/vinyl	Metal	2.10	1.87	1.82	2.10	1.87	1.82											
	Insulated	n/a	1.82	1.48	n/a	1.82	1.48											
Structural glazing	Metal												10.22	7.21	5.91	10.33	7.27	5.96
	Insulated												n/a	5.79	4.26	n/a	5.79	4.26

Note: This table should only be used as an estimating tool for early phases of design.
[a] Depends strongly on width of thermal break. Value given is for 9.5 mm.
[b] Single glazing corresponds to individual glazing unit thickness of 3 mm (nominal).
[c] Double glazing corresponds to individual glazing unit thickness of 19 mm (nominal).
[d] Triple glazing corresponds to individual glazing unit thickness of 34.9 mm (nominal).
[e] Glass thickness in curtain wall and sloped/overhead glazing is 6.4 mm.
[f] Single glazing corresponds to individual glazing unit thickness of 6.4 mm (nominal).
[g] Double glazing corresponds to individual glazing unit thickness of 25.4 mm. (nominal).
[h] Triple glazing corresponds to individual glazing unit thickness of 44.4 mm. (nominal).
n/a Not applicable
Source: From ASHRAE Handbook: Fundamentals 2017.

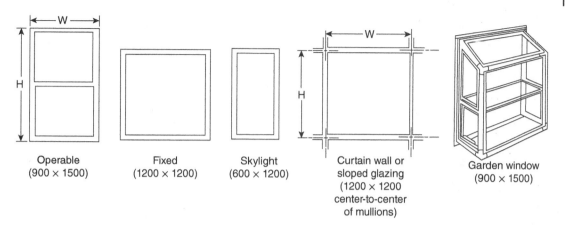

Operable
(900 × 1500)

Fixed
(1200 × 1200)

Skylight
(600 × 1200)

Curtain wall or
sloped glazing
(1200 × 1200
center-to-center
of mullions)

Garden window
(900 × 1500)

Figure 9.17 Representative fenestration frame types as used in Table 9.2.

ASHRAE fundamental handbook also provides the SHGC (and other property) values for various glazings and windows as demonstrated in Table 9.3.

Example 9.4

Estimate the U-factor of a wood-framed swinging French door of 970×2080 mm. The door has eight 280×400 mm clear double-glazing true-divided panes with a 6.4 cm air space and a metal spacer.

Solution

$$A_t = 970 \text{ mm} \times 2080 \text{ mm} = 2.0176 \text{ m}^2$$

$$A_g = 8 \times 280 \text{ mm} \times 400 \text{ mm} = 0.896 \text{ m}^2$$

$$A_f = A_t - A_g = 1.1216 \text{ m}^2$$

$$A_{cg} = 8 \times (280 - 2 \times 65) \text{ mm} \times (400 - 2 \times 65) \text{ mm} = 8 \times 150 \text{ mm} \times 270 \text{ mm} = 0.324 \text{ m}^2$$

$$A_{eg} = A_g - A_{cg} = 0.572 \text{ m}^2$$

Check Table 9.1, find: U_{eg}=3.63 W/m²·°C, U_{cg}=3.12 W/m²·°C
Check Table 9.2, find: U_f=2.90 W/m²·°C

$$U = \left(U_{cg}A_{cg} + U_{eg}A_{eg} + U_fA_f\right)/A_t = (3.12 \times 0.324 + 3.63 \times 0.572 + 2.90 \times 1.1216)/2.0176 = 3.14 \text{ W/m}^2 \cdot°C$$

Table 9.3 SHGC and other property values for several glazings and windows.

| Glazing system | | | | Center-of-glazing properties | | | | | | | Total window SHGC at normal incidence | | | | Total window T$_v$ at normal incidence | | | |
| | | | | Incidence angles | | | | | | Hemis., | Aluminum | | Other frames | | Aluminum | | Other frames | |
ID	Glass thick, mm	Center glazing T$_v$		Normal 0.00	40.00	50.00	60.00	70.00	80.00	Diffuse	Operable	Fixed	Operable	Fixed	Operable	Fixed	Operable	Fixed
Uncoated single glazing																		
1a	3	CLR 0.90	SHGC	0.86	0.84	0.82	0.78	0.67	0.42	0.78	0.75	0.78	0.64	0.75	0.77	0.80	0.66	0.78
			T	0.83	0.82	0.80	0.75	0.64	0.39	0.75								
			R^f	0.08	0.08	0.10	0.14	0.25	0.51	0.14								
			R^b	0.08	0.08	0.10	0.14	0.25	0.51	0.14								
			A$_i^f$	0.09	0.10	0.11	0.11	0.11	0.11	0.10								
1b	6	CLR 0.88	SHGC	0.81	0.80	0.78	0.73	0.62	0.39	0.73	0.71	0.74	0.60	0.71	0.75	0.79	0.64	0.77
			T	0.88	0.87	0.85	0.80	0.69	0.43	0.80								
			R^f	0.08	0.09	0.11	0.15	0.27	0.53	0.14								
			R^b	0.08	0.09	0.11	0.15	0.27	0.53	0.14								
			A$_i^f$	0.16	0.17	0.18	0.19	0.19	0.17	0.17								
1c	3	BRZ 0.68	SHGC	0.73	0.71	0.68	0.64	0.55	0.34	0.65	0.64	0.67	0.54	0.64	0.58	0.61	0.50	0.59
			T	0.65	0.62	0.59	0.55	0.46	0.27	0.56								
			R^f	0.06	0.07	0.08	0.12	0.22	0.45	0.12								
			R^b	0.06	0.07	0.08	0.12	0.22	0.45	0.12								
			A$_i^f$	0.29	0.31	0.32	0.33	0.33	0.29	0.31								

Source: From ASHRAE Handbook: Fundamentals 2017.

Homework Problems

1 Please use the following chart to find the azimuth and altitude angles at 9:30 a.m. on 31 May, and the sunrise and sunset time of the day. If the city is Las Vegas, NV, what is the standard/local time of the solar noon on 31 May. What is the altitude of the solar noon on 31 May (use both hand calculation and chart reading)?

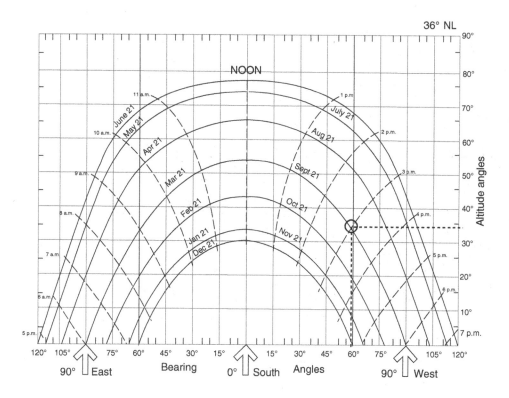

2 Please use the following chart to find the azimuth and altitude angles at 4:00 p.m. on 5 October and the sunrise and sunset time of the day. If the city is Monterey, CA, what are the standard/local times of the sunrise and sunset of the day?

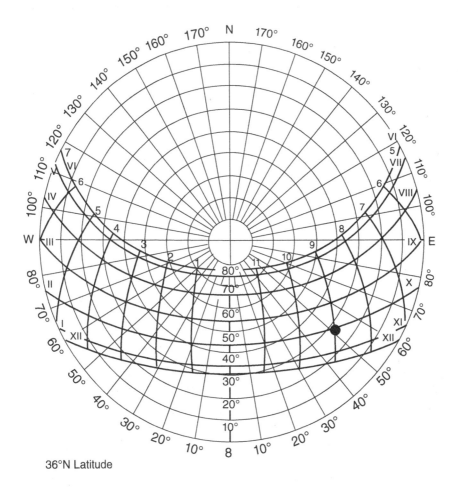

36°N Latitude

3 Please project the neighboring building of 50 m × 50 m × 100 m (tall), located at 45° southwest from the site and at direct distance of 50 m, on the horizontal sun-path diagram below. Please comment on the potential shading influence on the site.

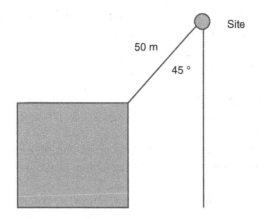

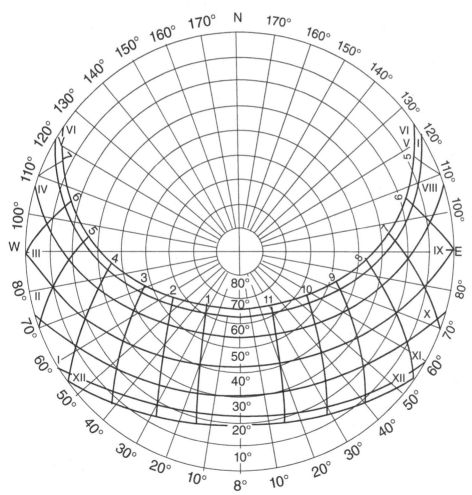

48 °N Latitude

4 Determine the heat transfer per unit area (W/m^2) through a fixed double pane (both are 3-mm clear glasses) window with wood frame and vertical installation. The indoor and outdoor air temperatures are, respectively, 20 and 5 °C and the panes are 6.4 mm apart. The incident solar radiation on 21 October 2 p.m. (solar time) for Boulder, CO is 650 W/m^2.

5 Staged Project Assignment (Teamwork)

   (a) Check relevant standards and guidelines for suggested minimum and maximum window-wall-ratios (WWR) at different envelope orientations.

   (b) Improve and use your developed Excel tool to justify and determine proper WWR for your building (considering both with solar and without solar conditions – the balance of heating and cooling loads).

   (c) Use the obstruction mask (on sun-path) to evaluate if any wind reduction or enhancement approaches considered may provide any shading effect on the façade of your building.

10

Passive Building Systems

10.1 Introduction

Buildings have significant impacts on the environment and natural resources. The construction, maintenance, and demolition of buildings consume tremendous quantities of natural resources and energy and produce significant amounts of environmental pollutants. The emerging world energy and environmental crises demand a substantial revolution of building design philosophies, strategies, technologies, and construction and management methods. Historically, humans were smart enough to utilize natural forces and resources to cool, heat, and ventilate spaces. Over the course of time, dwellings have evolved to respond to challenges in their climate, building materials, and cultural expectations in a given place. This evolution has occurred through a long period of trial and error and through the ingenuity of local builders who possess specific knowledge about their place on the planet. Passive architecture varies widely around the world with its vast array of local climate conditions, culture, and terrain. These architectures contain significant information and knowledge on how to optimize the energy performance of buildings at a low cost using local materials.

The Sun, Earth, sky, and atmosphere are adequate, fundamental, and renewable resources that can be taken advantage of to increase building comfort while reducing building energy use (Table 10.1). A good understanding of heat transfer mechanisms (conduction, convection, radiation, evaporation, and condensation) between the Earth, environment, and building elements will facilitate proper designs that fully use the cooling and heating capacity of ambient conditions (solar, air, sky, etc.).

In addition to saving energy, passive cooling and heating techniques typically demonstrate significant comfort and health benefits, as well as human preference. For instance, research has shown that the acceptable thermal comfort range for naturally ventilated buildings is noticeably larger than for buildings with standard mechanical HVAC systems. The generally high level of occupant control associated with passive buildings is also thought to contribute to the acceptance of warmer indoor temperatures. Furthermore, passive systems have been shown to consistently outperform mechanical systems with respect to complaints of sick building syndrome (SBS) and its associated symptoms. Generally, when building occupants are more satisfied with their working environment, their productivity and job satisfaction also increase. The advantages of using passive techniques are, therefore, significant beyond solely the benefit of energy savings.

This chapter first introduces the general principles and considerations of utilizing passive cooling and heating techniques for sustainable building development. A brief check list is presented to prescreen the feasibility of using passive cooling and heating techniques for specific projects.

Energy Efficient Buildings: Fundamentals of Building Science and Thermal Systems, First Edition. Zhiqiang (John) Zhai.
© 2023 John Wiley & Sons, Inc. Published 2023 by John Wiley & Sons, Inc.

Table 10.1 Heat transfer mechanisms and sources for passive cooling and heating.

	Conduction	Convection	Radiation	Evaporation
Passive cooling	Adjust insulation	Reduce infiltration	Reduce solar gain	Use evaporative cooling
	Use Earth cooling	Use ventilation	Use radiant cooling	
Passive heating	Increase insulation	Reduce infiltration	Increase solar gain	
		Reduce external airflow		
Source	Atmosphere; Earth	Atmosphere	Sun; sky	Atmosphere

Seven commonly used passive cooling and heating techniques are then discussed in detail. The chapter focuses on presenting general principles, thermal and energy performance, and key design considerations of these techniques. The introduced passive techniques are classified into three categories based on function:

1) **Passive cooling techniques:** including natural ventilation; night cooling and thermal mass; and evaporative cooling.
2) **Passive heating techniques:** including Trombe wall and sunspace.
3) **Combined passive cooling and heating techniques:** including double skin façade (DSF) and phase change material (PCM).

10.2 Overview of Passive Cooling

Before the technology of refrigeration, people used many natural methods to keep themselves cool such as wind through windows, water evaporation from pools and fountains, and stone or Earth for storing extra heat. People developed these ideas thousands of years ago which are called "passive cooling" in today's terminology. Passive cooling is defined as "a building design approach that focuses on heat gain control and heat dissipation in a building in order to improve indoor thermal comfort with low or none energy consumption. This approach works either by preventing heat from entering the interior (heat gain prevention) or by passively removing heat from the building (natural cooling). Natural cooling utilizes on-site energy, available from the natural environment, combined with the architectural design of building components (e.g. building windows), rather than mechanical systems, to dissipate heat." (*https://en.wikipedia.org/wiki/Passive_cooling*). People are able to reduce the capital and operational costs of the equipment and even eliminate the cooling energy by adopting passive cooling.

Passive cooling can be implemented in a number of approaches with different heat sinks (Figure 10.1): natural ventilation, DSF, night precooling, Earth cooling, and evaporative cooling. Proper natural cooling design must rely on a detailed study of the local climate, integrated building design strategies, and often requires advanced computational analysis. A successful passive design is intrinsically integrated with the architectural design making an integrated building design process necessary. Natural ventilation is one of the most popular and important passive cooling strategies that can be generally divided into two categories: wind-driven natural ventilation and buoyancy-driven natural ventilation. Wind-driven natural ventilation uses the air pressure differential across the building, caused by wind, to drive ventilated

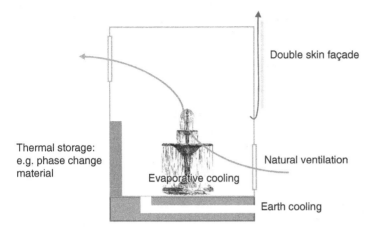

Double skin façade

Thermal storage:
e.g. phase change
material

Natural ventilation

Evaporative cooling

Earth cooling

Figure 10.1 Illustration of passive cooling techniques.

air through the building. This type of design is highly dependent on the external site weather conditions. Buoyancy-driven natural ventilation, also called the stack effect, uses the principle of thermal stratification in buildings to induce airflow through the building. This technique is different from wind-driven natural ventilation in that it does not rely on the ambient wind for motivating ventilation air. Buoyancy-driven flow is typically used with atriums, clerestories, solar chimneys, and courtyards.

Night cooling and Earth cooling are two other important and widely used passive cooling techniques that use night-time cool ambient air and the Earth as heat sinks to remove the heat stored during the daytime. Evaporative cooling is most appropriate in dry, semi-arid to arid climates and uses the latent heat of evaporation of water to cool incoming air. Evaporative cooling systems are used quite often in the Southwestern United States. It should be noted that passive cooling systems can be combined with conventional mechanical ventilation technologies, e.g. solar-assisted ventilation systems. The adoption of passive systems can largely reduce the size and energy use of active ventilation systems.

10.3 Overview of Passive Heating

Passive heating utilizes solar and thermal energy to maintain occupants' comfort without the use of mechanical systems. Passive solar heating is one of the primary passive heating methods focusing on the effective use of solar energy. When integrated properly, these passive strategies can collectively contribute to the heating, cooling, and daylighting of nearly any building. The types of buildings that benefit from the application of passive heating may range from barracks to large maintenance facilities.

Typically, passive solar heating involves (https://www.wbdg.org/resources/passive-solar-heating),

- collection of solar energy through properly-oriented (e.g. south-facing) glazing;
- storage of this energy in "thermal mass," comprised of building materials with a high heat capacity such as concrete slabs, brick walls, and tile floors;
- distribution of the stored energy back to the living space, when required, through the mechanisms of natural convection and radiation.

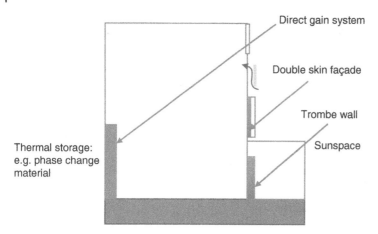

Figure 10.2 Illustration of passive heating techniques.

Passive solar design is the act of designing a building envelope, primarily the orientation and fenestration, to capture solar to heat a building. Some of the first examples of passive solar design were in ancient Greece, 2500 years ago. Passive solar heating systems usually have a low initial cost and short-term payback period, both of which are common with many active solar heating systems. Passive solar heating will increase user comfort which is another benefit. If properly designed, passive solar buildings are able to provide a delightful place to live and work in, contributing to increased satisfaction and user productivity. A variety of different passive solar techniques are currently used in building design, including: direct gain systems, thermal storage walls, and sunspaces (Figure 10.2).

Direct gain systems allow the direct solar radiation to strike surfaces in a building. The heat from solar radiation is stored in these surfaces and then released later during the day. The main disadvantage of this approach is that the large amount of direct solar radiation in living spaces can cause UV deterioration and sun fading of fabrics and dyes. It also causes glare and thermal discomfort for the occupants. Thermal storage walls that are typically located at the south façade of a building are thus proposed, which receive direct solar radiation and transfer the heat through the wall by conduction at a later time period. The conducted heat is distributed to the space by convection and radiation. This approach is a response to the drawbacks of the direct gain system and can provide far greater comfort in the room. Integrated design of these two systems can ensure that heat is introduced to the space at a desired time of a day. Another direct heating design with passive solar is the sunspace. While a sunspace will generally not provide all of the required heating, it can make a positive contribution. Sunspaces have the added benefit of connecting the outdoor and indoor environments, as well as providing additional living space.

10.4   Prescreening Feasibility of Passive Cooling and Heating Techniques

Passive cooling and heating techniques use natural resources to condition building living spaces and thus are highly sensitive to environmental conditions. Similar design strategies may be applicable for one project but not another, even at the same location under the same climate conditions. Surrounding buildings and environments (e.g. water body, terrain, landscaping, etc.) will all play an important role in determining appropriate passive cooling and heating strategies and techniques for a specific project. In addition, due to the large variation (or instability) of climatic conditions with time, it is difficult to maintain strictly specified indoor environmental conditions (e.g. temperature, humidity, acoustics, lighting, etc.) using passive techniques to the level that most mechanical systems can achieve. Buildings with rigorous environmental control needs such as hospitals, clean-rooms, and laboratories may not be suitable for using passive cooling, heating, and ventilation techniques. Table 10.2 presents a checklist to prescreen the

Table 10.2 Prescreening checklist for feasibility of using passive strategies.

	Passive cooling			Passive heating		
	Low	Medium	High	Low	Medium	High
Climate						
Temperature						
Relative humidity						
Solar						
Precipitation						
Micro-climate						
Terrain						
Landscape						
Shadow						
Air pollution						
Noise						
Indoor environment requirement						
Thermal comfort						
Air quality						
Lighting quality						
Acoustic quality						
Building function						
Activity type/intensity						
Schedule variation						
Occupant age						
Space availability for systems						
Budget						
First cost						
Utility rate						
Maintenance cost						

feasibility of using passive strategies for specific building projects. The shaded blocks represent the feasibility of project environmental conditions (e.g. climate) and design conditions (e.g. function, indoor environment quality need, and budget) for passive technique applications. Note that a "low" or "high" value for some conditions may be good for one passive technique while unideal for another passive technique. As a result, multiple appropriate values (low, medium, high) are selected for various possible passive solutions for different heating, cooling, and ventilation needs.

A host of computational tools are developed to assist the design of passively cooled and heated buildings at different design stages. Simple tools, such as Ecotect and Climate Consultant, can determine the viability and potential savings of specific passive techniques by analyzing climate data (Figure 10.3). This conclusion, however, does not consider specific project features (e.g. orientation, geometry, indoor design criteria, human activities, etc.) and surrounding environmental conditions (e.g. surrounding buildings, atmospheric quality, etc.). Passive design is truly case-dependent. Although rules-of-thumb can be found in literature for various passive techniques and these rules are of great value for preliminary design, an in-depth design and optimization of passive cooling, heating, and ventilation techniques requires advanced simulation tools to quantify and compare the performance of different solutions. It is important, however, to note that advanced models are only as accurate as the inputs, which can be difficult in the case of passive designs with many unknowns or uncertainties. For example, weather data that most simulation programs require may not account for the local climatic variations (e.g. wind speed and direction). A proper passive design will only occur through a detailed analysis of the microclimate, site conditions, and building design.

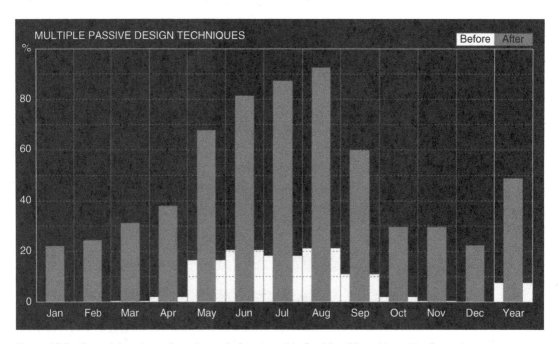

Figure 10.3 Potential savings of passive techniques used in Boulder, CO, estimated by Ecotect.

10.5   Natural Ventilation

10.5.1   Principle

With an increased awareness of human comfort and the environmental impacts of energy use, natural ventilation has become an attractive approach for reducing energy use and for providing acceptable indoor environmental quality. It plays an important role in maintaining a healthy, comfortable, and productive indoor climate rather than the more prevailing approach of using mechanical ventilation (https://www.wbdg.org/resources/natural-ventilation?r=wbdg_approach).

Natural ventilation is the process of supplying and removing air through an indoor space by natural means, i.e. without the use of a fan or other mechanical system. Air is driven by pressure differences between the building and its surroundings or among different building zones. Natural wind and buoyancy effects caused by temperature differences and humidity differences are the two main reasons for pressure differences. The amount of ventilation depends on the size and placement of openings in the building. Transom windows, louvers, grills, or open plans are techniques to complete the airflow circuit through a building. Smoke and fire transfer requirements challenge the designer when designing a natural ventilation system.

10.5.2   Performance

1) Energy consumption

There is a lot of research focusing on energy usage reduction by taking advantage of natural ventilation. In favorable climates and building types, natural ventilation can be used as an alternative to air-conditioning systems, saving 10–30% of total energy consumption. A 30% reduction of the cooling energy consumption, and a 40% reduction of the installed cooling capacity were predicted for a UK low energy office building with a night stack ventilation air exchange of 10 per hour (Kolokotroni and Aronis 1999). A 40% reduction in the daily cooling demand was simulated for a high thermal mass office building in Belgium (Gratia and Herde 2004). Night ventilation air changes of 8 per hour can reduce cooling requirements by 12–54%, depending on the temperature set point (Blondeau et al. 1997). The primary energy consumption of naturally ventilated office buildings in Denmark was compared with mechanical ventilation systems. The naturally ventilated buildings consumed 40 kWh/m^2 per year, whereas the mechanical systems consumption varied from 50 kWh/m^2 per year (VAV system) to 90 kWh/m^2 per year (CAV). The primary energy conservation for naturally ventilated office buildings in Belgium was calculated to be 8 kWh/m^2 per year (Schulze and Eicker 2013). Studies conducted for the 23-story Liberty Tower of Meiji University in Tokyo showed that about 17% of energy consumption for cooling is saved by using the natural ventilation system (Kolokotroni and Aronis 1999).

2) Thermal Comfort and Health

Besides offering benefits of energy usage reductions, many research and field experiments have shown that the occupants' thermal response in naturally ventilated spaces depends in part on the outdoor climate and may differ from thermal responses in buildings with centralized HVAC systems because of different thermal experiences, changes in clothing, availability of control, and shifts in occupant expectations. The acceptable thermal comfort range for naturally ventilated buildings is

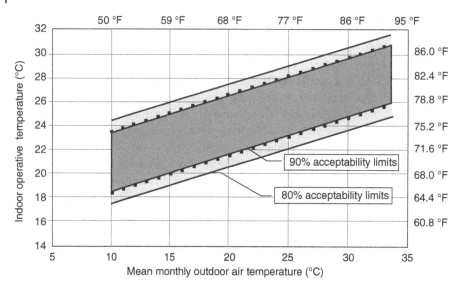

Figure 10.4 Comfort zone in naturally ventilated spaces. *Source:* ASHRAE (2020).

significantly larger than for buildings with standard mechanical HVAC systems. Based on this, ASHRAE Standard 55 has been revised to include an adaptive thermal comfort standard specifically for naturally ventilated buildings, which allows for increasing indoor temperatures as outdoor temperature increases. Figure 10.4 is the acceptable operative temperature ranges for naturally conditioned spaces from ASHRAE Standard 55-2020 "Thermal Environmental Conditions for Human Occupancy" (ASHRAE 2020). This adaptive model relates acceptable indoor operative temperature ranges to mean monthly outdoor temperature (defined as arithmetic average of mean monthly minimum and maximum air temperature). This information can be found easily by examining readily available climate data. Natural ventilation systems have been shown to consistently outperform mechanical systems with respect to complaints of SBS and its associated symptoms. Sepannen and Fisk (2002), in a review of 18 different studies on SBS and ventilation systems, found that the prevalence of SBS symptoms was 30–200% higher in air-conditioned buildings in comparison to naturally ventilated ones.

10.5.3 Design Considerations

Natural ventilation can be divided into: wind-driven vs buoyancy-driven (stack) natural ventilation according to driving forces and cross vs single-sided natural ventilation according to opening settings. Figure 10.5 illustrates the rule of thumb for the space depths penetrable by cross and single-sided ventilation. This rule of thumb indeed matches the prescriptive requirements for natural ventilation in ASHRAE 62.1 standard "Ventilation for Acceptable Indoor Air Quality" (ASHRAE 2019). ASHRAE 62.1 further requires that the openable area of wall openings directly to the outdoors is a minimum of 4% of the net occupiable floor area being naturally ventilated.

Wind-driven natural ventilation: when wind blows across a building, the wind hits the windward wall causing a direct positive pressure and after the wind moves around the building and leaves the

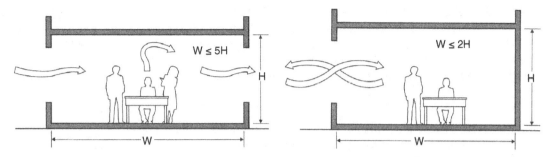

Figure 10.5 Rule of thumb for the maximum space depths for cross and single-sided.

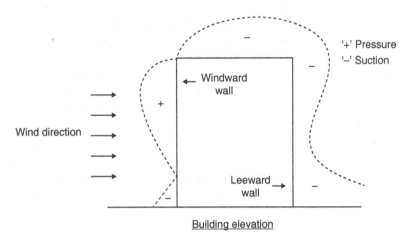

Figure 10.6 Diagram of wind-driven ventilation.

leeward wall it leaves a negative pressure (Figure 10.6). If there are any openings on the windward and leeward walls of the building at the same time, fresh air will rush in the windward wall opening and exit the leeward wall opening. Thus, the pressures on the windward and leeward walls can be balanced. In the summer and/or shoulder seasons, when the outdoor air temperature is appropriate, wind can be used to remove indoor heat and provide fresh air, while in winter, ventilation is normally reduced to the level that is sufficient to remove excess moisture and pollutants.

The air volume flow rate induced by wind can be calculated via:

$$Q_{wind} = K \times A \times V \tag{10.1}$$

where Q_{wind} is the volume of airflow (m^3/h); A is the area of smaller opening (m^2); V is the outdoor wind speed (m/h); K is the coefficient of effectiveness. The coefficient of effectiveness depends on the angle of the wind and the relative size of entry and exit openings. It ranges from about 0.4 for wind hitting an opening at a 45° angle of incidence to 0.8 for wind hitting directly at a 90° angle.

Buoyancy-driven natural ventilation: both a temperature gradient (known as stack ventilation) and humidity disparity (known as cool tower) can induce airflow, due to the air density difference caused by temperature and humidity. Stack ventilation is the most commonly used ventilation strategy when outdoor wind is not sufficient. In stack ventilation, the temperature difference between the warm air inside the building and the cool air outside can move the air in the building to rise and exit at upper openings, while air is supplemented from outside via lower openings.

The airflow rate induced by the stack effect can be estimated as:

$$Q_{stack} = C_d * A * [2gh(T_i - T_o)/T_i]^{1/2} \tag{10.2}$$

where Q_{stack} is the volume of ventilation rate (m^3/s); C_d is the discharge coefficient (e.g. 0.65); A is the net area of inlet opening (m^2), which equals area of outlet opening; g is the acceleration due to gravity (9.8 m/s^2); h is the vertical distance between inlet and outlet midpoints (m); T_i = average temperature of indoor air (K); T_o = average temperature of outdoor air (K).

Similarly, buoyancy caused by differences in humidity can allow a pressurized column of wet and cool air to naturally enter a space and repel the dry and warm indoor air from the top. Cool tower ventilation is most effective where outdoor humidity is low and evaporation can be directly applied. Prediction of airflow rate induced by a humidity difference can be somewhat involved. Equation (10.3) for estimating the airflow rate in humidity-based ventilation is based on the expression developed by Thompson, with the coefficient from the data measured at the Zion National Park Visitor Center that has a cool tower of 7.4 m tall, 2.4 m^2 cross-section, and 3.1 m^2 opening (http://www.wbdg.org/resources/naturalventilation.php?r=wbdg_approach).

$$Q_{cool\ tower} = 0.49 * A * [2gh\ (T_{db} - T_{wb})/T_{db}]^{1/2} \tag{10.3}$$

where $Q_{cool\ tower}$ is volume of ventilation rate (m^3/s); 0.49 is an empirical coefficient calculated with data from Zion Visitor Center, UT, which includes humidity density correction, friction effects, and evaporative pad effectiveness; A is the free area of inlet opening (m^2), which equals area of outlet opening; g is the acceleration due to gravity (9.8 m/s^2); h is the vertical distance between inlet and outlet midpoints (m); T_{db} = dry-bulb temperature of outdoor air (K); T_{wb} = wet-bulb temperature of outdoor air (K).

In design practice, stack ventilation and cool tower can be combined by having a cool tower deliver evaporative air to a space and then relying on the increased buoyancy of the warm air to exhaust heat through upper stacks. In the design of a naturally ventilated building, climate conditions, location topography and vegetation, and surrounding objects all need to be carefully evaluated. Wind data collected at local airports may not say much about microclimates that can be heavily altered by natural and man-made obstructions around the building site. Window openings should be accessible and operable by the occupants. A naturally ventilated structure often includes an articulated plan and large window and door openings, while an artificially conditioned building is sometimes best served by a compact plan with sealed windows.

When designing ***wind-driven natural ventilation***, several important factors need be considered during design process.

1) Building orientation is the most important factor that determines how much air can flow through a building. Generally, a building should be oriented so that the windward wall is perpendicular to the summer wind to maximize the ventilation.
2) Building shape is one crucial factor to capture wind and bring ventilation into a building. A good building shape design can create proper wind pressures around the building that can effectively drive the airflow through the openings of the building. Naturally ventilated buildings should not be too deep as

it will be difficult to distribute fresh air to all portions of the building. It is important to avoid obstructions between the windward inlets and leeward exhaust openings. Avoid partitions in a space oriented perpendicular to the airflow. Buildings that rely on natural ventilation often have an articulated open floor plan.

3) Window typology and operation are critical to enhance and balance air-conditioning needs. Proper window design can achieve the same amount of ventilation but with less openings (and thus costs). When wind flow does not ideally prevail perpendicular to a building wall, optimal location, type, and opening direction of windows can act as a scoop to direct the wind into the space.

When designing **buoyancy-driven natural ventilation**, the use and creation of an indoor–outdoor temperature difference (indoors are warmer than outdoors) is the critical factor. Several design hints are:

1) Design low inlets and high outlets;
2) Increase vertical distance between inlets and outlets;
3) Use skylights or ridge vents as exhaust;
4) Provide adequate up-flow channel (e.g. atrium);
5) Use solar chimney to increase indoor–outdoor temperature gradient;
6) Do not compromise the function as fire exits of enclosed staircases.

Natural ventilation is typically supplemented with mechanical ventilation, called hybrid ventilation, due to the variation and unreliability of the climate and stack effect. Hybrid ventilation allows a building to capitalize on the benefits of natural ventilation, with assurance that the building can operate at a desirable thermal comfort level during extreme outdoor conditions. Thus, hybrid ventilation can be implemented in a wider range of climates where purely natural ventilation is insufficient to provide all of the cooling needs of a space. Other methods to enhance natural ventilation performance include solar chimneys and wind towers. A solar chimney may be an effective solution where prevailing breezes are not dependable enough to rely on wind-induced ventilation and where keeping indoor temperature sufficiently above outdoor temperature to drive buoyant flow would be unacceptably warm. The chimney is isolated from the occupied space and can be heated as much as possible by the sun or other means. Air is simply exhausted out the top of the chimney creating a suction at the bottom which is used to extract stale air. Wind towers, often topped with fabric sails that direct wind into the building, are a common feature in historic Arabic architecture. The incoming air is often routed past a fountain to achieve evaporative cooling as well as ventilation. At night, the process is reversed, and the wind tower acts as a chimney to ventilate room air. More design strategies and parameters for natural ventilation can be found in AM10 Natural Ventilation in Non-Domestic Buildings (CIBSE 2005).

10.6 Night Cooling with Thermal Mass

10.6.1 Principle

Night cooling with thermal mass is an effective application of natural (and hybrid) ventilation strategy, especially for areas with dramatic diurnal temperature swings. The night cooling method ventilates a building during the cool night and removes the heat stored in a building's structural elements (e.g. walls and floors) during the day. Thermal mass is necessary for this approach to store heat (from solar, outdoors, internal heats) and release it at a later time when it is cool outside and ventilation (opening) is on either by a passive approach or by the use of a mechanical ventilation system. The thermal mass of a building is cooled mainly by convection at night and then the cooled/discharged mass can act as a heat sink and reduces the rate of rising indoor temperatures the following day. Night cooling can affect internal conditions during the day in several ways, such as reducing the peak air temperatures and reducing air temperature throughout the day especially during morning hours. It can also flatten the diurnal air temperature swing. The energy impact of this technique can cause a reduction in the cooling load both in air-conditioned and naturally ventilated buildings.

10.6.2 Performance

Setting controls for night cooling can save a significant amount of energy, depending on location. Studies indicate cost savings range from 5% in Phoenix, AZ, to 18% in Denver, CO, for a typical office building. Night precooling also reduces peak demand. Simulation analyses show that precooling a 100 000 ft^2 three-story building in Sacramento, California, would reduce energy use by 12.6% and cause a peak demand reduction of 31.3%. Most of the studies conclude that the use of night ventilation in free floating buildings may decrease the next day peak indoor temperature up to 3 K. Research on night cooling applications for residential buildings shows that night ventilation may decrease the cooling load up to 40 kWh/m^2/year with an average contribution close to 12 kWh/m^2/year (Santamouris et al. 2010).

Given that the free energy offered by night ventilation techniques increases as a function of the initial cooling needs of the buildings, those with high cooling loads benefit much higher absolute contribution than buildings presenting a low cooling demand. By investigating the correlation between the cooling needs of buildings and the energy contributions of night ventilation, researchers found that the correlation is almost linear. In parallel, the uncertainty associated with the evaluation of the energy contribution of night ventilation decreases largely for higher airflow rates. Given the dissimilarity of the amount of energy stored in buildings and the variability of the night ventilation capability for each individual building, the percentage energy contribution of night ventilation is independent of the initial cooling load of buildings.

10.6.3 Design Considerations

Night cooling with thermal mass is most appropriate and effective for climates with a large diurnal air temperature range, where the cool night air can be used to remove the heat accumulated and stored during the hot day in the building mass. For a high-mass, well-insulated, and shaded building, closed during the daytime and ventilated only during the night, a drop in the indoor maximum below the outdoor maximum of about 45–55% of the outdoor range is possible. At night, the indoor temperatures are higher than outdoors. Figure 10.7 shows the climatic boundaries, in terms of the outdoor maximum daily temperatures, under which night cooling is applicable.

Generally, night cooling is very efficient in arid and desert regions where the maximum temperature is below about 36 °C. In desert regions with daytime temperatures above 36 °C, night ventilation alone would not maintain the indoor daytime temperature at an acceptable level and other passive cooling systems should be applied during the hot hours, such as evaporative cooling, compression, or absorption

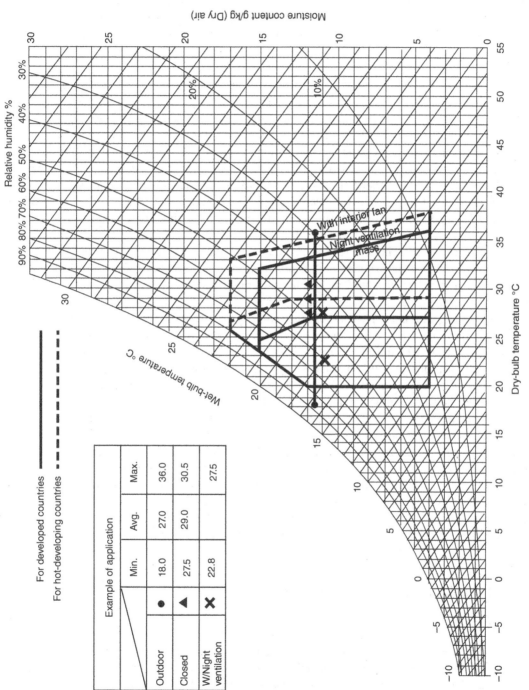

Figure 10.7 Climate boundaries for using night cooling technique. *Source:* Givoni (1992).

air-conditioning. Even in this case, the application of night ventilation can significantly reduce the length of the periods and duration of the time when the additional cooling systems will be used.

In reviewing the distribution of temperature ranges of many regions in the world, Givoni (1976) found a quantitative relationship between the vapor pressure (vp) and the diurnal ranges (T_{swing} (K)):

$$T_{swing} (K) = 26 - 0.83 \times (vp) (mmHG) \tag{10.4}$$

He gave some examples in his paper that in a very dry desert with a summer vapor pressure of about 7–8 mmHg, the expected diurnal temperature range would be about 20 K; in arid regions with a vapor pressure of about 12 mmHg, the expected range would be about 16 K, while in a humid region with vapor pressure of 22 mmHg, the expected range would be only about 8 K. It should be noted, however, that this is only a statistical relationship and in any given location the actual temperature range should be considered. As a general principle, it can be estimated that in arid and desert regions, with a summer diurnal temperature range of 15–20 K, the expected reduction of the indoor maximum temperature can be about 6–8 K below the outdoor maximum. On very hot days, which usually have a larger diurnal range, the drop in the indoor temperature, during the time of the outdoor maximum, may be up to about 10 K (Givoni 1992).

Thermal mass is an inevitable element in night cooling design which indeed determines the performance of night ventilation. A building can maintain daytime temperatures lower than the outdoor level only when it has sufficient thermal mass and thermal resistance and is protected from absorbed and penetrating solar radiation. In such a building, the diurnal temperature swing is very small, about 10–20% of the outdoor temperature range. A high thermal mass and materials with high thermal resistance are recommended along with the application of this technique.

Night cooling will not work without proper control and operation strategies. In principle, a building should be cooled down 1–2 K during the night below the lowest comfort level that can be accepted during the daytime. If this is not accomplished, and it is controlled at the same comfort level as applied during the daytime, then the benefits of night cooling will be reduced. To get effective night cooling, a building should be cooled down deliberately to lower temperatures.

It should be noted that the acceptable temperature range is reduced with higher humidity, reflecting the corresponding decrease in the outdoor range with higher humidity. Increasing the indoor air speed by internal fans can extend the indoor comfort range without elevating the indoor temperature. It is also possible to use an air handler and economizer to flush the building with night air to cool down the building mass. In general, the use of mechanical ventilation systems resolves a number of problems associated with natural systems. They require much smaller openings, can be easier to control, and provide sound absorption and security. However, they consume electricity and heat the air (systems can have fan heat gains of up to 2 °C). Therefore, careful monitoring is required when used for night ventilation.

10.7 Direct/Indirect Evaporative Cooling

10.7.1 Principle

Evaporative cooling is an effective, economical, environmentally friendly, and healthy cooling approach that is widely used in residential and commercial buildings. Evaporative cooling is different from traditional air-conditioning systems that use vapor-compression or absorption refrigeration cycles with chemical refrigerant. Instead, it uses water as the working fluid and provides excellent cooling by utilizing water's large enthalpy of vaporization. The temperature of dry air can be reduced significantly by passing through water. Evaporative cooling has the additional attraction of low energy consumption, easy maintenance, installation, and operation. Evaporative cooling is commonly used in cooling water towers, air washes, evaporative condensers, and fluid cooling, and is also used to uniformize the temperature in places where several heat sources are present (Camargo et al. 2006).

There are basically two types of evaporative cooling: direct evaporative cooling (DEC) and indirect evaporative cooling (IEC). The combination of DEC and IEC techniques is also very popular in practical applications. In **DEC**, water evaporates directly into the air thus bringing down the dry-bulb air temperature as well as adding the moisture to the airstream until the air is close to saturation. Therefore, the ideal status of air after a direct evaporative cooler is that the dry-bulb temperature can reach the wet-bulb temperature and the humidity of the air may come close to 100%. The air can be cooled either by direct contact with a liquid surface or a wet solid surface or else with the use of spray systems. **IEC** removes the heat from the primary air by crossing the secondary airstream in a heat exchanger. The secondary air is cooled by water in either a direct approach or indirect approach. In the direct approach, the secondary airstream passes through a cooling tower and circulates through one side of the heat exchanger. For the indirect approach, the secondary air, which is always the return air from the rooms, can be wetted directly and removes the heat from the conditioned supply air on the dry side in an air-to-air heat exchanger. IEC does not add moisture to the primary air but both the dry-bulb and wet-bulb air temperatures of the primary air are reduced. IEC is much flexible in applications than DEC, as it is also applicable even in a climate with a high wet-bulb temperature. In practice, a combination of IEC and DEC, which is also called two-phase cooling, is becoming more and more popular and is proven to be an efficient passive cooling strategy. The primary air is sensibly cooled first by IEC in phase one and then cooled by directly adding the moisture to the main airstream in phase two which is DEC (Figure 10.8).

10.7.2 Performance

The effectiveness of evaporative cooling is defined as the rate between the real decrease of dry-bulb temperature and the maximum theoretical decrease that dry-bulb temperature could have if the cooling were

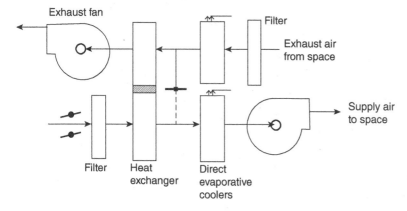

Figure 10.8   Sketch of an indirect–direct evaporative cooling system. *Source:* ASHRAE (2012).

100% efficient. In DEC, saturation will be a key factor in determining evaporative cooler performance. The theoretical temperature that the air can reach is the wet-bulb temperature of the entering air. Therefore, the direct saturation efficiency can be expressed as

$$\varepsilon_e = 100\frac{t_1 - t_2}{t_1 - t_s'} \tag{10.5}$$

where

ε_e = direct evaporative cooling saturation efficiency, %
t_1 = dry-bulb temperature of entering air, °F
t_2 = dry-bulb temperature of leaving air, °F
t_s' = thermodynamic wet-bulb temperature of entering air, °F

In IEC, the effectiveness of the heat exchanger is important. The performance is defined as wet-bulb depression efficiency (WBDE) as follows:

$$\text{WBDE} = 100\frac{(t_1 - t_2)}{(t_1 - t_s')} \tag{10.6}$$

where

WBDE = wet-bulb depression efficiency, %
t_1 = dry-bulb temperature of entering primary air, °F
t_2 = dry-bulb temperature of leaving primary air, °F
t_s' = wet-bulb temperature of entering secondary air, °F

The performance charts of DEC and IEC are shown in Figure 10.9.

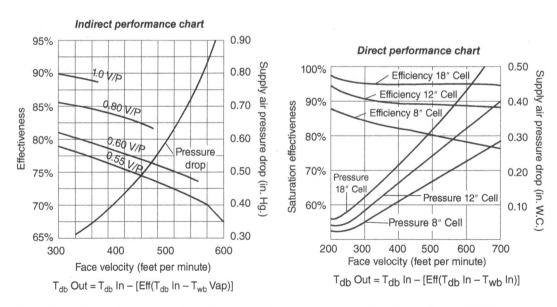

Figure 10.9   The performance charts of direct and indirect evaporative cooling. *Source:* Palmer (2002).

10.7.3 Design Considerations

Evaporative cooling varies in applicability and efficiency with the relative humidity of the outside air. It is much more suitable in a dry climate. This constraint should be especially paid attention to when using DEC, as the performance is directly determined by the wet-bulb temperature of the air. IEC can be somewhat flexible in climates where it can be used. The following three climate types are mostly suitable for the application of evaporative cooling:

- **Desert:** comfort during the entire cooling season
- **Steppe:** comfort during the dry period of the hot season and moderate relief cooling during more humid periods
- **Savanna:** only relief cooling during the hot season

The most commonly used evaporative cooling equipment or units are packaged direct evaporative air coolers, air washers, indirect evaporative air coolers, evaporative condensers, vacuum cooling apparatus, and cooling towers that exchange sensible heat for latent heat. The details of these pieces of equipment or units can be referred to ASHRAE Handbook – HVAC application (ASHRAE 2011).

Air passing through a direct evaporative cooler's wetted media is free of pollutants, minerals, and bacteria. A properly designed system, operating within the recommended face velocities, will not have water carryover. For example, the maximum face velocity for a rigid-type media is about 700 fpm while the systems are often designed for around 500 fpm. The primary air passing through an indirect evaporative cooler will not take any additional moisture. Besides acting as a scrubber for removing water-soluble gases, the wetted media has some particulate removal capability. That capability depends on media type and thickness. A 12 in rigid media has an ASHRAE Standard 52 dust spot efficiency of about 16%, which is equivalent to a fiber furnace filter. Water added in the evaporative systems also acts as an air washer by cleaning the inlet airstream of airborne particulates and soluble gases, which could be a significant benefit to downstream filter elements. Studies show that evaporative cooling also reduces NO_x emissions because of the increase in moisture added to the air.

It is recommended to supply the evaporative cooler makeup water from a quality drinking water source. Untreated surface water sources should be avoided due to the risk of a high concentration of nutrients and other pollutants. Two criteria for evaporative coolers are 500 mg/l of TDS and a pH in the range of 6.5–8.5. By using drinking quality water as the makeup water, the outside air to be conditioned will be the most likely source of nutrients. Depending on efficiency, upstream filtration can significantly reduce the concentration of these potential nutrients. A water bleed, required for scale control, will also keep nutrients that might penetrate upstream filtration from concentrating in the recirculating water. Besides moisture and nutrients, algae need light to grow. Evaporative coolers that have the wetted media attached or in close proximity to inlet louvers can be particularly susceptible to algae. Providing a barrier like a shade cloth or awnings will inhibit algae growth. The prevention of legionellosis is a health and safety concern for owners and operators of facilities and those who manage water-based evaporative cooling systems. ASHRAE 188 "Prevention of Legionellosis Associated with Building Water Systems" should be used for compliance.

Evaporative cooling is not considered/recognized as a viable humidification technology, yet it has quality and economic advantages when compared to other options such as steam injection, infrared, air injection (atomizing), or ultrasonic. Maintenance and equipment costs are moderate compared to the other four options. In a 100% outside air-makeup system, the thermal energy cost to humidify is essentially equal for all technologies. However, when humidification by wetted media is incorporated in an economizer system using a mixture of outside and return air, the energy content of the return air provides a significant portion and sometimes all of the energy required to humidify.

The operating costs for evaporative condensers are normally less than a cooling tower designed to dissipate the same heat load because the airflow for evaporative condensers will be about equal for equivalent service and the recirculating water pumping cost will usually be less. The fan static pressure for the evaporative cooler will usually be between 0.2 and 0.8 in. of water which is also true for the cooling tower. Comparison of brake horsepower requirements for equivalent service indicate that the evaporative cooler will generally be equal to or lower than the cooling tower. The rate of recirculating water in an evaporative cooler is less than half of the recirculating rate required for a cooling tower exchanger combination designed for equivalent service.

Maintenance requirements for evaporative coolers are comparable with shell and tube heat exchangers. If proper fouling resistances are used to determine the cooling surface, the frequency of cleaning can be reduced to practically any level desired. Even cleaning the outside of the tube surface can be accomplished either chemically or mechanically without shutting down the unit.

10.8 Trombe Wall

10.8.1 Principle

Thick adobe or stone walls can absorb and store sun heat during the day and release it slowly and evenly at night to heat a building inside. The performance of this thermal storage wall can be largely improved by attaching a layer of glazing outside the wall – an enhanced method for passive heating. Named after the French inventor Felix Trombe in the late 1950s, this energy storage Trombe wall generally is a very thick, solar-facing wall, that is typically painted a dark color and is made of a material with a large density and specific heat (for energy storage). A pane of glass or plastic glazing, installed a few inches in front of the wall, helps hold in the heat. The wall heats up slowly during the day and cools gradually during the night, transferring heat to the inside at a delayed time (Figure 10.10). Trombe walls take advantage of the unique transmittance properties of regular glass at different radiation wavelengths: they transmit most solar energy at low wavelengths but block most thermal energy at high wavelengths. As a result, the transmitted solar energy is trapped in the cavity between the glazing and the wall, gradually transporting it into

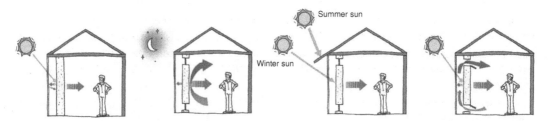

Figure 10.10 Trombe wall design and operation.

the indoor space via the storage wall. Trombe walls are a very useful passive heating system, particularly for buildings with large heating loads and for building retrofit. They require little or no effort to operate and are ideal for spaces where silence and privacy are desirable. They are often easily built from readily available materials and are very reliable. Rooms heated by a Trombe wall often feel more comfortable than those heated by forced-air systems, even at lower air temperatures, because of the radiantly warm surface of the wall. A successful Trombe wall with properly designed window systems can optimize heat gain and minimize heat loss during cold months, and avoids excess heat gain in hot months.

10.8.2 Performance

Trombe wall performance varies mostly with wall materials. For an 8-inch-thick (20-cm) concrete wall, heat may take about 8 hours to reach the interior of the building. This implies that rooms receive slow heating for many hours after the sun sets, greatly reducing the need for conventional heating (during a no-sun period). The performance of Trombe walls is diminished if the wall interior is not directly open to the interior zones. Based on previous experience with Trombe walls, the heat delivered by a Trombe wall in a residence was reduced by over 40% because kitchen cabinets were placed on the interior of the wall (Torcellini and Pless 2004). National Renewable Energy Lab of the United States monitored and analyzed the energy performance of the Trombe wall in a visitor center over a two-year period. The analysis consisted of measured electrical end uses, Trombe wall temperature profiles, and thermographic pictures to determine the performance of the Trombe wall. They found that the thermal distribution of the Trombe wall at 8:30 p.m. on 16 December 2000 is generally homogeneous, with the internal wall surface temperature ranging from 90 to 96 °F (32 to 36 °C). The wall temperature typically peaks between 8 and 9 p.m. The visitor center's electric radiant heating system used 22 680 kWh (81.6 GJ) over the year, with the Trombe wall contributing 20% of the total heating to the building. The Trombe wall imposed a heating load on the building for only 2 of the 151 days of the 2001–2002 heating season. For the other 149 heating days, the wall was a net positive. The peak heat flux through the wall was 11.2 W/ft^2 (89 W/m^2) or 8.3 kW over the entire Trombe wall area. The average efficiency of the wall (defined as the heat delivered to the building divided by the total solar radiation incident on the exterior of the wall) was 13%.

10.8.3 Design Considerations

Trombe walls are particularly well suited for sunny climates that have high diurnal (day–night) temperature swings, such as the mountain-west. They do not work well in either cloudy climates or where there isn't a large diurnal temperature swing. For instance, in New Mexico, where homes have been built out of adobe (dried mud) bricks for hundreds of years, even an unglazed south wall will deliver some heat into the house – if one adds a frame and layer of glazing on the outside of the wall, the performance improves dramatically.

A typical unvented Trombe wall consists of a 4–16 inch (10–41 cm) thick, solar-facing masonry wall with a dark, heat-absorbing material on the exterior surface and faced with a single or double layer of glass. The glass is placed ¾ to 2 inch (2–5 cm) from the masonry wall to create a small airspace. Heat from sunlight passing through the glass is absorbed by the dark surface, stored in the wall, and conducted slowly inward through the masonry. High transmission glass maximizes solar gains to the masonry wall. As an architectural detail, patterned glass can limit the exterior visibility of the dark concrete wall without sacrificing transmissivity. Using low-E glazing can prevent heat from re-radiating out through the glass of a Trombe wall and greatly reduce the amount of heat lost. Applying a spectrally selective surface or low-E coating to the wall itself can also improve performance by reducing the amount of infrared energy radiated toward the glass.

Trombe walls typically do not have additional insulation. The system works with material that is both very heavy (high heat capacity) and fairly conductive (low R-value per inch). The trick is to choose the right material and size the wall thickness so that the solar heat penetrates through it to the inner surface by a desired later time. In general, 6–10 inch are good for an adobe wall, while 10–16 may be needed for a concrete or brick wall. Water may also be used as the thermal storage medium but it usually requires a container (e.g. tube). When a water tank is used, eight or more inches of thickness are often suggested.

To avoid overheating at hot times of day or hot seasons of the year, Trombe walls may be used in conjunction with overhangs, eaves, and other building design elements to evenly balance solar heat delivery. An overhang is typically built that extends out over and above the Trombe wall. This will shade the wall from direct sun during the summer but will allow full solar exposure in the winter.

Top and bottom vents can be installed through the masonry wall to deliver more heat into the house during the daytime hours. Warm air in the space between the glazing and wall surface rises and enters the room, being replaced by air from the house entering through the lower vents in a convective loop. These vents should be closed at night so that the air circulation doesn't reverse, with air next to the glazing cooling off and pulling in warm air from the room through the upper vents and delivering chilled air to the room through lower vents. Vents through the glazing can also be installed and seasonally opened and closed. In the summer months – when one does not need the Trombe wall to deliver heat into the house – these vents are left open. Screens on the vents keep out insects and other unwanted visitors.

10.9 Sunspace

10.9.1 Principle

A sunspace is one popular passive solar system, due to its potential as an energy collecting system, its multiple functions, and pleasant appearance. Sunspaces represent additional space (and thus value) with positive architectural qualities. Sunspaces, as a retrofitting technique produced by glazing in balconies or as newly built extensions, can be one of the most useful techniques to store solar energy and are therefore extensively used.

The main principle of south-facing sunspaces and conservatories is similar to the Trombe wall but with a large and functional air "cavity." The greenhouse effect of glass allows solar beams to penetrate through the external glazing but keep the heat inside. Part of the heat is stored in the thermal storage media (e.g. wall and floor), which will be redistributed back to the space at a later time (e.g. at night). Part of the heat can be directly carried into the space via the air convection through the vent openings at the wall when heating is needed immediately (Figure 10.11).

10.9.2 Performance

Sunspaces are generally less efficient in collecting solar heat than direct solar gain via windows and Trombe walls, mostly due to the larger convection and radiation heat loss in larger air spaces between the glazing and the wall. Increasing the thermal resistance of a sunspace glazing by using multi-pane glasses (either air-filled or inert gas filled) can reduce the heat loss although it does slightly decrease the solar transmittance.

Figure 10.11 Sunspace design and operation.

A good orientation of a sunspace is critical. Ideally, the sunspace should face due south, but 30° east or west of due south will provide about 90% of the maximum static solar collection potential. The optimum orientation will depend on site-specific and local landscape features. Research compared the performance of sunspace with south, north, east, and west orientations in the cold period of a year and found that orientation alteration from north to south can improve the results almost up to 7 °C (Mihalakakou and Ferrante 2000; Mihalakakou 2002).

10.9.3 Design Considerations

Sunspace performance heavily depends on climate conditions and system designs. In southern latitudes, passive solar design can provide space heating in winter, but overheating problems in summer should be overcome by using effective solar control and passive cooling systems, such as night ventilation and Earth-to-air heat exchangers. In temperate and northern climates, although the conditions are more favorable, one may still need to pay attention to the climate variations by using effective shading devices, ventilation, and sufficient thermal insulation to ensure a comfortable built environment. In very cold climates, double glazing is needed to reduce heat losses through the glass to the outside. For all climates, except those with very cool summers, operable or mechanized windows should be considered at top and bottom. These allow the sunspace to avoid overheating by passively venting hot air out the top of the glazing and pulling cool air in through the bottom of the glazing.

The angle and type of sunspace window glazing determine the performance of sunspace. Although tilted glazing collects more heat in winter, it also loses more heat at night and may be covered with snow in winter, as well as causing overheating in warmer weather. Vertical glazing is commonly used that maximizes heat gain in winter when the solar angle is low and heat is needed most. It is also easy to shade and produces less heat gain in the summer. Compared with tilted glazing, vertical glazing is less expensive, easier to install and insulate, and not as prone to leaking, fogging, breaking, and other glazing failures. Vertical glazing is also often more aesthetically compatible with the retrofit of existing homes. Glazing sometimes is also implemented on the common wall that separates the sunspace from the indoors, allowing daylighting to the indoor spaces. Movable insulation is usually desired, especially at night at cold climates, to reduce the heat loss from the inside to the sunspace.

Massive materials (e.g. masonry or concrete) are required to flatten temperature swings and store energy for a later use. Walls and floors are typically the best locations for thermal mass which have direct exposures to the solar. In general, for insulated and uninsulated masonry common walls, the suggested wall thicknesses are 8–12 inch and 4–6 inch, respectively. If water is used as thermal storage, the rule of the thumb is 2 gallons of water for each square foot of glazing.

Operable vents are often designed at the top of sunspaces where temperature is the highest and at the bottom where temperature is the lowest. The vents can circulate air between the sunspace and indoors or directly exhaust hot air to the outdoors. They can be operated manually or with thermostatically controlled motors. Mechanical fans (e.g. ceiling fan) can be used to accelerate the air circulation to the entire house when natural convection is not adequate.

10.10 Double Skin Façade

10.10.1 Principle

Modern architecture is dominated by transparent buildings. The large, glazed areas result in high building heating and cooling loads, leading to high levels of energy consumption and therefore significant financial and environmental burdens. The DSF is one potential response to these problems. Compared to a single-skin façade, a DSF consists of an external glazing offset from an internal glazing integrated into a curtain wall, often with a controllable shading system located in the cavity between the two glazing systems. Typically, the external glazing is a single layer of heat-strengthened safety or laminated safety glass, while the interior layer consists of single- or double-pane glass with or without operable windows.

Air can flow through the cavity via natural or mechanical ventilation and is used to help moderate building thermal loads. Most commonly, outdoor airflows into the bottom of the cavity and exhausts from the top of the cavity to outdoors. During the cooling season, throttling flaps at the inlet and outlet of the cavity remain open to allow airflow through the cavity to exhaust heat that builds up in the cavity. If the DSF includes dynamic sunshades in the cavity, they are usually deployed during the cooling season. Consequently, instead of heating the building interior, the solar radiation heats the air in the cavity, causing the air to buoyantly rise out of the cavity. Thus, the DSF reduces the heat gain of the building. Because the buoyancy of the heated air increases with cavity height, DSFs are typically used in multistory buildings on one or more sides of the building that receive appreciable Sun. In addition, when combined with operable windows that open into the air cavity, a DSF can provide natural ventilation to cool spaces adjacent to the DSF without (or with much lower levels of) mechanical cooling when the outdoor air enthalpy is less than the indoor air enthalpy. During the heating season, the DSF cavity inlet and outlet vents close to prevent airflow, trapping solar gains in the cavity and reducing daytime heat losses through the façade. In addition, the sunshades remain open during the day to allow more solar radiation to enter the building and are closed at night to retain heat (i.e. the reverse of the operational scheme during cooling season).

10.10.2 Performance

A large amount of research has been done on the thermal performance of DSF. Generally, it is believed that DSF can achieve better energy performance than a single-skin façade. This is due to (i) an additional thermal buffer zone/layer created between the inside and outside and (ii) proper airflow controls depending on indoor and outdoor conditions. However, the actual performance of DSF buildings vary significantly with building type, scale, and location.

The following validated formulae provide a quick estimate on the most important parameters for predicting energy performance of DSF buildings.

1) The airflow rate as a function of opening size and h/d ratio for a single-story high cavity (3 m) with a roller shading device for buoyancy-driven airflow DSF:

$$\dot{V} = (-6.2 * {}^h\!/\!_d + 647.5) A_{opening} \sqrt{\Delta T_{Average}} \tag{10.7}$$

For a five-story (15 m) high cavity:

$$\dot{V} = 1375 * A_{opening} \sqrt{\Delta T_{Average}} \tag{10.8}$$

where h/d is the cavity height/depth ratio; $A_{opening}$ is the area in m^2 of one opening for cavities with two equal size openings or the smaller of two openings for cavities with different size openings; ΔT is the average cavity temperature minus outdoor air temperature in °C; and $\dot{V}$ is the airflow rate through the cavity in m^3/hr. Note that $A_{opening}$ is the equivalent opening area that has the same resistance performance as a physical opening.

2) The peak cavity air temperature to the average cavity air temperature:

$$T_{Peak\ cavity\ air - 1 - story} = 2.43 \left(T_{Average\ cavity\ air} \right)^{0.78} \tag{10.9}$$

$$T_{Peak\ cavity\ air - 5 - stories} = 3.68 \left(T_{Average\ cavity\ air} \right)^{0.69} \tag{10.10}$$

It is interesting to notice that the difference between peak cavity air temperature and outdoor air temperature is about 1.87 of that between average cavity air temperature and outdoor air temperature, independent of cavity height.

10.10.3 Design Considerations

A DSF consists of an external façade and an internal façade that form an internal cavity. Apart from this basic design, the cavity dimensions, openings, materiality, construction types, airflow paths, and control systems vary considerably. Different DSF designs are better suited to specific climates and design goals. Figure 10.12 shows a typical DSF configuration that was tested in Belgium. The primary parameters of a DSF design include:

a) Cavity Depth

DSF cavities have been built with depths from a few inches up to four or five feet. Most studies that have considered a variety of cavity widths have determined that this variable does not have a large effect on the overall thermal performance of the DSF. Some considerations in deciding on the cavity depth include: (i) making the cavity volume large enough to allow for sufficient airflow through operable windows to meet ventilation requirements; (ii) providing adequate space for the shading device and other structural elements; (iii) allowing for easy access to the cavity interior for maintenance and cleanness purposes; etc.

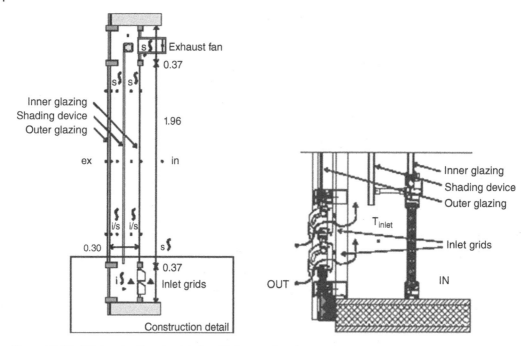

Figure 10.12 Vliet test cell cavity details with louvered cavity opening.

b) Cavity Width

DSF exist both with cavities extending the whole width of a building and with cavities stratified into 1- to 2-m-wide sections. The effect of cavity width on airflow has not been studied extensively, but it is likely that structural and aesthetic considerations would largely determine this design decision. It has been suggested, however, that creating narrow cavities will produce shafts that will allow fire and contaminants to propagate more quickly throughout a building.

c) Cavity Height and Opening Locations

Another design consideration is the height of the cavity: whether it extends the whole height of a building or is divided at each floor. Most existing DSFs have cavities that are divided along the height of the façade, although some have perforated catwalks at each floor for maintenance access that still allow air to flow through them. A taller cavity will produce a stronger buoyancy force, creating a greater airflow rate. However, the fire hazard resulting from a large vertical cavity with no divisions might be prohibitive. Also, air contaminants and noise pollution could flow readily from floor to floor. If the cavity is divided at each floor, there are typically air inlets and outlets near the top and bottom of each section. For a full building height cavity, there could be a single bottom inlet and a single top outlet. This configuration creates the strongest buoyancy-driven airflow due to the tall cavity height. Alternatively, there could be a number of inlets along the exterior façade with one main outlet at the cavity's peak. This will increase the airflow into the cavity, which is desirable in the case with operable windows.

d) Opening Structure

The cavity openings could be quite simple or complex with controlled louvers and wind shields meant to reduce the effect of wind pressure on airflow through the cavity. If no wind shields exist, the airflow within the cavity will often be driven by wind pressure, forcing the air to flow predominantly in the downward direction. Two sets of automatically controlled louvers allow one or two grids to be open, providing precise control over the airflow rate through the cavity.

e) Cavity Materiality

Most DSFs for new buildings are designed with the goal of maximizing the transparency of the building, in which case both the interior and exterior façades will be mostly glazed. Typically, the interior façade is the thermal barrier, while the exterior facade is a higher transmittance single-pane glazing. This is the logical configuration if outdoor air is drawn into the cavity through openings on the exterior façade. If indoor air is cycled through the cavity, conversely, the exterior façade should be the thermal barrier. In the renovation of an existing masonry building, a glazed exterior façade could be placed as a screen in front of the original building. In such a case, the exterior glazing and cavity could protect the original building and operable windows from pollutants, noise, wind, insolation, rain, and vandalism. In either case, the glazing could be clear, allowing as much insolation in as possible, or tinted or etched to filter out the sun as desired.

f) Shading Device

One frequently cited reason for the implementation of a DSF is the potential for positioning a shading device within the cavity. This reduces maintenance costs over locating it on the exterior of the building and more effectively prevents solar gains than when it is located within the building. Generally, all the options exist as for conventionally located shading devices: roller shades, louvered blinds, fixed vs manually or automatically controlled, horizontal vs vertical fins, etc. The designer should consider how the device will affect airflow within the cavity and how solar gains absorbed by the shade will be radiated relative to the interior façade. The roller shade positioned in the center of the cavity, shown in Figure 10.1, will allow air to flow freely against the interior façade. Daylight might be more difficult to control precisely with this design, however.

g) Airflow Control

The airflow through the cavity can be naturally driven (with buoyancy and wind pressure), mechanically driven, or a combination of both. Operable windows could allow airflow into the building, or the cavity air could be completely closed off from the building's interior. The airflow could be bi-directional or it might always flow in the same direction. Additionally, depending on the application seasons and purposes, the cavity inlet could draw in outdoor air, indoor air, or alternate between the two, and the cavity outlet could discharge the air to the outdoors, to the building's interior, or to preheat for the HVAC system.

10.11   Phase Change Material

10.11.1   Principle

Shifting part of the peak load to off-peak time by adding building thermal mass is a promising approach to managing the building load demand. However, as promising as this approach may be, when common building materials are used, large masses of these materials are required. PCMs, on the other hand, can absorb large amounts of heat during the phase change process without being so massive. For example, Benard et al. (1985) found that for the same thermal performance, a wall outfitted with PCMs would only need about one-twelfth of the weight of a concrete wall. When PCMs are placed in building walls, they absorb a major part of the heat transferred from hot outside environment in the daytime and release the absorbed heat during the night and early morning hours. As a result, part of the peak space-cooling load is reduced and shifted to off-peak hours. For real applications, PCMs should be able to work in daily cycles, which means the PCM melted in the daytime should solidify at night and/or during early morning hours and then be ready for the next melting process. PCMs can be generally categorized as organic, inorganic, etc. Table 10.3 lists the main characteristics of inorganic and organic PCMs (Fang 2009).

According to the methods of energy storage, there are two kinds of phase change energy storage building envelopes. One is the passive phase change energy storage building envelope, which is based on indoor air temperature or solar radiation changes; the other is the active phase change energy storage building envelope, which is combined with heating or cooling equipment and the heat transfer can be controlled actively. Table 10.4 lists some applications of PCM in building envelopes.

10.11.2   Performance

PCM is an effective passive technique to reduce the effect of heat waves, resulting in a reduced and delayed thermal zonal load. Several studies show promising conceptual and real examples of PCM integration into building applications (e.g. Khudhair and Farid 2004; Baetens et al. 2010). Several

Table 10.3   Classification and properties of PCMs.

Classification	Inorganic	Organic
Category	Crystallin hydrate, molten salt, metal or alloy	High aliphatic hydrocarbon, acid/esters or salts, alcohols, aromatic hydrocarbons, aromatic ketone, lactam, freon, multi-carbonated category, polymers
Advantages	Higher energy storage density, higher thermal conductivity, nonflammable, inexpensive	Physical and chemical stability, good thermal behavior, adjustable transition zone
Disadvantages	Supercooling, phase segregation, corrosive	Low thermal conductivity, low density, low melting point, highly volatile, flammable, volume change
Methods for improvement	Mixed with nucleating and thickening agents, thin layer arranged horizontally, mechanical stir	High thermal conductivity additives, fire-retardant additives

Source: Based on Fang (2009).

Table 10.4 Illustration of a few passive and active PCMs applications in building.

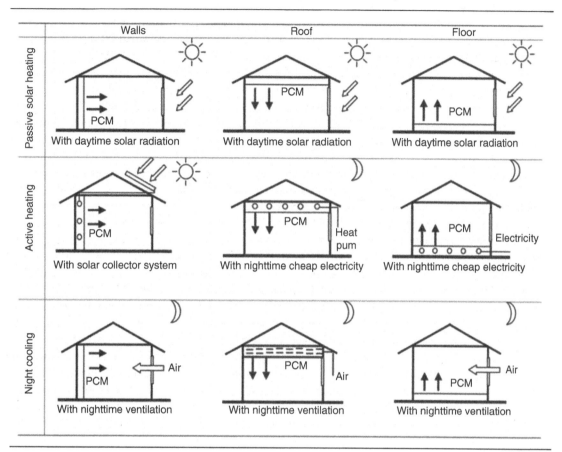

Source: Wang et al. (2009)/Springer Nature.

experimental studies (e.g. Zhang et al. 2005; Ko 2013) conducted over summer or winter days at the University of Kansas revealed that PCM has significant reduction in peak and daily loads. The average reduction in peak heating load for all walls ranges from as low as 5.7% to as high as 15% depending on the thermal properties of PCM, PCM design configurations, PCM concentrations, and the time period of the study. Experimental testing at Arizona State University on the PCM panels installed on every surface of the test shed on the interior side showed a reduction of 27% in peak cooling load and a reduction of 19% in annual cooling energy in Phoenix, AZ. Since no internal heat gain are imposed, except the solar radiation from a single east window, the high surface area of PCM has to exclusively handle the cooling load generated by the heat transfer from outside climate through exterior envelope. At the Natural Exposure Testing facility located in Charleston in Southern California, two testing campaigns in the summer and winter have been conducted to evaluate the performance of different PCM panels divided into two wall groups under real climatic conditions (Givoni 1976). Results from one wall group show a reduction in heat gain that ranges between 21.8 and 22.9% in the summer. In the winter, reductions in heat gain range from 5.7 to 15.4% and reductions in heat loss from 25.5 to 27.7% have been reported.

Simulation tools have also been used to evaluate the energy performance of PCM in buildings. The US National Renewable Energy Laboratory (NREL) research team has used EnergyPlus to simulate the PCM integrations into different envelope systems using a typical house as per America benchmark protocols under Phoenix, AZ, weather file (Tabares-Velasco et al. 2012). The results show that PCM has minor effect on reducing the peak cooling load in the cooling season of Phoenix. For the best PCM application in the wall, a maximum reduction of around 8% in peak cooling was achieved in the month of May with only 4% peak cooling reduction in July. A couple of hospital spaces including an administration office space and group treatment and patient rooms have been individually simulated using EnergyPlus for the Oregon State Hospital in Junction City. PCM layers with different thermal properties are integrated into these spaces for three envelope options: (i) external walls only, (ii) external walls and ceiling, and (iii) all surfaces. The charging occurs naturally during the day but discharging is performed using night flush via integrated economizer with HVAC system. Average reduction of 15, 17, and 28% in annual cooling energy and 9.5, 11, 12% reduction in peak cooling loads are achieved for external walls only, an external walls and ceiling design option, and an all surfaces option, respectively.

In general, these various studies indicated that the PCM's performance is highly dependent on many factors including the PCM's thermal properties (i.e. latent heat, melting temperature, and melting range), zonal thermostat set points, PCM's design configurations and integration mechanism, the insulation level of the wall assembly, PCM's surface areas, exposure to internal heat and solar gains, charging and discharging strategies and the climate. Recent studies at the University of Colorado at Boulder indicated that PCM performs relatively better when it is located in direct contact with the controlled indoor environment. On annual bases, the maximum saving for the tested latent heat case is 6 and 2.4% for cooling and heating loads, respectively. PCM shows improved performance during transition season when compared to winter or summer seasons. For the month of May (a transition month in Colorado), a maximum savings of 10 and 23% are achieved for heating and cooling loads, respectively. A closer look at PCM layer level indicated that the diurnal cycle during this month was the main reason for the improved PCM's performance. As a result, the charging and discharging cycles occur on a daily bases, driven mainly by the climate. Multiple PCM layers – one for cooling and another for heating – were found to provide slightly better performance. For this design, the order of PCM layers relative to the indoor environment was not sensitive.

PCM shows improved performance when integrated with other passive strategies such as natural ventilation, night cooling, direct solar gain techniques and PCM-enhanced Trombe wall, etc. The Colorado study showed that the reduction in annual heating loads for a south wall with PCM is only 0.59% compared to 7.2% when a cavity is used (Trombe wall). Ideally, the charging and discharging process should occur on a daily basis for maximum benefits. In mild climates, the PCM should be flushed with outside cooled air via natural ventilation during night to prepare the PCM for the next hot day cycle. In cold winter climates, solar harvesting techniques should be fully exploited to store the solar heat for later night use.

10.11.3 Design Considerations

Selecting proper PCM properties is the first important step for designing a successful PCM-enhanced building. In general, the optimal melting temperature of PCM is found to hover around the thermostat set points: 1 °C below the cooling set point for maximum savings in annual cooling load while at heating set point of 22 °C for maximum savings in annual heating load. The corresponding melting range should be 0.1–1 and 2 °C for maximum savings in annual cooling and heating loads, respectively.

The following general guidelines for choosing desirable thermo-physical, kinetic, and chemical properties of PCMs can be followed:

1) **Thermo-physical properties:**
   - melting temperature in the desired operating temperature range (temperature range of application),
   - high latent heat of fusion per unit volume so that the required volume of the container to store a given amount of energy is smaller,
   - high specific heat to provide additional significant sensible heat storage,
   - high thermal conductivity of both solid and liquid phases to assist the charging and discharging energy of the storage system,
   - small volume change on phase transformation and small vapor pressure at operating temperature to reduce the containment problem,
   - congruent melting of the PCM for a constant storage capacity of the material with each freezing/melting cycle.
2) **Kinetic properties:**
   - high nucleation rate to avoid supercooling of the liquid phase,
   - high rate of crystal growth, so that the system can meet demand of heat recovery from the storage system.
3) **Chemical properties:**
   - complete reversible freeze/melt cycle,
   - no degradation after a large number of freeze/melt cycle,
   - no corrosiveness to the construction materials,
   - nontoxic, nonflammable, and nonexplosive material for safety.

PCM can be integrated into different building enclosures, ceilings, floors, or walls. The PCM can be thermally activated (i.e. charging or discharging) using passive strategies such as solar radiation from the sun, internal heat gain, or using outside cooled air via natural ventilation. Active systems can also be used to thermally activate the PCM especially when the energy is available at low cost. To be fully exploited, PCM should go through a charging and discharging cycle at least one time a day. Multiple cycles can be achieved through active systems which demand tuned control algorithms for optimizing the charging and discharging process to meet the zone demand.

PCM can be integrated into either passive or active building's systems. Passive systems perform their intended function using explicit inherited properties with no or negligible external aid. On the other hand, active systems are those associated with using mechanical, electrical, and electronic equipment to perform their intended function. One of the simplest and easiest solutions is when PCM is incorporated into drop ceiling of a zone. Hybrid (automatic or manually) controlled windows are installed to recharge the PCM when outside air is favorable. PCM can also be embedded into floor systems. The PCM-enhanced concrete floor is charged by the direct exposure to the solar radiation during the day. At night, the PCM is then naturally discharged to meet the heating demand.

A wall system is perhaps one of the most common integration mechanisms for PCM in buildings. In addition to PCM-enhanced multilayer wall systems, PCM can be integrated into ventilated cavity walls (similar to a Trombe wall with top and bottom vents). PCM can be utilized in the south wall for solar collection and consequently discharged when heat demand is required. Different designs, operational strategies, and controlled mechanisms can be used to passively or actively charge and discharge the PCM.

Air is commonly used as a heat transfer medium but embedded pipes can be integrated into the PCM-enhanced wall. Using the embedded pipes, water can be utilized to quickly charge and discharge the PCM.

Homework Problems

1

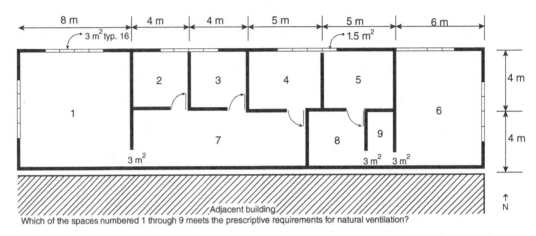

Which of the spaces numbered 1 through 9 meets the prescriptive requirements for natural ventilation?

The gross window area (per window) in Rooms 1–3 is 1 m (w) × 3 m (h) = 3 m² with an openable area of 20%. The gross window area (per window) in Rooms 4–6 is 1 m (w) × 1.5 m (h) = 1.5 m² with an openable area of 40%. The building ceiling height is 4 m. Please refer to ASHRAE 62.1.

2

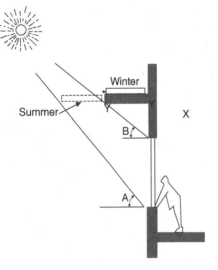

Design appropriate X and Y to meet both full shade line and full sun line for Denver, CO. Please search and study the definition of "full shade line" and "full sun line".

3 Staged Project Assignment (Teamwork)

   (a) Assess if natural ventilation can be used (considering weather), and whether and how the natural ventilation design can comply with ASHRAE Standard 62.1.

   (b) Design proper shadings for the windows with justification.

References

ASHRAE (2011). *ASHRAE Handbook – HVAC Application*. ASHRAE.

ASHRAE (2012). *ASHARE Handbook – HVAC Systems and Equipment*. ASHRAE.

ASHRAE (2019). *ASHRAE Standard 62.1: Ventilation for Acceptable Indoor Air Quality*. Atlanta: ASHRAE.

ASHRAE (2020). *ASHRAE Standard 55: Thermal Environmental Conditions for Human Occupancy*. Atlanta: ASHRAE.

Baetens, R., PetterJelle, B., and Arild Gustavsen, A. (2010). Phase change materials for building applications: a state-of-the-art review. *Energy and Buildings* 42: 1361–1368.

Benard, C., Body, Y., and Zanoli, A. (1985). Experimental comparison of latent and sensible heat thermal walls. *Solar Energy* 34: 475–487.

Blondeau, P., Sperandio, M., and Allard, F. (1997). Night ventilation for building cooling in summer. *Solar Energy* 61 (5): 327–335.

Camargo, J.R., Ebinuma, C.D., and Cardoso, S. (2006). Three methods to evaluate the use of evaporative cooling for human thermal comfort. *Revista de Engenharia Térmica* 5 (2): 9–15.

CIBSE (2005). *AM10 Natural Ventilation in Non0Domestic Buildings*. CIBSE.

Fang, Y. (2009). A comprehensive study of phase change materials for building walls applications. PhD thesis. University of Kansas. [29]

Givoni, B. (1976). *Climate and Architecture*. London: Applied Science Publishers.

Givoni, B. (1992). Comfort, climate analysis and building design guidelines. *Energy and Buildings* 18: 11–23.

Gratia, E. and Herde, A.D. (2004). Natural cooling strategies efficiency in an office building with a double-skin façade. *Energy and Buildings* 36 (11): 1139–1152.

Khudhair, A.M. and Farid, M.M. (2004). A review on energy conservation in building applications with thermal storage by latent heat using phase change materials. *Energy Conversion and Management* 45: 263–275.

Ko, L. (2013). Using hydrated salt phase change materials for residential air conditioning peak demand reduction and energy conservation in coastal and transitional climates in the state of California. MS thesis. University of Kansas.

Kolokotroni, M. and Aronis, A. (1999). Cooling-energy reduction in air-conditioned offices by using night ventilation. *Applied Energy* 63 (4): 241–253.

Mihalakakou, G. (2002). On the use of sunspace for space heating/cooling in Europe. *Renewable Energy* 26: 415–429.

Mihalakakou, G. and Ferrante, A. (2000). Energy conservation and potential of a sunspace: sensitivity analysis. *Energy Conversion and Management* 41: 1247–1264.

Palmer, J.D. (2002). Evaporative cooling design guidelines manual for New Mexico schools, and commercial buildings. *NRG Engineering* 2626: 1–99.

Santamouris, M., Sfakianaki, A., and Pavlou, K. (2010). On the efficiency of night ventilation techniques applied to residential buildings. *Energy and Buildings* 42: 1309–1313.

Schulze, T. and Eicker, U. (2013). Controlled natural ventilation for energy efficient buildings. *Energy and Building* 56: 221–232.

Sepannen, O. and Fisk, W. (2002). Association of ventilation type with SBS symptoms in office workers. *International Journal of Indoor Environment and Health* 12 (2): 98–112.

Tabares-Velasco, P.C., Christensen, C., and Bianchi, M. (2012). Verification and validation of EnergyPlus phase change material model for opaque wall assemblies. *Building and Environment* 54: 186–196.

Torcellini, P. and Pless, S. (2004). Trombe walls in low-energy buildings: practical experiences. The World Renewable Energy Congress VIII and Expo, Denver, Colorado, August 29–September 3, 2004.

Wang, X., Zhang, Y., Xiao, W. et al. (2009). Review on thermal performance of phase change energy storage building envelope. *Chinese Science Bulletin* 54 (6): 920–928.

Zhang, M., Medina, M.A., and King, J.B. (2005). Development of a thermally enhanced frame wall with phase-change materials for on-peak air conditioning demand reduction and energy savings in residential buildings. *International Journal of Energy Research* 29: 795–809.

11

Building Load Calculation

The heating and cooling of buildings account for the majority of energy consumption in buildings. Designing an energy efficient building requires an accurate prediction of a building's heating and cooling load. This load estimate can help determine proper building HVAC system types and reasonable system sizes. This chapter will introduce the principles and methods of basic building load calculations. Once the HVAC systems are identified, the annual energy use of the systems for a typical or actual year can be computed, which will be introduced in Chapter 13. Energy efficient design solutions, from shape alternatives to envelope properties to system parameters, can then be tested, compared, and optimized in both building load and energy consumption calculations.

Figure 11.1 illustrates different methods for estimating building loads, with various complexity and accuracy levels. Simple methods such as the instantaneous Q = UAdT method and the ASHRAE CLTD/CLF method are easy to use and can be performed by hand-calculations but have a lower prediction accuracy. Advanced methods such as the ASHRAE Radiant Time-Series (RTS) method and the ASHRAE Transfer Function method, as well as the heat balance method are commonly programed in software, which consider more heat transfer physics. Thus, these advanced methods are more accurate but are also more time-consuming for both case development and simulation. In general, advanced methods are more accurate than simple methods, but they require more inputs. Uncertainties of these input parameters can significantly decrease the calculation accuracy. This is especially true for the early design stage when most of the design parameters remain unknown. For these cases, simple methods can provide reliable initial estimates of building loads, while advanced methods can be applied at a later stage when the designs are more mature. This chapter will introduce the principles of the instantaneous Q = UAdT method, the ASHRAE CLTD/CLF method, and the heat balance method.

11.1 Residential and Light Commercial Buildings

The instantaneous Q = UAdT method is fairly straightforward, especially for calculating the heating load, to be introduced below. The major deficiency of this method is that it does not consider the thermal storage effects, which can defer (and reduce) the heat transfer depending on the properties of the envelope materials (as discussed in Chapter 8). For residential and light commercial buildings, whose envelopes have much less storage capacity than commercial buildings, the thermal storage impacts can thus be ignored and simply approximated. This is especially true for the heating load calculation, where solar radiation is not

Energy Efficient Buildings: Fundamentals of Building Science and Thermal Systems, First Edition. Zhiqiang (John) Zhai.
© 2023 John Wiley & Sons, Inc. Published 2023 by John Wiley & Sons, Inc.

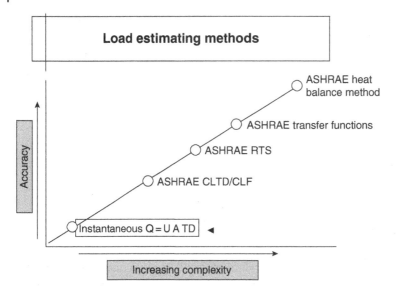

Figure 11.1 Building load estimating methods, with different complexity and accuracy levels.

included. The load calculation should consider the worst scenario (e.g. cloudy, raining, or snowing days) so that the designed systems can provide adequate heating and cooling to ensure the spaces remain at the designed indoor conditions on these days. Solar radiation is the major driving force and impact factor on thermal storage, much more so than the indoor–outdoor air temperature differences.

Residential buildings commonly share these characteristics:

- 24-hour conditioned (buildings have less thermal storage impacts);
- small internal loads (building loads are determined by indoor and outdoor air conditions);
- single thermal zone (with one controller/thermostat, without zone cross-airflow interference);
- thermostat control (with sensible heat control only)
- small capacity unit (with easy-to-identify specs)
- dehumidification for cooling only (or none at all)

These characteristics ensure the effectiveness of a simple calculation method for such buildings. As typically there is no latent load control for these buildings, the following sections focus on the calculation of sensible heating and cooling loads.

11.1.1 Heating Load Calculation

Since the load calculation is used to select and size appropriate mechanical systems, it requires consideration of the worst (most unfavorable) outdoor and operating conditions. As a result, the heat contributed by irregular or uncontrollable factors, such as internal loads (from occupants, appliances, lightings, and equipment) and outdoor solar radiation, cannot be counted in the calculation of the space heating load. If these were considered, the resultant systems might not provide desired indoor environments when these conditions were not available.

During the heating season, heat is provided to compensate for the heat losses through two mechanisms:

11.1.1.1 Through Envelope Structures and Windows

$$\dot{Q}_e = U \times A \times (T_o - T_i) \tag{11.1}$$

where $\dot{Q}_e$ is the heat transfer rate through envelopes (in W or Btu/h); U is the total thermal conductance of the building (in $W/(m^2 \cdot °C)$, or $Btu/(h \cdot ft^2 \cdot °F)$); A is the total building envelope surface area (in m^2 or ft^2); T_o and T_i are, respectively, the outdoor and indoor design air temperatures identified according to Chapters 5 and 6. The UA value of the building can be calculated according to Example 8.10.

11.1.1.2 Through Infiltration

$$\dot{Q}_{inf} = \rho \times \dot{V}_{infiltration} \times C_p \times (T_o - T_i) \tag{11.2}$$

where $\dot{Q}_{inf}$ is the heat transfer rate through infiltration (in W or Btu/h); ρ is the air density (in kg/m^3 or lb/ft^3); $\dot{V}_{infiltration}$ is the infiltration flow rate (in m^3/s, or ft^3/h); C_p is the specific heat of air (in $J/(kg \cdot °C)$ or $Btu/(lb \cdot °F)$); T_o and T_i are, respectively, the outdoor and indoor design air temperature identified according to Chapters 5 and 6. The $\dot{V}_{infiltration}$ value of the building can be either estimated or measured according to Chapter 7.

Combining Equations (11.1) and (11.2) yields:

$$\dot{Q}_{heat} = \left(U \times A + \rho \times \dot{V}_{infiltration} \times C_p\right) \times (T_o - T_i) = BLC \times (T_o - T_i) \tag{11.3}$$

where $\dot{Q}_{heat}$ is the total sensible heating load of the building (in W or Btu/h); and BLC is the total building load coefficient (in $W/°C$, or $Btu/(h \cdot °F)$). Example 8.10 demonstrates the calculation of the BLC and $\dot{Q}$. If the space has constantly operated internal loads $\dot{Q}_{int}$, such as those from appliances (e.g. refrigerator), the total sensible heat load $\dot{Q}$ can subtract this load $\dot{Q}_{int}$ to avoid oversizing the heating systems.

11.1.2 Cooling Load Calculation

The calculation of the sensible cooling load is similar to that for the heating load. The major difference is that the cooling load has to consider all of the worst conditions for cooling, such as those from solar radiation and internal heat gains. Therefore, in the cooling season, coolness is provided to compensate for the heat gains through four mechanisms:

11.1.2.1 Through Envelope Structures

$$\dot{Q}_e = U \times A \times CLTD \tag{11.4}$$

where $\dot{Q}_e$ is the heat transfer rate through envelopes (in W or Btu/h); U is the thermal conductance of the building envelope (in $W/(m^2 \cdot °C)$, or $Btu/(h \cdot ft^2 \cdot °F)$); A is the building envelope surface area (in m^2 or ft^2); CLTD stands for the cooling load temperature difference that combines the indoor–outdoor temperature difference and solar radiation, as well as considering thermal capacity of building envelope. Table 11.1 presents CLTD values for single-family detached residences from the ASHRAE fundamentals handbook. In order to

Table 11.1 CLTD values for single-family detached residences.[a]

Daily temperature range[b]	Design temperature, °C											
	29		32			35			38		41	43
	L	M	L	M	H	L	M	H	M	H	M	H
All walls and doors												
North	4	2	7	4	2	10	7	4	10	7	10	13
NE and NW	8	5	11	8	5	13	11	8	13	11	13	16
East and West	10	7	13	10	7	16	13	10	16	13	16	18
SE and SW	9	6	12	9	6	14	12	9	14	12	14	17
South	6	3	9	6	3	12	9	6	12	9	12	14
Roofs and ceilings												
Attic or flat built-up	23	21	26	23	21	28	26	23	28	26	28	31
Floors and ceilings												
Under conditioned space, over unconditioned room, over crawl space	5	2	7	5	2	8	7	5	8	7	8	11
Partitions												
Inside or shaded	5	2	7	5	2	8	7	5	8	7	8	11

[a] Cooling load temperature differences (CLTDs) for single-family detached houses, duplexes, or multifamily, with both east and west exposed walls or only north and south exposed walls, K.
[b] L denotes low daily range, less than 9 K; M denotes medium daily range, 9–14 K; and H denotes high daily range, greater than 14 K.

use this table, the outdoor air design temperature and the daily temperature range need to be identified. The daily temperature swing affects the effectiveness of envelope thermal storage, where a range less than 9 °K is defined as a low daily range, between 9 and 14 °K as a medium daily range, and greater than 14 °K as a high daily range. The CLTD values vary with envelope type (e.g. roof and walls) and the orientation of the envelope (e.g. south vs north). The CLTD method is superior to the UAdT method as it considers the thermal storage impacts, which are dominant under solar radiation and are thus critical for the cooling load calculation.

11.1.2.2 Through Envelope Glasses

$$\dot{Q}_g = A \times GLF \tag{11.5}$$

where $\dot{Q}_g$ is the heat transfer rate through building windows/glass (in W or Btu/h); A is the building window/glass surface area (in m^2 or ft^2); GLF stands for glass load factor and includes the effects of both transmission and solar radiation. The transmitted solar radiation is an important cooling load that distinguishes the windows/glasses from other opaque envelopes. The ASHRAE fundamentals handbook also provides the window glass load factors for single-family detached residences, as demonstrated in Table 11.2. With a given outdoor air design temperature and the window glass type, the GLF values vary with window orientation and the type of shading devices.

Table 11.2 Window glass load factors for single-family detached residences.[a]

Design temperature, °C	Regular single glass						Regular double glass						Heat-absorbing double glass						Clear triple glass		
	29	32	35	38	41	43	29	32	35	38	41	43	29	32	35	38	41	43	29	32	35
No inside shading																					
North	107	114	129	148	151	158	95	95	107	117	120	129	63	63	73	79	82	88	85	85	95
NE and NW	199	205	221	237	243	262	173	177	186	196	199	208	114	117	123	132	139	139	158	158	167
East and West	278	284	300	315	322	337	243	246	255	265	268	278	161	161	170	177	186	186	221	221	230
SE and SW[b]	249	255	271	287	290	309	218	221	230	240	243	252	142	145	155	161	170	170	196	199	205
South[b]	167	173	189	205	211	227	145	148	158	167	170	180	98	98	107	114	123	123	132	132	142
Horizontal skylight	492	492	508	524	527	539	432	435	442	451	454	464	284	287	293	300	303	309	391	394	401
Draperies, venetian blinds, translucent roller shades, fully drawn																					
North	57	60	73	85	91	104	50	50	60	69	73	82	41	44	50	57	60	66	47	50	57
NE and NW	101	104	120	132	136	148	91	95	101	110	114	123	76	76	85	91	91	101	88	88	95
East and West	142	145	158	170	173	186	126	129	139	145	148	158	104	104	114	120	120	129	123	123	129
SE and SW[b]	126	129	145	155	161	173	114	117	123	132	136	145	91	95	101	107	110	117	110	114	120
South[b]	85	88	104	117	120	132	76	79	88	98	98	107	63	66	73	79	82	88	73	76	82
Horizontal skylight	246	249	262	271	274	284	224	224	233	240	243	249	183	186	192	199	199	205	218	218	224
Opaque roller shades, fully drawn																					
North	44	47	63	73	79	91	41	44	54	60	63	73	38	38	47	54	54	63	41	41	47
NE and NW	79	82	98	107	114	126	73	76	85	95	95	104	66	69	76	82	85	91	73	73	82
East and West	107	114	126	139	142	155	101	104	114	120	123	132	91	95	101	107	110	117	101	101	no
SE and SW[b]	98	101	114	126	132	145	91	95	104	110	114	123	82	85	91	98	101	107	91	91	98
South[b]	66	69	85	95	101	114	63	63	73	82	85	95	57	60	66	73	76	82	60	63	69
Horizontal skylight	189	192	202	214	218	227	180	180	189	196	199	205	164	164	173	180	180	186	177	180	186

[a] Glass load factors (GLFs) for single-family detached houses, duplexes, or multifamily residences, with both east and west exposed walls or only north and south exposed walls, W/m².

[b] Correct by +30% for latitude of 48° and by −30% for latitude of 32°. Use linear interpolation for latitude from 40 to 48 and from 40 to 32°.

11.1.2.3 Through Infiltration

$$\dot{Q}_{inf} = \rho \times \dot{V}_{infiltration} \times C_p \times (T_o - T_i) \tag{11.6}$$

When ACH (air change rate per hour) is used to represent the infiltration, Equation (11.6) becomes:

$$\dot{Q}_{inf} = \rho \times ACH \times V_{room} \times C_p \times (T_o - T_i) \tag{11.7}$$

where V_{room} is the space volume. Equation (11.7) works for IP units (e.g. Btu/h). It needs to be divided by 3600 sec/h if SI units are used (e.g. W).

As indicated in Chapter 7, for a modern house, the mean ACH is about 0.5 ACH. For a healthy building, the infiltration rate could be maintained at a level of 0.5–1.5 ACH depending on the building functions. State-of-the-art construction can, at best, reduce infiltration to approximately 0.2 ACH (where the indoor air quality should be watched closely though). At the other extreme, an older, leaky building may be expected to suffer from infiltration rates from 3 to 8 ACH, with 1 ACH as the mean value for most old houses.

11.1.2.4 Due to Occupants and Appliances

Internal heat gains from occupants, lighting, equipment, and appliances in residential buildings may not be as significant as those in commercial buildings but can still play an important role in increasing indoor air temperature and humidity. To estimate the internal heat gains for a cooling load calculation, one can simply count the number of people and appliances, as well as their heat/power capacities in the space of design. For instance, a typical person releases 67 W heat; an average home refrigerator has a power of 100–200 W; and a 65-in LED TV uses 120 W. Since most of the equipment and appliances may not operate at the same time, a lump sum estimate based on empirical observations may also serve the same purpose. For instance, $\dot{Q}_{int} = 470$ W for both the kitchen and laundry room for single-family residences and $\dot{Q}_{int} = 350$ W for multifamily residences.

A recent study (https://www.nrel.gov/docs/fy14osti/60266.pdf) on plug and process load (PPL) capacity for 14 office buildings and 7 higher education buildings indicates:

- On average, the peak PPL energy use intensities for offices (without laboratories or data centers) is 0.50 W/ft^2 and 0.64 W/ft^2 for higher education buildings;
- On average, the average PPL energy use intensity for offices (without laboratories or data centers) is around 0.28 W/ft^2, and 0.27 W/ft^2 for higher education buildings.

These numbers, multiplied by the floor area, can also help estimate the internal heat gains from the PPLs for typical office and education buildings.

Adding all the heat gains above yields the total sensible cooling load:

$$\dot{Q}_{cool} = \dot{Q}_e + \dot{Q}_g + \dot{Q}_{inf} + \dot{Q}_{int} \tag{11.8}$$

where $\dot{Q}_{cool}$ is the total sensible cooling load of the building (in W or Btu/h).

For buildings that require humidity controls, the latent cooling load can be either calculated by counting individual components that release or absorb water vapor or estimated as 30% of the sensible load – a reasonable approximation for residential and light commercial buildings.

11.2 Commercial Buildings

In nonresidential building, it is important to include the time lag in conductive heat gain through opaque exteriors and the thermal storage in converting radiant heat gain to cooling load. Advanced methods such as the RTS method, the Transfer Function method, and the heat balance method are commonly used. The CLTD/CLF method would require the use of various sophisticated coefficients from quite a few tables. The following equation illustrates the impacts of the internal heat gain from people on the cooling load, as an example of requiring dynamic coefficients for commercial building load calculation:

$$Q_{total,int} = N \times Q_{person\text{-}sensible} \times CLF \tag{11.9}$$

where N is the number of occupants; CLF is the cooling load factor, by hour of occupancy, to be determined from Table 11.3. Depending on the type of space/zone and the total hours that the people stay in the space, the CFL values vary in every hour after the person enters the space. For a person staying in a space of Zone Type A for 2 hours, the heat impact of this person can last for 10 hours due to the thermal storage and time lag of the heat. Similar sophistication would be required for other heating and cooling elements, leading to the inconvenience of applying this approach.

The heat balance method is thus commonly adopted to calculate the heating and cooling load for commercial buildings, which becomes the foundation of most commercial energy simulation software. The heat balance method applies the thermal network model (i.e. the RC network model) as introduced in Chapter 8, "Heat Transfer in Building." For a building envelope as illustrated in Figure 11.2, the associated thermal RC network can be built below. Figure 11.3 shows a simplified wall thermal network with three wall thermal storage layers, and R_1 and R_4 are the surface heat resistances.

The energy balance on the first node T_1 can be expressed as:

$$\frac{T_0 - T_1}{R_1} - \frac{T_1 - T_2}{R_2} = C_1 \frac{dT_1}{dt} \tag{11.10}$$

and further as:

$$\frac{dT_1}{dt} = \frac{T_0 - T_1}{R_1 C_1} - \frac{T_1 - T_2}{R_2 C_1} \tag{11.11}$$

The general representation for each capacitance node can thus be obtained:

$$\frac{dT_i}{dt} = \frac{T_{i-1} - T_i}{R_i C_i} - \frac{T_i - T_{i+1}}{R_{i+1} C_i} \tag{11.12}$$

With the given R_i and C_i as well as the boundary conditions T_0 and T_4, the three equations for T_1, T_2, and T_3 can produce the values of T_1, T_2, and T_3. Equation (11.12) can be solved numerically or iteratively:

$$\frac{dT_i}{dt} = \frac{\Delta T_i}{\Delta t} \approx \frac{T_i^+ - T_i}{\Delta t} = \frac{T_{i-1} - T_i}{R_i C_i} - \frac{T_i - T_{i+1}}{R_{i+1} C_i} \tag{11.13}$$

where T_i^+ is the temperature at the next time step, T_i is the temperature at the current time step, and Δt is the time step.

Table 11.3 Cooling load factors for people and unhooded equipment (ASHRAE Fundamentals).

Hours in space	Number of hours after entry into space or equipment turned on																							
	1	2	3	4	5	6	7	8	9	10	11	12	13	14	15	16	17	18	19	20	21	22	23	24
Zone type A																								
2	0.75	0.88	0.18	0.08	0.08	0.04	0.02	0.01	0.01	0.01	0.00	0.00	0.00	0.00	0.00	0.00	0.00	0.00	0.00	0.00	0.00	0.00	0.00	0.00
4	0.75	0.88	0.93	0.95	0.95	0.22	0.10	0.05	0.03	0.02	0.01	0.01	0.01	0.01	0.00	0.00	0.00	0.00	0.00	0.00	0.00	0.00	0.00	0.00
6	0.75	0.88	0.93	0.95	0.97	0.97	0.23	0.11	0.06	0.04	0.03	0.02	0.02	0.01	0.01	0.01	0.01	0.00	0.00	0.00	0.00	0.00	0.00	0.00
8	0.75	0.88	0.93	0.95	0.97	0.97	0.97	0.98	0.24	0.11	0.06	0.04	0.03	0.02	0.02	0.01	0.01	0.01	0.01	0.01	0.00	0.00	0.00	0.00
10	0.75	0.88	0.93	0.95	0.97	0.97	0.98	0.98	0.99	0.99	0.24	0.12	0.07	0.04	0.03	0.02	0.02	0.01	0.01	0.01	0.01	0.01	0.00	0.00
12	0.75	0.88	0.93	0.96	0.97	0.98	0.98	0.98	0.99	0.99	0.99	0.99	0.25	0.12	0.07	0.04	0.03	0.02	0.02	0.02	0.01	0.01	0.01	0.01
14	0.76	0.88	0.93	0.96	0.97	0.98	0.98	0.99	0.99	0.99	0.99	0.99	1.00	1.00	0.25	0.12	0.07	0.05	0.03	0.03	0.02	0.02	0.01	0.01
16	0.76	0.89	0.94	0.96	0.97	0.98	0.98	0.99	0.99	0.99	0.99	0.99	1.00	1.00	1.00	1.00	0.25	0.12	0.07	0.05	0.03	0.03	0.02	0.02
18	0.77	0.89	0.94	0.96	0.97	0.98	0.98	0.99	0.99	0.99	0.99	1.00	1.00	1.00	1.00	1.00	1.00	1.00	0.25	0.12	0.07	0.05	0.03	0.03
Zone type B																								
2	0.65	0.74	0.16	0.11	0.08	0.06	0.05	0.04	0.03	0.02	0.02	0.01	0.01	0.01	0.01	0.00	0.00	0.00	0.00	0.00	0.00	0.00	0.00	0.00
4	0.65	0.75	0.81	0.85	0.85	0.24	0.17	0.13	0.10	0.06	0.04	0.03	0.03	0.02	0.02	0.01	0.01	0.01	0.01	0.00	0.00	0.00	0.00	0.00
6	0.65	0.75	0.81	0.85	0.89	0.91	0.29	0.20	0.15	0.12	0.09	0.07	0.05	0.04	0.03	0.02	0.02	0.01	0.01	0.01	0.01	0.01	0.00	0.00
8	0.65	0.75	0.81	0.85	0.89	0.91	0.91	0.93	0.31	0.22	0.17	0.13	0.10	0.08	0.06	0.05	0.04	0.03	0.02	0.02	0.01	0.01	0.01	0.01
10	0.65	0.75	0.81	0.85	0.89	0.91	0.93	0.95	0.96	0.97	0.33	0.24	0.18	0.14	0.11	0.08	0.06	0.05	0.04	0.03	0.02	0.02	0.01	0.01
12	0.66	0.76	0.81	0.86	0.89	0.92	0.94	0.95	0.96	0.97	0.98	0.98	0.34	0.24	0.19	0.14	0.11	0.08	0.06	0.05	0.04	0.03	0.02	0.02
14	0.67	0.76	0.82	0.86	0.89	0.92	0.94	0.95	0.96	0.97	0.98	0.98	0.99	0.99	0.35	0.25	0.19	0.15	0.11	0.09	0.07	0.05	0.04	0.03
16	0.69	0.78	0.83	0.87	0.90	0.92	0.94	0.95	0.96	0.97	0.98	0.98	0.99	0.99	0.99	0.99	0.35	0.25	0.19	0.15	0.11	0.09	0.07	0.05
18	0.71	0.80	0.85	0.88	0.91	0.93	0.95	0.96	0.97	0.98	0.98	0.99	0.99	0.99	0.99	0.99	1.00	1.00	0.35	0.25	0.19	0.15	0.11	0.09

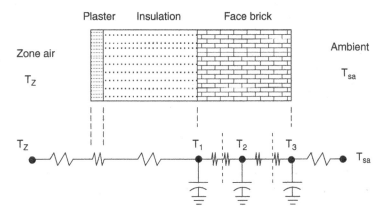

Figure 11.2 Thermal RC network for a building envelope.

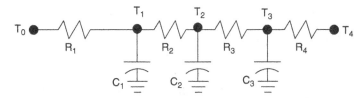

Figure 11.3 A simplified wall thermal network with three wall thermal storage layers (C_1, C_2, C_3) and two surface heat resistances R_1 and R_4.

The heat flux through the wall can then be calculated as:

$$Q_{w,1} = \frac{T_0 - T_1}{R_1} \tag{11.14}$$

$$Q_{w,4} = \frac{T_3 - T_4}{R_4} \tag{11.15}$$

Due to the thermal storage, $Q_{w,1}$ may not be equal to $Q_{w,4}$.

Building envelope heat transfer can be linked to zone energy conservation (Figure 11.4). Figure 11.4 only illustrates the link between the zone and one envelope for simplification. Linking to all envelopes will be discussed later. The resulting thermal network is shown in Figure 11.5.

The energy conservation equations for nodes T_1, T_2, T_3, T_m, T_z in Figure 11.5 can be expressed, respectively, as:

$$\frac{dT_1}{dt} = \frac{T_{sa} - T_1}{R_1 C_1} - \frac{T_1 - T_2}{R_2 C_1} \tag{11.16}$$

$$\frac{dT_2}{dt} = \frac{T_1 - T_2}{R_2 C_2} - \frac{T_2 - T_3}{R_3 C_2} \tag{11.17}$$

$$\frac{dT_3}{dt} = \frac{T_2 - T_3}{R_3 C_3} - \frac{T_3 - T_z}{R_c C_3} - \frac{T_3 - T_m}{R_r C_3} \tag{11.18}$$

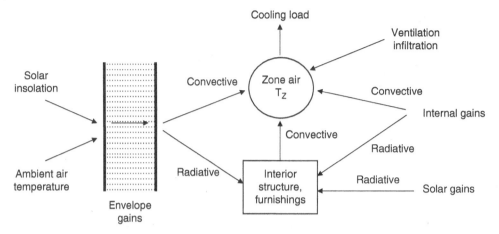

Figure 11.4 The heat transfer link between the zone and one building envelope.

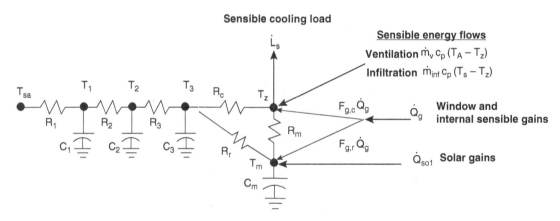

Figure 11.5 The thermal network lining the zone with one building envelope.

$$\frac{dT_m}{dt} = \frac{T_z - T_m}{R_m C_m} + \frac{T_3 - T_m}{R_r C_m} + \frac{F_{g,r}\dot{Q}_g + \dot{Q}_s}{C_m} \tag{11.19}$$

$$\frac{dT_z}{dt} = \frac{T_m - T_z}{R_m C_z} + \frac{T_3 - T_z}{R_c C_z} + \frac{F_{g,c}\dot{Q}_g + \dot{m}_v C_p(T_A - T_z) + \dot{m}_{inf}C_p(T_{sa} - T_z)}{C_z} + \dot{L}_s \tag{11.20}$$

There are five equations for five variables to be solved simultaneously. Note that in Equation (11.20), if $\frac{dT_z}{dt} = 0$, which means the indoor air temperature T_z is constant (e.g. as the indoor design temperature), $\dot{L}_s$ is the required heating and cooling load to maintain the indoor temperature. If $\dot{L}_s = 0$, which means there is no active heating and cooling supplied to the zone, the zone air temperature will freely swing with the ambient and room conditions. The ventilation $\dot{m}_v$ may come from other systems than the heating and cooling systems (e.g. the demand-controlled ventilation (DCV) systems) with its own supply air temperature T_A. If natural ventilation is used, $T_A = T_{sa}$.

When multiple envelopes are considered in the thermal network, the envelope-i is linked not only with the zone air but also with the other envelopes through radiation. Equations (11.16)–(11.20) will thus change to:

$$\frac{dT_{i,1}}{dt} = \frac{T_{sa} - T_{i,1}}{R_{i,1}C_{i,1}} - \frac{T_{i,1} - T_{i,2}}{R_{i,2}C_{i,1}} \tag{11.21}$$

$$\frac{dT_{i,2}}{dt} = \frac{T_{i,1} - T_{i,2}}{R_{i,2}C_{i,2}} - \frac{T_{i,2} - T_{i,3}}{R_{i,3}C_{i,2}} \tag{11.22}$$

$$\frac{dT_{i,3}}{dt} = \frac{T_{i,2} - T_{i,3}}{R_{i,3}C_{i,3}} - \frac{T_{i,3} - T_z}{R_{i,c}C_{i,3}} - \frac{T_{i,3} - T_m}{R_{i,r}C_{i,3}} - \sum_{j=1}^{n} \frac{T_{i,3} - T_{j,3}}{R_{i,j,r}C_{i,3}} \tag{11.23}$$

$$\frac{dT_m}{dt} = \frac{T_z - T_m}{R_m C_m} + \sum_{j=1}^{n} \frac{T_{j,3} - T_m}{R_{j,r}C_m} + \frac{F_{g,r}\dot{Q}_g + \dot{Q}_s}{C_m} \tag{11.24}$$

$$\frac{dT_z}{dt} = \frac{T_m - T_z}{R_m C_z} + \sum_{j=1}^{n} \frac{T_{j,3} - T_z}{R_{j,c}C_z} + \frac{F_{g,c}\dot{Q}_g + \dot{m}_v C_p(T_A - T_z) + \dot{m}_{inf}C_p(T_{sa} - T_z)}{C_z} + \dot{L}_s \tag{11.25}$$

where $R_{i,j,r}$ is the radiant thermal resistance between envelope-i and envelope-j; $R_{j,r}$ is the radiant thermal resistance between envelope-j and interior structure and furnishings; $R_{j,c}$ is the convective thermal resistance between envelope-j and the zone; n is the total number of the envelopes. Equations (11.21)–(11.25) have a total of $(3 \times n + 2)$ equations for $(3 \times n + 2)$ variables: $T_{i,1}$, $T_{i,2}$, $T_{i,3}$ $(i = 1, 2, ..., n)$, T_m, and T_z (or $\dot{L}_s$ if T_z is fixed as an indoor design air temperature). Equations (11.21)–(11.25) are the foundation for most building load calculations and energy simulation programs.

Example 11.1

Calculate the heating load for a commercial building with one external wall that has a total R and C. The interior surface temperatures of the other walls are identical to the room air temperature. There is no interior structure and furnishings and no active ventilation.

Solution

Equations (11.21)–(11.25) become:

$$\frac{dT_{i,1}}{dt} = \frac{T_{sa} - T_{i,1}}{R_{i,1}C_{i,1}} - \frac{T_{i,1} - T_{i,2}}{R_{i,2}C_{i,1}} \tag{11.26}$$

$$\frac{dT_{i,2}}{dt} = \frac{T_{i,1} - T_{i,2}}{R_{i,2}C_{i,2}} - \frac{T_{i,2} - T_{i,3}}{R_{i,3}C_{i,2}} \tag{11.27}$$

$$\frac{dT_{i,3}}{dt} = \frac{T_{i,2} - T_{i,3}}{R_{i,3}C_{i,3}} - \frac{T_{i,3} - T_z}{R_{i,c}C_{i,3}} - \frac{T_{i,3} - T_z}{R_{i,j,r}C_{i,3}} \tag{11.28}$$

$$\frac{dT_z}{dt} = \frac{T_{i,3} - T_z}{R_{i,c}C_z} + \frac{\dot{m}_{inf}C_p(T_{sa} - T_z)}{C_z} + \dot{L}_s \tag{11.29}$$

Homework Problems

1 Compare the UAdT method with the CLTD method to calculate the cooling load for the following space:

A large room in Denver, CO, with $60 \times 60 \, \text{ft}^2$ floor area and 12 ft ceiling high in a building has a large south window-wall as shown below. The ceiling, floor and other walls are next to rooms of similar conditions (and thus can be assumed adiabatic). Infiltration rate of 0.25 ACH, and the total internal heat gains from people/equipment is 10 000 Btu/h. No inside window shading. Only consider the direct/beam solar radiation impacts on the wall and window, $I_{beam} = 974 \, \text{W/m}^2 = 309 \, \text{Btu/h} \cdot \text{ft}^2$ at the solar noon of 21 June. Assume the wall solar absorptivity is 0.1 (white). Make other assumptions as necessary! *Comment on the difference of the calculated results!*

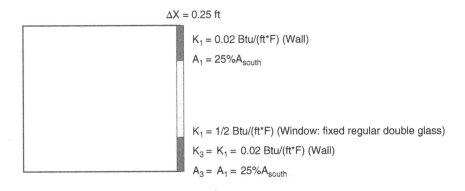

2 An office with $25 \times 25 \, \text{ft}^2$ floor area and 12 ft ceiling high in a large building has a large south envelope with 50% area for a R30 ($30 \, \text{ft}^2\text{-h-}^\circ\text{F/Btu}$) wall and 50% area for a R2 ($2 \, \text{ft}^2\text{-h-}^\circ\text{F/Btu}$) glass window. Assuming the indoor and outdoor $h_{(C+R)}$ are, respectively, 2 and 5 $\text{Btu/(ft}^2\text{-h-}^\circ\text{F)}$. The ceiling, floor, and other walls are next to rooms of similar conditions (and thus can be assumed adiabatic). The outdoor air temperature is $-10 \, ^\circ\text{F}$

(a) At solar noon of 27 January, if the direct normal irradiance (i.e., direct beam solar on a surface normal to the sun) is 40 Btu/h-ft^2, only consider the solar normally transmitted through the window and the transmittance of the glass is 0.9, calculate the indoor air temperature when considering this passive solar heat gain (with 1000 Btu/h internal load but without supplying air from ceiling).

(b) In order to keep the indoor air temperature at 72°F, what will be the supply air temperature if the total supply air flow rate is 0.5 ACH?

(c) In real operation, what measures can be used to adjust the room temperature to be around 72°F without turning on the active air system?

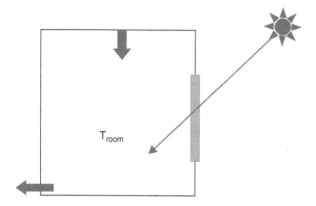

3 Staged Project Assignment (Teamwork)

(a) Estimate the heating and cooling load of your designed building using hand/manual calculation.

(b) Adjust design parameters to minimize the heating and cooling loads (with reasonable assumptions and justifications).

12

Heating, Cooling, and Ventilation Systems

12.1 Basics of Heating and Cooling Systems

A wide variety of HVAC systems (also called building mechanical systems) are available in practice, ranging from simple window units to advanced central systems. These various systems, however, share the same principles and mechanisms. This section will introduce the fundamentals of heating and cooling systems, starting from the most traditional products and systems, upon which new systems are developed incrementally.

12.1.1 Heating Systems

12.1.1.1 Fire Pit and Fireplace

Fire pits and fireplaces are probably the oldest heating solutions, which may use wood, oil, or gas as fuel. Radiation is the primary heat transfer mechanism when using fire pits or fireplaces. As a result, the distance between a person and a heat source significantly determines the amount of heat received. The shorter the distance, the more heat one will feel due to the larger view factor. In order to avoid the contaminants released from the combustion entering the occupied spaces, a chimney is used to exhaust the combustion air. The traditional radiation-based fireplace has a low heating efficiency (around 10%). A newer design of fireplaces creates a separate air channel around the combustion chamber to passively suck in cool room air and exhaust heated air to the indoors (Figure 12.1). This convection and radiation combined fireplace can increase the efficiency to 30%. Although this number may look low compared to traditional furnaces or water boilers (usually above 80%), fireplaces are an aesthetic design feature. In fact, they are very effective as a localized heating solution for a space where the rest of the space may not need simultaneous heat. Such examples include family/living rooms and bedrooms (whereas other spaces may not need the same level of heating at the same time).

12.1.1.2 Hot Water Heating Systems

Hot water heating systems are the second oldest heating approach, which provide efficient and quiet heating to many large spaces at a time, with a lower space requirement for the system. The hot water can be produced locally in the building (e.g. in the basement or garage) via water boilers. It can also be acquired from the district's heating pipe networks, which supply hot water or steam from the central plant (with larger boilers) to individual buildings. The supplied hot water may be used directly through the zone

Energy Efficient Buildings: Fundamentals of Building Science and Thermal Systems, First Edition. Zhiqiang (John) Zhai.
© 2023 John Wiley & Sons, Inc. Published 2023 by John Wiley & Sons, Inc.

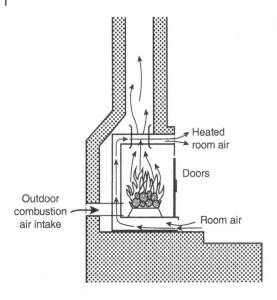

Figure 12.1 Illustration of a fireplace.

water pipes or it will go through a heat exchanger to deliver the heat to a sperate and closed water loop in each building (this is especially true for steam due to the high temperature and pressure of hot steam). Figure 12.2 shows the configuration of a water heating system with a basement boiler. Figure 12.3 illustrates the working mechanism inside the boiler. Similar to a fireplace, fuel (e.g. wood, oil, gas) is burned in the combustion chamber to heat the water flowing in the water pipes through the chamber. Ambient air is supplied to sustain the combustion, while combustion air is exhausted through the chimney. A water

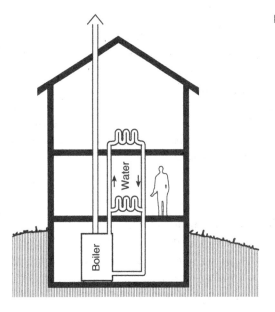

Figure 12.2 Water heating system with a basement boiler.

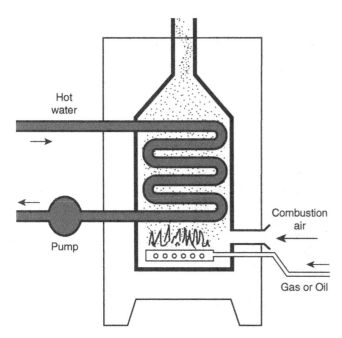

Figure 12.3 Principle of a boiler for water heating system.

pump is used to circulate water through the entire pipe network and deliver water to the spaces. Large commercial boilers (for central plants) share the same principles but have larger sizes. Figure 12.4 demonstrates the packaged fire-tube hot water boiler, where the tank is filled with water and the hot combustion gas flows through the tubes that are immerged in the water.

Hot water is delivered through the water pipes and eventually through the distribution terminals to heat the indoor environment. Hot water pipe materials may range from chlorinated polyvinyl chloride (CPVC) to cross-linked polyethylene (PEX) to steel to copper. Insulation may be used to avoid heat loss during transportation. Distribution terminals work as a heat exchanger in the working zone to deliver the heat from the hot water to the cool room air. As a result, distribution terminals are mostly made of conductive materials such as iron and aluminum, and external fins are commonly used to increase the heat contact area, as shown in Figure 12.5.

Recently, radiation floor heating sytems, which provide more thermal comfort (and less noise) to indoor spaces, have been receiving a lot of attention. The same hot water system is used to deliver hot water, while the distribution terrminals are changed to small hot pipes (ususally made of copper or PVC) embeded into the floor as illustrated in Figure 12.6. Proper layout of the distributed floor water pipes is critical to deliver uniform heating to the floor and the space.

12.1.1.3 Hot Air Heating Systems

Hot water heating is an effective method for space heating, with a long history of practical applications and a recent re-boom attributed to interest in a radiation heating concept. However, hot water heating has some inherent shortcomings such as not providing ventilation and a slow response to heating demand.

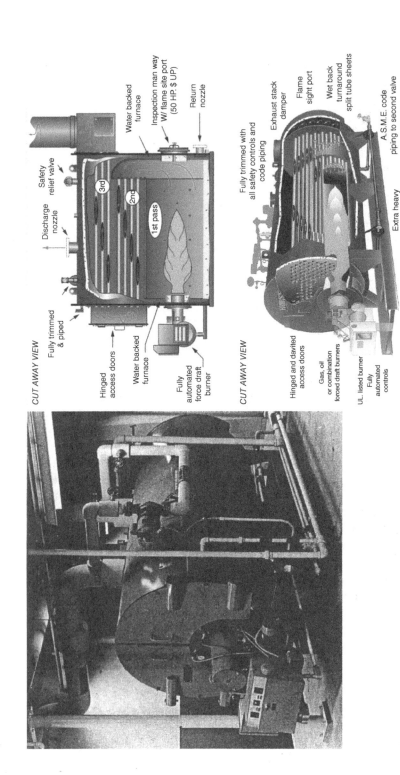

CUT AWAY VIEW

Fully trimmed & piped

Water backed furnace

Inspection man way W/ flame site port (50 HP. $ UP)

Return nozzle

Safety relief valve

Discharge nozzle

3rd

2nd

1st pass

Hinged access doors

Water backed furnace

Fully automated force draft burner

CUT AWAY VIEW

Fully trimmed with all safety controls and code piping

Exhaust stack damper

Flame sight port

Wet back turnaround split tube sheets

A.S.M.E. code piping to second valve

Extra heavy skids and supports

Hinged and davited access doors

Gas, oil or combination forced draft burners

UL listed burner

Fully automated controls

Figure 12.4 Packaged fire-tube hot water boiler.

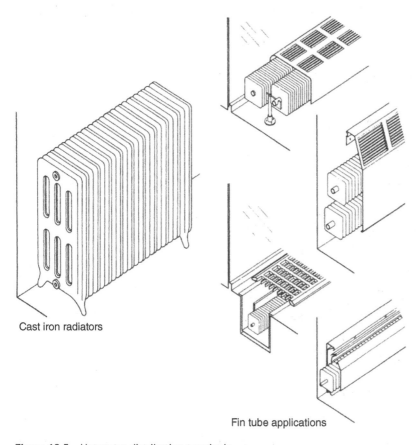

Cast iron radiators

Fin tube applications

Figure 12.5 Hot water distribution terminals.

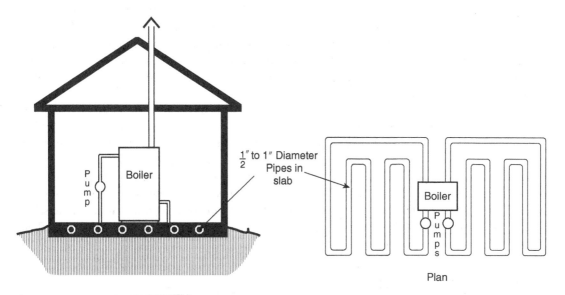

$\frac{1}{2}''$ to $1''$ Diameter
Pipes in
slab

Boiler

Plan

Figure 12.6 Hot water floor heating system.

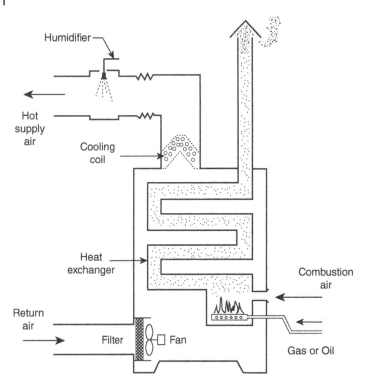

Figure 12.7 Principle of a hot-air furnace (with cooling and humidification).

Air-based heating systems gained their popularity over the past few decades. Fundamentally similar to hot water heating systems, hot air systems burn the combustion chamber (fueled with gas or oil) to heat the air circulated by the fan in the duct system (Figure 12.7). The heated air can be returned from spaces, supplied from the outdoors, or a mixture of outdoor and room air (most cases). A filter is installed before the air enters the heating coil to remove the contaminants in the air such as dust and pollen. The heated air is supplied to spaces through the deliberately designed duct networks, which provide not only the needed heat but also the ventilation. The same duct and fan system can be used to cool and/or humidify the air (as illustrated in Figure 12.7).

Figure 12.8 shows the configuration and a picture of an actual hot-air furnace typically used in residential and small commercial buildings. Similar configurations can be found for large furnaces for commercial buildings. Hot and cool air is delivered to the spaces via a wide variety of distribution terminals as shown in Figure 12.9. Different terminals may have different air-mixing performances and ventilation efficiencies. Selecting property air terminals depends on many factors such as the function and geometry of the space, available space for air ducts and terminal control boxes, and indoor environment control requirements (e.g. thermal comfort, air quality, noise, etc.). This often requires a comprehensive study of indoor air distribution, usually by applying engineering simulation tools such as computational fluid dynamics techniques.

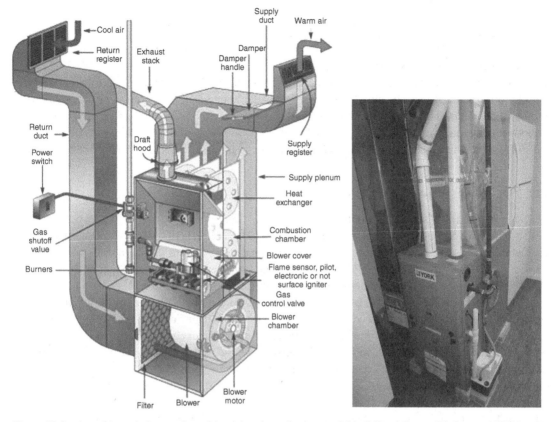

Figure 12.8 Actual hot-air furnace in residential and small commercial buildings. *Source:* Wtshymanski/Wikimedia Commons.

Figure 12.9 Air distribution terminals.

12.1.1.4 Electrical Heating Systems

Electrical heating presents a most convenient and flexible solution for space heating. While it is least efficient in terms of energy use, its plug-in and "use as needed" character allows the system to be installed and applied whenever and wherever there is heating demand, with the least infrastructure investment. Various conduction, convection, and radiation based electrical heating units and systems can be found off the shelf (as seen in Figure 12.10). Table 12.1 compares the pros and cons of various heating systems.

12.1.2 Cooling Systems

12.1.2.1 Principles of Compressive Refrigeration

The first modern air conditioner (AC) was invented in 1902 by Willis Haviland Carrier, an electrical engineer who tried to solve a humidity control problem for a printing plant in Brooklyn, NY. At that stage it was not even called air conditioning. The term was coined by a gentleman named Stuart W. Cramer in a May 1906 speech in Asheville, NC, before the American Cotton Manufacturers Association. Initial ACs used flammable and toxic gases like ammonia, propane, and methyl chloride and thus were mostly applied for industrial environment controls. In 1928, Thomas Midgley discovered Freon, which led to the rise of better, safer refrigerants that could be applied to domestic air conditioning. For decades, Freon, also known as R-22 and HCFC-22, was the main refrigerant used in AC units. However, new AC systems made after 2010 no longer rely on Freon, instead using a refrigerant called R410A, or Puron, that has been shown not to harm the ozone layer. Starting from 1 January 2020, Freon will no longer be made in or imported to the United States.

Figure 12.11 shows a typical cycling process of compressive refrigeration, where the cold and low-pressure liquid-phase refrigerant absorbs heat when it evaporates, and the hot and high-pressure gas-phase refrigerant releases heat when it condenses. The compressor increases the temperature and pressure of the gas-phase refrigerant so that it can release heat at a higher ambient temperature. The

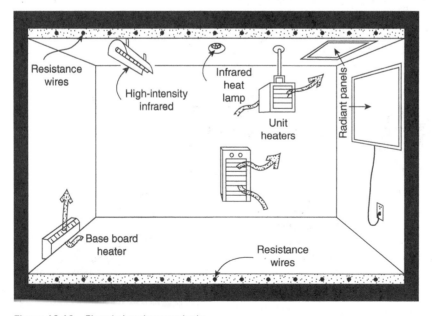

Figure 12.10 Electric heating terminals.

Table 12.1 Advantages and disadvantages of various heating systems.

System	Advantages	Disadvantages
Air	Can also perform other functions such as ventilation, cooling, humidity control, and filtering	Very bulky ducts require careful planning and space allocation
	Prevents stratification and uneven temperatures by mixing air	Can be noisy if not designed properly
		Very difficult to use in renovations
	Very quick response to changes of temperature	Zones are not easy to create. Cold floors result if air outlets are high in the room
	No equipment required in rooms being heated	
Water	Compact pipes are easily hidden within walls and floor	For the most part can only heat and not cool (exceptions: fan-coil units and valance units)
	Can be combined with domestic hot water system	No ventilation
	Good for radiant floor heating	No humidity control
		No filtering of air
		Leaks can be a problem
		Slightly bulky equipment in spaces being heated (base-board and cabinet convectors)
		Radiant floors are slow to respond to temperature changes
Electricity	Most compact	Very expensive to operate (except heat pump)
	Quick response to temperature changes	Wasteful, unenvironmental
	Very easily zoned	Cannot cool (except for heat pump)
	Low initial cost	

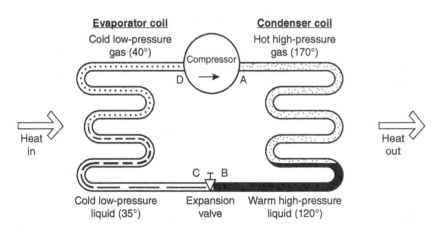

Figure 12.11 Typical cycling process of compressive refrigeration (in °F).

expansion valve reduces the temperature and pressure of the warm and high-pressure liquid-phase refrigerant so that it can return to a cold and low-pressure state for the next cycle of cooling. As a result, an AC is made of four primary elements: an evaporator coil, a condenser coil, a compressor, and an expansion valve. Figure 12.12 shows the theoretic cooling processes on the temperature-entropy diagram and the pressure-enthalpy diagram, respectively.

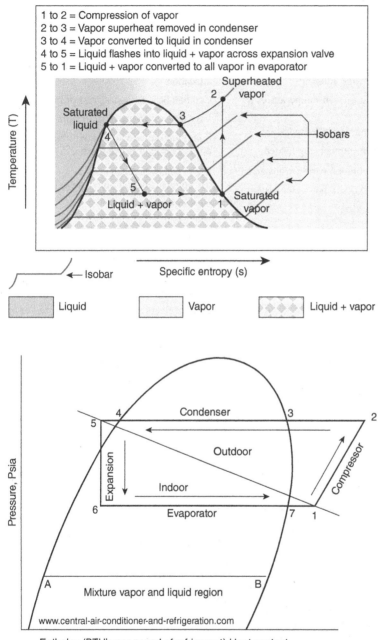

1 to 2 = Compression of vapor
2 to 3 = Vapor superheat removed in condenser
3 to 4 = Vapor converted to liquid in condenser
4 to 5 = Liquid flashes into liquid + vapor across expansion valve
5 to 1 = Liquid + vapor converted to all vapor in evaporator

Figure 12.12 Theoretic cooling processes on the temperature-entropy diagram and the pressure-enthalpy diagram, respectively.

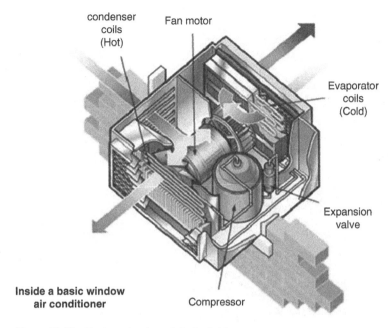

condenser coils (Hot)

Fan motor

Evaporator coils (Cold)

Expansion valve

Inside a basic window air conditioner

Compressor

Figure 12.13 Packaged unitary (window) AC.

12.1.2.2 Various Air-Conditioning Systems

The principles of AC are shared to build various AC products ranging from a packaged unitary AC (Figure 12.13) to a residential whole house AC unit (Figure 12.14) to a rooftop unit (Figure 12.15) for small commercial buildings to a packaged central AC system (Figure 12.16) for midsize commercial buildings and further to larger central AC systems.

A few common features should be noted among these systems of different capacities:

1) Outdoor air can be used to mix with return air to provide the required fresh air for ventilation and reduce the cooling energy consumption (due to the lower room air temperature compared to the outdoor air temperature).
2) Evaporator can be installed in the same duct and fan system for air heating.
3) Supply air fan can be installed either before or after the cooling coil (evaporator).
4) Cooling coil (evaporator) can be designed in a V-shape to increase the heat exchange contact area.
5) Compressor is often placed outdoors to avoid operation noise and heat.
6) Condensation collector pan is required under the condenser to collect the condensed water.

12.2 Basics of Heating and Cooling Distribution Systems

Except the unitary systems that can deliver heat and coldness directly to the spaces, a distributed system is often required to deliver the conditioned media from the building mechanical systems to various compartments in a building. Mechanical systems are installed in the mechanical rooms, usually located in the

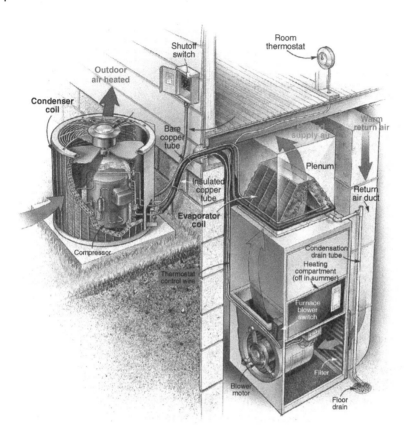

Figure 12.14   Residential whole house AC.

basement, the top floor, or a dedicated location on a certain floor. A rule-of-thumb estimate is that every 12 floors require 1 floor of mechanical rooms, while the actual size requirement varies with system type and capacity choices, etc. Three different types of distribution systems are generally used.

12.2.1   All Air System

All air systems deliver conditioned air from the mechanical equipment room to the occupied rooms (as showed in Figure 12.17). The mechanical room houses the full heating and AC systems, which heat and cool the return air (as well as the outdoor fresh air). The air system can thus provide ventilation in addition to heating and cooling. This becomes extremely useful in mild environments (e.g. spring and fall as well as cool nights) when the heating and cooling can be turned off, whereas the outdoor air can still be supplied into the spaces to remove internal heat and moisture. This is called mechanical ventilation cooling.

Figure 12.15 Rooftop AC unit. *Source:* Taya/Adobe Stock.

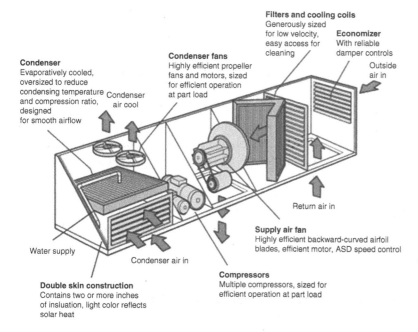

Filters and cooling coils
Generously sized
for low velocity,
easy access for
cleaning

Economizer
With reliable
damper controls

Outside
air in

Condenser fans
Highly efficient propeller
fans and motors, sized
for efficient operation
at part load

Condenser
Evaporatively cooled,
oversized to reduce
condensing temperature
and compression ratio,
designed
for smooth airflow

Condenser
air cool

Return air in

Water supply

Condenser air in

Supply air fan
Highly efficient backward-curved airfoil
blades, efficient motor, ASD speed control

Double skin construction
Contains two or more inches
of insluation, light color reflects
solar heat

Compressors
Multiple compressors, sized for
efficient operation at part load

Figure 12.16 Packaged central system for midsize buildings. *Source:* E source.

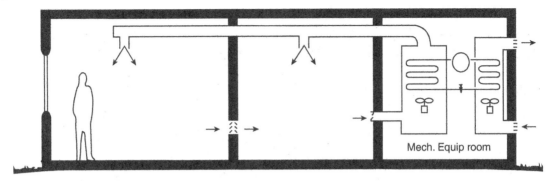

Figure 12.17 All air systems.

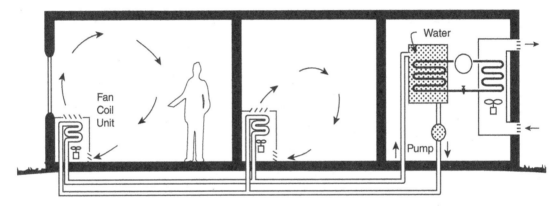

Figure 12.18 All water systems.

12.2.2 All Water System

All water systems deliver conditioned water from the mechanical equipment room to the occupied rooms (as showed in Figure 12.18). The conditioned water releases the heat and coldness through the room terminals with convection, radiation, or a combination of the two. As water has a much higher thermal capacity than air, the water pipe network system requires much less space than the air duct system to deliver the same amount of energy. As a result, the energy required for circulating media throughout a building is smaller for water (pump) than for air (fan). Electricity used by fan can take up to 40% of the entire building's electricity usage. In addition, water heating and cooling provide better thermal comfort and less noise compared to the all-air system. However, the air system can deliver the heating and cooling to the spaces, more rapidly and uniformly, due to the strong convection and mixing. The water system may take a fairly long period of time to warm-up and cool-down and thus is not ideal for spaces that require frequent on and off control.

12.2.3 Air Water System

Air water systems attempt to adopt the advantages of both air and water systems while removing the disadvantages of both. As shown in Figure 12.19, the mechanical room will condition both air and water as

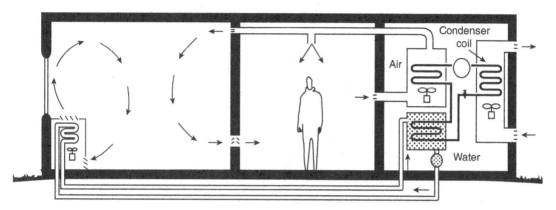

Figure 12.19 Air water systems.

the transport medias and supply them through their own duct/pipe systems to the spaces in need. For a space with both air and water supply, water is the primary source for heating and cooling while air is mostly for ventilation purpose with supplementary heating and cooling functions.

A fan-coil unit is one of such examples that is commonly used for hotel and office rooms (Figure 12.20). Each Fan-coil unit holds one or two water coils. Two coils can run heating and cooling at the same time, while a seasonal valve is used if one coil is used for both hot and cold water. A room thermostat controls the water flow. Ventilation is met with either local outdoor air or central air systems. Another promising air water system technology is the active chilled beam system as shown in Figure 12.21, where the chilled water in the copper pipes removes 90% indoor heat while the cool air provides the other 10% cooling and the ventilation. As a result, the fan energy is largely reduced. While an air water system presents more energy efficiency, it results in many more complications from design to operation to maintenance due to the dual systems. Table 12.2 compares the advantages and disadvantages of various distribution systems.

12.3 Heating and Cooling on Psychrometric Chart

The heating and cooling processes can be described, quantitatively, on the psychrometric chart, upon which the system capacities can be properly determined.

12.3.1 Change of Sensible Heat

Figure 12.22 shows the processes of sensible heating and cooling on the psychrometric chart and the associated drawing symbols. The sensible heat transfer without water vapor content (W) changes are only a function of dry-bulb temperature changes, as calculated below:

$$Q = \rho \times V \times C_p \times (T_2 - T_1) \tag{12.1}$$

Or

$$Q = \rho \times V \times (h_2 - h_1) \tag{12.2}$$

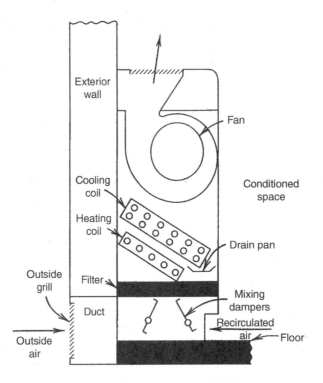

Exterior wall

Fan

Conditioned space

Cooling coil

Heating coil

Drain pan

Outside grill

Filter

Duct

Mixing dampers

Recirculated air

Floor

Outside air

Main fan-coil units

Cassette type fan-coil unit	Horizontal concealed fan-coil unit 1	Horizontal concealed fan-coil unit 2	Vertical concealed fan-coil unit
Vertical exposed fan-coil unit	Horizontal exposed fan-coil unit	Horizontal mounted fan-coil unit	Cabinet type fan-coil unit

Figure 12.20 Fan-coil unit.

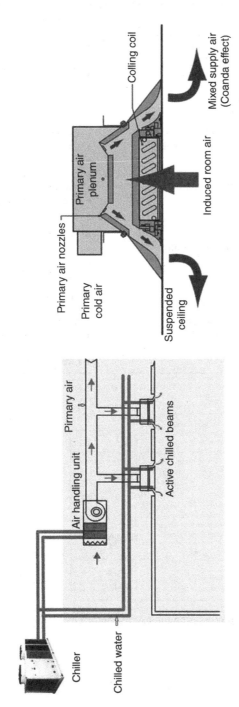

Primary air nozzles

Primary
cold air

Primary air
plenum

Colling coil

Mixed supply air
(Coanda effect)

Induced room air

Suspended
ceiling

Primary air

Air handling unit

Active chilled beams

Chiller

Chilled water

Figure 12.21   Active chilled beam system.

Table 12.2 Advantages and disadvantages of various distribution systems.

System	Advantages	Disadvantages
All air	• Central equipment location • No piping in occupied area • Use of outside air (free cooling) • Easy seasonal change • Heat recovery possible • Closest operating conditions	• Duct clearance • Large ducts – space • Air balancing difficulties
Air water	• Individual room control • Separate secondary heating/cooling • Less space for ducts • Smaller HVAC central equipment • Central filter, humidification	• Changeover if only two pipes • Operating complex if two pipes • Control is numerous • Fan-coil clearance problem • No-shut off for primary air • High pressure for induction • Four pipe system is too expensive
All water	• Less space • Locally shutoff (individual control) • Quick pull down • Good for existing buildings	• More maintenance in occupied area • Coil cleaning difficulties • Filter • Open window for IAQ
Unitary	• Individual room control • Simple and inexpensive • Independent of other buildings • Manufacturer made it ready	• Limited performance • No humidity control (general) • More energy (low efficiency) • Control of air distribution • Filter • Overall appearance

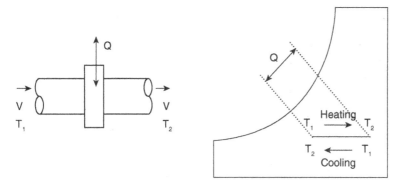

Figure 12.22 Sensible heating and cooling processes on the psychrometric chart and the associated drawing symbols.

where Q is the total heat transfer (J or Btu); ρ is the air density; V is the air volume (in m^3 or ft^3); C_p is the air specific heat; T_2 and T_1 are the air temperatures; and h_2 and h_1 are the air enthalpy per unit air mass that can be read from the psychrometric chart. Note that the chart does not include any information about the air volume or volume flow rate. If V in Equations (12.1) and (12.2) is replaced with the air volume flow rate $\dot{V}$ (in m^3/s or ft^3/h), Q becomes $\dot{Q}$ (in W or Btu/h).

Example 12.1

Determine the energy required by a baseboard heater to heat 1 lb of cold air at 30 °F and 50% RH to 70 °F. Also find the relative humidity at State 2.

Solution

From the psychrometric chart, relative humidity = 12%, $h_1 = 8.9$ Btu/lb$_a$, $h_2 = 18.7$ Btu/lb$_a$.

$$Q = \rho \times V \times (h_2 - h_1) = 1\ lb_a \times (18.7 - 8.9)\ Btu/lb_a = 9.6\ Btu$$

or

$$Q = \rho \times V \times C_p \times (T_2 - T_1) = 1\ lb_a \times 0.24\ Btu/lb_a \cdot °F \times (70°F - 30°F) = 9.6\ Btu$$

12.3.2 Humidification and Dehumidification

This is a process of latent heat transfer, where the air temperature is kept constant, and the water vapor content (humidity ratio) is increased (humidification) or decreased (dehumidification) as shown in Figure 12.23. This implies a direct supply or removal of water vapor at the room air temperature. The total enthalpy change is thus:

$$Q = \rho \times V \times h_g \times (W_2 - W_1) = \rho \times V \times (h_2 - h_1) \tag{12.3}$$

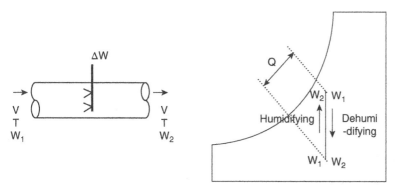

Figure 12.23 Humidifying and dehumidifying processes on the psychrometric chart and the associated drawing symbols.

where h_g is the enthalpy of saturated water vapor and $h_g = 2501.3$ kJ/kg (at 0 °C) $= 1061.2$ Btu/lb (at 32 °F) (or $= 2442$ kJ/kg at 24 °C $= 1050$ Btu/lb at 75 °F).

Example 12.2

Determine the energy required by a steam humidifier to humidify 1 lb of dry air at 10% RH and 70 °F to 50% RH.

Solution

From the psychrometric chart, $h_1 = 18.7$ Btu/lb, $h_2 = 25.2$ Btu/lb.

$$Q = \rho \times V \times (h_2 - h_1) = 1 \times (25.2 - 18.7) = 6.5 \text{ Btu}$$

or

$$Q = \rho \times V \times h_g \times (W_2 - W_1) = 1 \text{ lb}_a \times 1050 \text{ Btu/lb}_a \times (0.0077 \text{ lb}_w/\text{lb}_a - 0.0015 \text{ lb}_w/\text{lb}_a) = 6.5 \text{ Btu}$$

12.3.3 Cooling and Dehumidification

When the cooling process reaches the saturation line (i.e. the dew point temperature), water vapor in the air starts to condense and humidity ratio reduces. Further cooling will lead to more condensation of water vapor in the air while keeping the air relative humidity at saturation ($\varphi = 100\%$). Figure 12.24 demonstrates the process.

The energy reduced during the process can be calculated:

$$Q = \rho \times V \times C_p \times (T_2 - T_1) + \rho \times V \times h_g \times (W_2 - W_1) = \rho \times V \, (h_2 - h_1) \tag{12.4}$$

It appears that the entire cooling process can be split into two imaginary sub-processes: latent (from T_1 to B) and sensible (from B to T_2). Note that the cooling process along the saturation line embeds both sensible and latent heat changes.

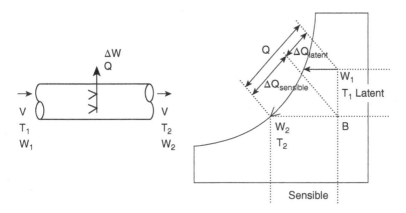

Figure 12.24 Cooling and dehumidifying processes on the psychrometric chart and the associated drawing symbols.

Example 12.3

An air conditioner is used to cool 1 lb of warm air at 70% RH and $T_{dry} = 90\,°F$ to $55\,°F$. Determine relative humidity at State 2 and sensible and latent heat removed.

Solution

From the psychrometric chart, $\phi_{new} = 100\%$

$$h_1 = 45.6 \text{ Btu/lb} \qquad h_2 = 23.3 \text{ Btu/lb} \qquad h_B = 32.0 \text{ Btu/lb}$$

$$Q_{sensible} = \rho \times V \times (h_2 - h_B) = 1 \times (23.3 - 32.0) = -8.7 \text{ Btu}$$

$$Q_{latent} = \rho \times V \times (h_B - h_1) = 1 \times (32.0 - 45.6) = -12.6 \text{ Btu}$$

or

$$Q_{sensible} = \rho \times V \times C_p \times (T_2 - T_1) = 1 \times 0.24 \times (55 - 90) = -8.4 \text{ Btu}$$

$$Q_{latent} = \rho \times V \times h_g \times (W_2 - W_1) = 1 \times 1050 \times (0.0093 - 0.0217) = -13.0 \text{ Btu}$$

12.3.4 Heating and Humidification

The heating process increases the capability of the air to hold water vapor (i.e. the increase of $P_{v,s}$). As a result, the relative humidity is decreased during the heating process although the humidity ratio stays constant. Hence, humidification is commonly used along with the heating process. Figure 12.25 illustrates the heating process with humidification. Depending on the media that is used for the humidification (e.g. water, water vapor, steam, etc.), the humidification process follows different paths as shown in Figure 12.26, which may include a pure latent heat transfer, or a combined sensible and latent heat transfer. When directly supplying the water vapor at the room air temperature, the humidity ratio is increased while the room air temperature is kept constant. Supplying warm/hot steam (water vapor) increases both the humidity ratio and the air temperature. When water at room air temperature is supplied, water becomes water vapor through evaporation that absorbs the heat from the air. As a result, the air temperature is reduced while the humidity ratio is increased, and the enthalpy during the process remains

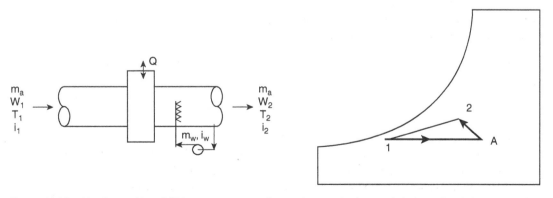

Figure 12.25 Heating and humidifying processes on the psychrometric chart and the associated drawing symbols.

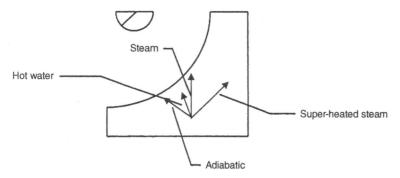

Figure 12.26 Different humidification processes depending on the used media.

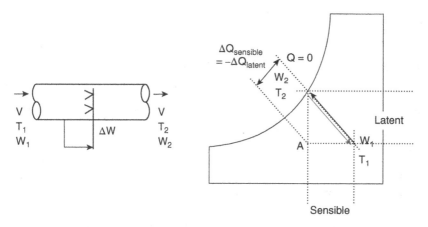

Figure 12.27 Adiabatic humidification and desiccant dehumidification processes on the psychrometric chart and the associated drawing symbols.

constant. This is called adiabatic humidification. The reverse process is called adiabatic/desiccant dehumidification, as shown in Figure 12.27.

Since the enthalpy during the adiabatic humidification and desiccant dehumidification process remains constant, this can be expressed as,

$$Q = \rho \times V \times C_p \times (T_2 - T_1) + \rho \times V \times h_g \times (W_2 - W_1) = \rho \times V \times (h_2 - h_1) = 0 \tag{12.5}$$

As a result,

$$Q_{sensible} = \rho \times V \times C_p \times (T_2 - T_1) = -Q_{latent} = -\rho \times V \times h_g \times (W_2 - W_1) \tag{12.6}$$

Example 12.4

In Phoenix, it is possible to use evaporative cooling in summer. In a room of $30' \times 20' \times 10'$, the infiltration rate is 2 ACH. The outdoor air temperature is 90 °F and relative humidity is 20%. Comfort standards allow the relative humidity to be increased to 70% by evaporative cooling. Determine the dry-bulb temperature in the room and the sensible and latent heat added or removed.

Solution

This is an adiabatic humidification process. The air process on a psychrometric chart is iso-enthalpy. From the psychrometric chart

$$T_2 = 67.5°F$$

$$W_1 = 0.006\ lb_w/lb_a \qquad W_2 = 0.0107\ lb_w/lb_a$$

$$h_1 = 28.4\ Btu/lb_a \qquad h_2 = 28.4\ Btu/lb_a \qquad h_A = 23.2\ Btu/lb_a$$

$$\rho \times \dot{V} = 0.075\ lb/ft^3 \times 2/h \times (20 \times 20 \times 10)\ ft^3 = 600\ lb/h$$

$$\dot{Q}_{sensible} = \rho \times V \times C_p \times (T_2 - T_1) = 600 \times 0.24 \times (69.5 - 90) = -2962\ Btu/h$$

$$\dot{Q}_{latent} = \rho \times V \times h_g \times (W_2 - W_1) = 600 \times 1050 \times (0.0107 - 0.006)$$
$$= 2961\ Btu/h\ (this\ is\ about\ 2.82\ lb_w/h)$$

or

$$Q_{sensible} = \rho \times V \times (h_A - h_1) = 600 \times (22.9 - 27.8) = -2940\ Btu/h$$

$$Q_{latent} = \rho \times V \times (h_2 - h_A) = 600 \times (27.8 - 22.9) = 2940\ Btu/h$$

12.3.5 Adiabatic Mixing of Air

Mixing two streams or bodies of air of different properties is very common in practice. One example is the mixing of the return air (from the space) with the outdoor air before the mixed air is conditioned through the heating, cooling, and/or humidifying/dehumidifying elements. Figure 12.28 demonstrates the mixing

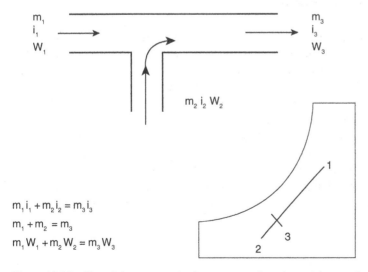

$$m_1 i_1 + m_2 i_2 = m_3 i_3$$
$$m_1 + m_2 = m_3$$
$$m_1 W_1 + m_2 W_2 = m_3 W_3$$

Figure 12.28 The mixing process in the system and on the psychrometric chart and three conservations (i is enthalpy).

process in the system and on the psychrometric chart. Three conservations are obeyed during the process: the energy of the mixture, the mass of the mixture, and the mass of the water vapor.

By solving the conservation equations in Figure 12.28 one can obtain:

$$h_3 = \frac{m_1 h_1 + m_2 h_2}{m_1 + m_2} \tag{12.7}$$

$$W_3 = \frac{m_1 W_1 + m_2 W_2}{m_1 + m_2} \tag{12.8}$$

Therefore, the resultant enthalpy (h_3) and humidity ratio (W_3) are linearly proportional to the mass contributions of the incoming flows (m_1 and m_2). The mixing air (3) on the psychrometric chart is thus on the line connecting the incoming airflow (1) and (2) and is closer to the air that contributes more mass according to their mass weighting factors.

Example 12.5

Return air at 25 °C, 50%, $V_r = 5\,\text{m}^3/\text{s}$ is mixed with outdoor air at 35 °C, 60%, $V_o = 1.25\,\text{m}^3/\text{s}$. Determine mixed air conditions and flow rate. Assume at 1 atm.

Solution

$$V_m = V_o + V_r = 6.25\,\text{m}^3/\text{s}$$

$$h_m = \frac{m_r h_r + m_o h_o}{m_r + m_o} = 0.8 \times h_r + 0.2 \times h_o = 0.8 \times 50.26 + 0.2 \times 90.11 = 58.23\,\text{kJ/kg}$$

$$W_m = \frac{m_r W_r + m_o W_o}{m_r + m_o} = 0.8 \times W_r + 0.2 \times W_o = 0.8 \times 9.78 + 0.2 \times 20.98 = 12.02\,\frac{\text{g}}{\text{kg}}$$

12.4 Central HVAC Systems on Psychrometric Chart

The entire heating and cooling processes through a central HVAC system can be presented on the psychrometric chart, upon which the system performance can be quantified, and the sizes of heating and cooling coils can be estimated. Figure 12.29 shows the flow diagram and states for a typical central HVAC system. Part of the return air from building spaces is recirculated and mixed with ventilation (outdoor) air. The mixed air is treated by heating and cooling coils (and humidifier and dehumidifier if any). The processed air can be reheated, if needed, to increase the relative humidity before it is supplied to the spaces to remove the sensible (L_S) and latent (L_L) heating and cooling loads. The room air is returned, and some is exhausted to outdoors and some is recirculated to save energy. Figure 12.30 illustrates the corresponding flow processes and air states on the psychrometric chart, where the hot outdoor air A is mixed with the zone air Z to yield the mixed air E that is conditioned to C and further reheated to S for the zone supply.

During the processes as shown in Figure 12.30, the indoor sensible and latent load removed by the supply air are, respectively:

$$\dot{L}_S = \dot{m}_s(h_X - h_S) \tag{12.9}$$

$$\dot{L}_L = \dot{m}_s(h_Z - h_X) \tag{12.10}$$

where $\dot{m}_s$ is the supply air mass flow rate and h is the air enthalpy at different states.

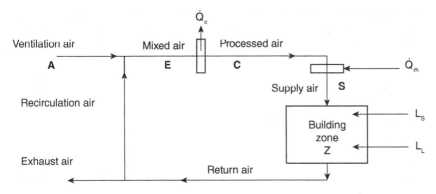

Figure 12.29 Flow diagram and states for a typical central HVAC system.

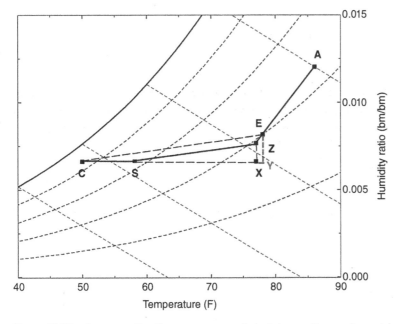

Figure 12.30 Corresponding flow processes and air states on the psychrometric chart.

The total indoor load is:

$$\dot{L}_T = \dot{m}_s(h_Z - h_S) \tag{12.11}$$

The ratio of $\dot{L}_S$ to $\dot{L}_T$ is called the sensible heat ratio (SHR) of the zone.

$$SHR_Z = \frac{\dot{L}_S}{\dot{L}_T} = \frac{\dot{L}_S}{\dot{L}_S + \dot{L}_L} \tag{12.12}$$

The loads imposed on the heating and cooling coils, however, are different from the loads removed from the indoor spaces. This is attributed to the ventilation/outdoor air that comes in with a different enthalpy from the zone air (e.g. hotter in summer and colder in winter). The total coil load for Figure 12.30 is:

$$\dot{Q}_{coil} = \dot{m}_s(h_E - h_C) \tag{12.13}$$

where the sensible and latent load of the cooling coil are, respectively,

$$\dot{Q}_{coil,S} = \dot{m}_s(h_Y - h_C) \tag{12.14}$$

$$\dot{Q}_{coil,L} = \dot{m}_s(h_E - h_Y) \tag{12.15}$$

Similarly, the ratio of $\dot{Q}_{coil,S}$ to $\dot{Q}_{coil}$ is called the sensible heat ratio (SHR) of the coil.

$$SHR_C = \frac{\dot{Q}_{coil,S}}{\dot{Q}_{coil}} = \frac{\dot{Q}_{coil,S}}{\dot{Q}_{coil,S} + \dot{Q}_{coil,L}} \tag{12.16}$$

The additional reheat energy at the zone terminal is:

$$\dot{Q}_{rh} = \dot{m}_s(h_S - h_C) \tag{12.17}$$

It may not sound wise to reheat air after cooling in order to increase the relative humidity of the supply air. Using passive approaches such as bypassing unconditioned airflow to mix with the cooled air may serve the same purpose. Figure 12.31 illustrates the principles of the bypass model for cooling coil.

The bypass factor is defined as, which can be adjusted in operation:

$$f_{bp} = \frac{\dot{m}_{bp}}{\dot{m}_s} \tag{12.18}$$

where $\dot{m}_s$ is the total supply air mass flow rate entering the coil and $\dot{m}_{bp}$ is the bypass flow rate. The total heat transfer through the coil is thus:

$$\dot{Q}_{coil} = \dot{m}_s(1 - f_{bp})(h_E - h_C) \tag{12.19}$$

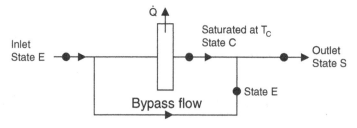

Figure 12.31 Principles of the bypass model for cooling coil. *Source:* ASHRAE (1992).

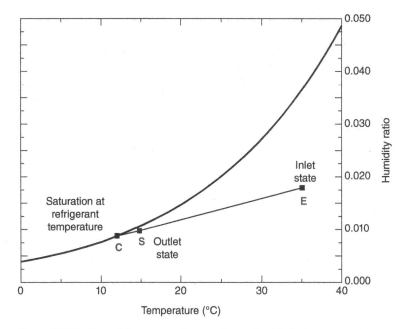

Figure 12.32 The mixing process in a cooling coil with the bypass model.

The supply air properties are:

$$W_S = (1 - f_{bp}) \times W_C + f_{bp} \times W_E \tag{12.20}$$

$$h_S = (1 - f_{bp}) \times h_C + f_{bp} \times h_E \tag{12.21}$$

Figure 12.32 shows the mixing process of the conditioned air at State C with the bypassed air at State E to create the supply air at State S. T_C is often called the apparatus dew point temperature. The coil effectiveness is defined and calculated as:

$$\varepsilon = \frac{\dot{m}_s(h_E - h_S)}{\dot{m}_s(h_E - h_C)} = \frac{h_E - (1 - f_{bp}) \times h_C - f_{bp} \times h_E}{h_E - h_C} = 1 - f_{bp} \tag{12.22}$$

12.5 Coil Sizing and Selection

Coil sizing starts with space conditions, including the indoor design conditions (e.g. the desired $T_{dry\text{-}air}$ and relative humidity ϕ) (Chapter 6) and the calculated indoor sensible and latent load (Chapter 11), as well as the ventilation air requirement (Chapter 7).

Step-1: propose a supply air temperature according to indoor environment requirements and system considerations. Draw a line from the room air condition (Z), following the SHR slope (which is provided at the top-left corner of the psychrometric chart as a guide, as shown in Figure 12.33). The intersection of the line with the supply air temperature (dash) line forms the required room supply air condition (S) (Figure 12.34).

Step-2: mix the room air (Z) with the outdoor air (A) according to the required ventilation rate to find the mixture air that enters the coil (E) (Figure 12.35).

Step-3: connect E and S and extend the line to reach the saturation curve. The intersection is the apparatus dewpoint temperature (C) (Figure 12.35). Based on the distances among E, C, and S, the bypass coefficient (f_{bp}) and the coil effectiveness (ε) can be calculated. The black arrows in Figure 12.36 reveal the actual airflow processes in a complete cooling cycle.

Step-4: determine the coil capacity by measuring the enthalpy difference of air at E and S (Figure 12.37) and using the following equation:

$$\dot{Q}_{coil} = \dot{m}_s(h_E - h_S) \tag{12.23}$$

Equation (12.23) is the same as Equation (12.19).

Example 12.6

Determine the minimum cooling coil load and required chilled water flow rate for a system serving an office space with a sensible cooling load of 30 800 Btu/h and a latent load of 8500 Btu/h. Assume the indoor design conditions are 75 °F DB and 50% RH, and the outdoor design conditions are 85 °F DB and 70 °F WB. The minimum ventilation air is 33% of the total supply air that is delivered at 55 °F. The fan selected is a 70% efficient "draw-through" type, which delivers air at 4.0 inch total pressure. No manufacturer's coil data is available. Assume a 10 °F chilled water temperature rise through the coil.

Solution

Calculate the total airflow rate based on the sensible cooling load:

$$CFM = \frac{30\,800}{1.10(75 - 55)} = 1400$$

Calculate the SHR based on the given sensible and latent cooling load:

$$SHR = \frac{30\,800}{30\,800 + 8500} = 0.80$$

Calculate air temperature rise due to the draw-through fan:

$$Q_{loss} = CFM \times \Delta P/eff = \rho \times CFM \times C_p \times \Delta t_{fan}$$

$$\Delta t_{fan} = \frac{0.363 \times 4}{0.7} = 1.95\,^{\circ}F$$

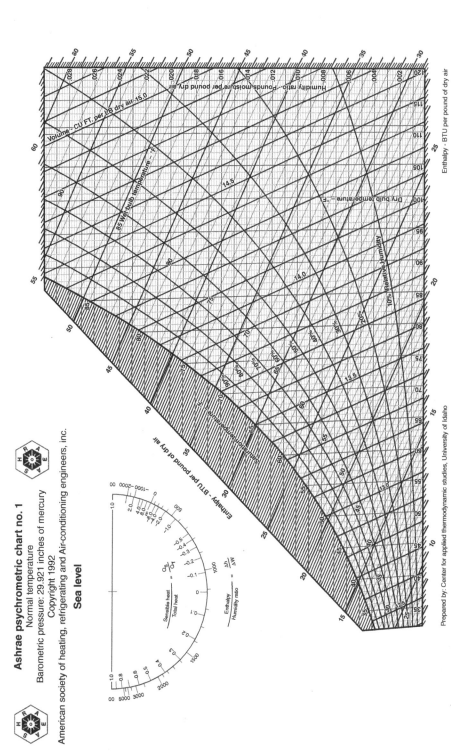

Figure 12.33 The complete psychrometric chart at the sea level. *Source:* ASHRAE (1992).

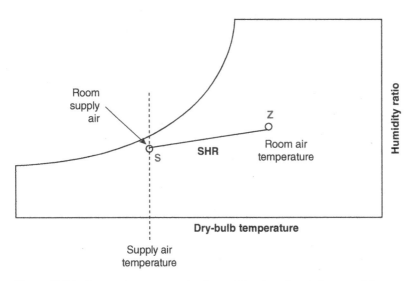

Figure 12.34 Locate the room supply air condition based on indoor conditions.

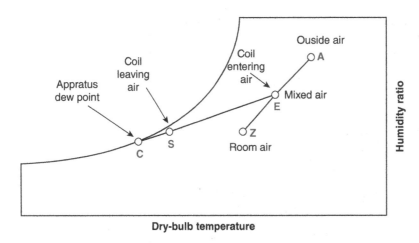

Figure 12.35 Locate the apparatus dewpoint temperature.

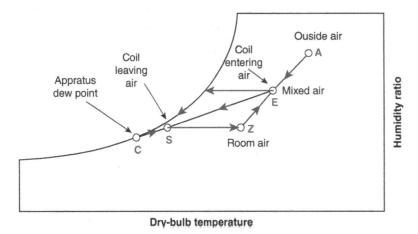

Figure 12.36 The complete cooling cycle.

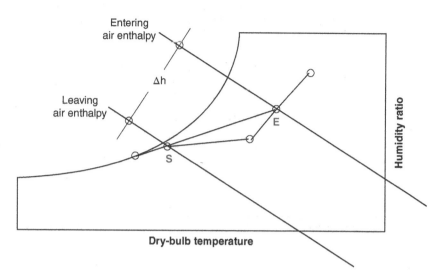

Figure 12.37 Determine the coil capacity by measuring the enthalpy difference of air at E and S.

Plot space conditions of 75 °F DB and 50% RH on the psychrometric chart, and find the intersection of the SHR line with the space temperature and supply air temperature:

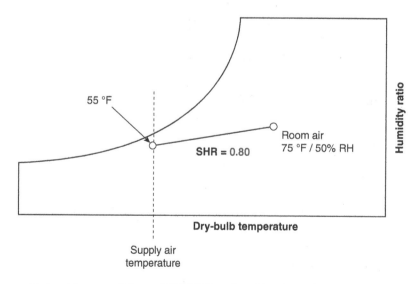

Plot outdoor conditions of 85 °F DB and 70 °F WB on the psychrometric chart, and find the mixed air condition at 33% outdoor air (OA). Account for the fan caused temperature rise Δt_{fan}:

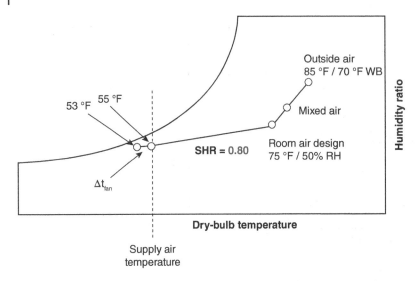

Determine the enthalpy for coil entering and leaving conditions:

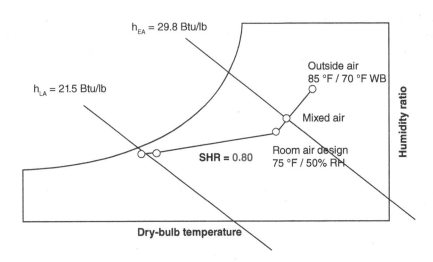

$$\dot{Q}_{coil} = \dot{m}_s(h_{EA} - h_{LA}) = 4.5 \times CFM \times (h_{EA} - h_{LA})$$

$$\dot{Q}_{coil} = 4.5 \times 1400 \times (29.8 - 21.5) = 52{,}290 \, Btu/h$$

Since 12,000 Btu/h=1 ton, 52,290 Btu/h=4.36 ton (round to 5 ton).
Determine the chilled water flow rate:

$$\dot{Q}_{coil} = 52{,}290 \frac{Btu}{h} = \dot{m}_w \times C_{p,w} \times \Delta T_w = 500 \times GPM \times \Delta T_w$$

$$GPM = \frac{52,290}{500 \times 10} = 10.5$$

Check manufacturers' brochures, to find a product with parameters of 4.36 ton and 10.5 GPM or the immediate above (e.g. 5 ton and 15 GPM).

Homework Problems

1 On a winter day at sea level, ventilation air for a building must be heated from the outdoor conditions of 40 °F and 70% relative humidity. If the outdoor air is heated to 70 °F, please calculate the resulting relative humidity. How much energy is needed to heat one pound of dry air? Verify the result with the chart reading.

2 On a winter day at sea level, the conditions of a room in a residence are measured to be 70 °F dry bulb and 40 °F dewpoint. The square room measures 15 ft on each side and 8 ft high. If the occupants require that the indoor relative humidity be 40% for comfort reasons, how many pounds of water vapor must be added to the room air to increase its relative humidity to 40% while maintaining the temperature at 70 °F?

3 In steady-state operation, a room air conditioner at the sea level takes in air at 25 °C and 60% relative humidity. The mass flow rate of dry air is 0.07 kg/s. The air passes over several rows of cold tubes. The air is cooled to 10 °C and returned to the room; there is not any reheat added to the air after it passes over the cold tubes. Moisture from the air condenses as liquid on the outside surface of the tubes. On the enclosed sea-level psychrometric chart, **locate** the states of the inlet and outlet air from the air conditioner. (You can use the psychrometric chart to complete the following questions.) (*air density: 1.2 kg/m³; specific heat: 1005 J/kg*C*).

(a) How much water vapor (in grams) was condensed from the air per second?

(b) What is sensible cooling capacity (in Watts) of the air conditioner?

Depending on the room conditions, the refrigerant compressor shuts off for a period of time. In that time period, there is not any cooling refrigerant within the tubes. In most air conditioner designs, the fan circulating the air over the tubes continues to run. The tubes have liquid water on their surface, and there is some re-evaporation of liquid water into the flowing air. Assume the air entering the air conditioner is still at 25 °C and 60% relative humidity.

(c) On the psychrometric chart shows the trace of possible states for the air after it passes over the wet tubes. Will the outlet air temperature increase or decrease (compared to the inlet at 25 °C)? Will the outlet air humidity ratio increase or decrease (compared to the inlet)? (Hint: this is an evaporation process.)

(d) Using part c, what is the maximum possible rate (in $\mathrm{gram_{vapor}}$/second) of water evaporation from the surface of the tubes to the air if the airflow rate is still 0.07 kg/s?

Room air conditioner

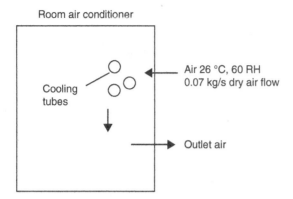

4 The figure shows a typical arrangement for mixing ventilation air with return air from the occupied space in a commercial building air handler. At sea level, the outdoor air at 95 °F and 40% RH mixes with return air at 80 °F and 50% RH in a mass flow ratio of 1 : 4 (20% outdoor air). Please calculate (you can use the psychrometric chart):

(a) What are the unit enthalpy, humidity ratio, temperature, and relative humidity of the mixed air upsteam of the filter?

(b) If the fan airflow rate is 10,000 lb/hr, how much net energy must be removed from the mixed air to deliver supply air to the building at 55 °F and 90% RH? What fraction of this load is due to sensible cooling rather than latent?

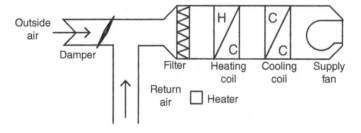

5 In summer, a single zone building in New York is cooled and dehumidified by the system shown below. The sensible load in the zone is 120 000 Btu/hr. Supply air is delivered to the zone at 53 °F and 80% RH (A desiccant dehumidifier – iso-enthalpy change – is used to decrease the supply air relative humidity from 100 to 80%). The indoor conditions are desired to be maintained at 75 °F and 50% RH. [air density: 0.075 lb/ft³; specific heat: 0.24 Btu/lb*F] (you can use the psychrometric chart):

(a) What are the humidity ratio and enthalpy of the supply air?

(b) What is the required supply air mass flow rate to meet the zone load?

(c) What is the latent heat removed by the system? What is the SHR?

(d) Plot the entire air handling process on attached psychrometric chart if the outdoor air is 95°F and 50% RH and the total supply air uses 25% return air to mix with outdoor air to save energy.

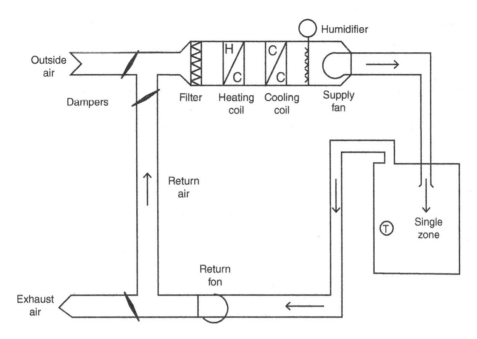

6 Draw the HVAC processes on the psychrometric chart and present all the air properties (i.e., T_{dry}, RH, W, and i) at critical steps for a CAV (constant air volume) system at design conditions for a commercial office building at sea level. The design ambient conditions are 92 °F dry bulb and 78 °F wet bulb. The building can be treated as a single zone with a total zone load of 500 000 Btu/hr and a sensible heat ratio of 0.8. The design occupancy is 80 people and the corresponding ventilation air flow rate is 1600 cfm, which is based on 20 cfm of outdoor air per person. The zone temperature thermostat setting is 77 °F, which is within the summer comfort zone. *The bypass factor for the coil is 0*, and the chilled water temperature is 50 °F (the lowest temperature air can achieve in the coil). Determine: (1) total supply volume flow rate; (2) recirculated volume flow rate; (2) coil load.

7 Draw the HVAC processes on the psychrometric chart and present all the air properties (i.e., T_{dry}, RH, W, and i) at critical steps for a CAV (constant air volume) system at design conditions for a commercial office building at sea level. The design ambient conditions are 92 °F dry bulb and 78 °F wet bulb. The building can be treated as a single zone with a total zone load of 500 000 Btu/hr and a sensible heat ratio of 0.8. The zone temperature thermostat setting is 77 °F and RH = 50%, which is within the summer comfort zone. The chilled water temperature is 50 °F and the supply air temperature is 55 °F. Determine: (1) the bypass factor for the coil; (2) ventilation air flow rate; (3) coil load.

8 Determine the coil load, reheat energy, and indoor RH for the CAV system of Problem 1, but operating at part-load with ambient conditions of 86 °F dry bulb and 70 °F wet bulb. The total load is now 350 000 Btu/hr with the same sensible heat ratio of 0.8, and it keeps the same zone temperature of 77 °F, the same total circulation flow rate and ventilation flow rate. The chilled water temperature is 50 °F, which is the same as design conditions. Reheat is needed for CAV system to meet varying internal loads with dynamic ambient conditions.

9 Staged Project Assignment (Teamwork)

Determine HVAC systems: Research and review prevalent **mechanical systems** used in a small commercial building of this size and climate:

(a) Compare pros and cons of them (at least three types).

(b) Conduct feasibility check in terms of (1) weather condition; (2) utility accessibility; (3) space availability; (4) soil condition; (5) regulations, etc.

(c) Obtain specifications (specs) of the desired systems/products from manufacturers.

(d) Size the system using your calculated heating and cooling loads.

Reference

ASHRAE, (1992). *Handbook: Heating, Ventilating, and Air-Conditioning Systems and Equipment*. AHRAE Transactions ISBN-13: 978-0910110860.

13

Building Energy Consumption

13.1 Manual Calculation

Designing an energy efficient building requires an estimate of the energy consumption in buildings. A building energy calculation is used to predict the total energy consumption (in kWh or Btu) of a building over a period of time (e.g. a day, month, quarter, season, or year). It considers the actual building operations (e.g. occupant activities and schedules) and actual/typical dynamic weather conditions. Building energy calculation methods can generally be divided into two categories: manual calculation methods and computer simulation methods. Manual calculation methods, such as the degree-day (DD) and bin methods (ASHRAE 2021), are widely used in practical design due to their simplicity and efficiency, although they are not precise.

13.1.1 The Degree-Day Method

The DD method is a simple and conventional approach to estimating the annual building energy consumption, widely used at the earlier stage of a building design, retrofit, or performance evaluation. The heating and cooling DDs are defined as below:

$$DD_h = \sum_{year}(T_{bal} - T_o)/24 \qquad \text{for heating} \tag{13.1}$$

$$DD_c = \sum_{year}(T_o - T_{bal})/24 \qquad \text{for cooling} \tag{13.2}$$

where DD_h and DD_c are, respectively, the heating and cooling DDs (unit: $°C \cdot day$ or $°F \cdot day$). T_o is the hourly outdoor air (dry-air) temperature (°C or °F) and T_{bal} is the balance point temperature (°C or °F). The balance point temperature is the outdoor air (dry-air) temperature at given thermal conditions that would be in equilibrium with the desired indoor air temperature without auxiliary heating/cooling. T_{bal} can be projected using the following equation:

$$Q_{load} = Q_{infiltration} + Q_{conduction} - Q_{people} - Q_{light} - Q_{equipment} = 0 \tag{13.3}$$

This implies that the internal heat gains (from people, lighting, and equipment) will be released to the outdoors through the envelope and infiltration, and therefore a constant/comfortable indoor air

Energy Efficient Buildings: Fundamentals of Building Science and Thermal Systems, First Edition. Zhiqiang (John) Zhai.
© 2023 John Wiley & Sons, Inc. Published 2023 by John Wiley & Sons, Inc.

temperature will be maintained (without using additional heating and cooling). For a given T_{indoor}, T_{bal} can be identified for a specific building. When $T_o > T_{bal}$, there is no heating required for the building; When $T_o < T_{bal}$, auxiliary heating is necessary to maintain the required T_{indoor}. The lower the T_o, the more heating needed, while the lower the T_{bal}, the less heating needed. A similar concept is used for cooling, but solar radiation must be included in determining the T_{bal}.

Generally, T_{bal} varies with buildings. Each building has a different T_{bal} because of the different internal heat gains (e.g. people, lighting, and equipment) and different building envelope performance (including infiltration behaviors). The balance point temperature may also be different for heating and cooling. Therefore, for a specific building, T_{bal} for both heating and cooling needs to be determined first, upon which the heating and cooling DDs for this building can be calculated. For a generic building, $T_{bal} = 65\,°F\,(18.3\,°C)$ is commonly used to pre-calculate the location-specific but building-independent heating and cooling DDs. These general heating and cooling degree days (DD_h and DD_c) for a specific location (city) can be applied for a rough estimate of building heating and cooling energy consumption at earlier design/study stages.

Example 13.1

Determine the balance point temperature of a classroom in Denver, CO. Assume the internal heat gains in the occupied hour to be 235 W. The heat loss through envelope at design conditions ($T_i = 20\,°C$ and $T_o = -14\,°C$) is $Q_{loss} = 4000\,W$.

Solution

The building load coefficient BLC $= Q_{loss}/(T_i - T_o) = 4000/(20 + 14) = 117.6\,W/°C$

Assume BLC is constant at different indoor and outdoor conditions:

$$Q = BLC \times (T_i - T_{bal}) = Q_{gain} = 235\,W$$

$$T_{bal} = T_i - Q_{gain}/BLC = 20 - 235/117.6 = 18°C$$

When $T_{out} = T_{bal} = 18\,°C$, the classroom needs neither heating nor cooling.

Once the balance point temperature is determined, the hourly outdoor air temperature from a typical meteorological year (TMY) weather file for the location of interest can be used to calculate the heating and cooling degree hours. The heating and cooling degree hours are accumulated by summarizing the difference of T_{out} and T_{bal} at each hour throughout the entire year, as demonstrated in Table 13.1 for 1 January (where $T_{bal} = 65\,°F$ for both heating and cooling).

The total degree hours for 1 January are thus the summation of the column of (T_{bal}-T_o), which is 582 °F · h. The calculation is repeated for all of the remaining days in the year. Only the hours when $T_o < T_{bal}$ will be accumulated for calculating the heating degree hours. The hours when $T_o > T_{bal}$ will be accumulated to calculate the cooling degree hours. The resultant heating and cooling degree hours throughout the year are often large numbers, and thus are converted into DDs by dividing the large numbers by 24 hours/day. DDs are mostly for convenience of expression, and degree hours are the actual numbers that are needed in the energy calculation.

The annual energy consumption can thus be computed as:

$$Q_{h,year} = BLC \times DD_h \times 24/\eta_h \tag{13.4}$$

Table 13.1 Degree hours for 1 January.

Time	T_o (°F)	T_{bal}-T_o (°F·h)	Time	T_o (°F)	T_{bal} – T_o (°F·h)
1:00	38	27	13:00	55	10
2:00	36	29	14:00	53	12
3:00	33	32	15:00	52	13
4:00	29	36	16:00	50	15
5:00	28	37	17:00	48	17
6:00	28	37	18:00	46	19
7:00	29	36	19:00	44	21
8:00	33	32	20:00	43	22
9:00	36	29	21:00	42	23
10:00	40	25	22:00	41	24
11:00	45	20	23:00	40	25
12:00	50	15	24:00	39	26

$$Q_{c,year} = BLC \times DD_c \times 24/\eta_c \tag{13.5}$$

where η is the efficiency of boiler or chiller. Equations (13.4) and (13.5) summarize every-hour energy use for heating and cooling:

$$Q_{h,year} = BLC \times \sum_{year}(T_{bal} - T_o)/\eta_h = \sum_{year}[BLC \times (T_{bal} - T_o)/\eta_h] \tag{13.6}$$

$$Q_{c,year} = BLC \times \sum_{year}(T_o - T_{bal})/\eta_c = \sum_{year}[BLC \times (T_o - T_{bal})/\eta_h] \tag{13.7}$$

where BLC and η are assumed as constant throughout the year.

Example 13.2

Continuing on Example 13.1, estimate the annual heating energy cost of the classroom. Suppose that the classroom is heated by a heat pump with a coefficient of performance (COP) value of 2.5.

Solution

Check the ASHRAE systems handbook: $DD_h = 6283$ °C · day for Denver, CO.

$$Q_{h,year} = BLC \times DD_h \times 24/COP = 117.6 \times 6283 \times 24/2.5 = 7093.3\ kWh$$

13.1.2   The Bin Method

In actual buildings, both BLC and η may vary with the indoor and outdoor environmental conditions. Equations (13.8) and (13.9) should be used for calculating energy use at every single hour:

$$Q_{h,year} = \sum_{year}[BLC \times (T_{bal} - T_o)/\eta_h] \tag{13.8}$$

$$Q_{c,year} = \sum_{year}[BLC \times (T_o - T_{bal})/\eta_h] \tag{13.9}$$

This would require an 8760-hour calculation. Since BLC and η may not differ significantly within a certain range of operational conditions, the outdoor air temperature can be grouped into a set of bins (usually in the temperature intervals of 5 °F or 2.8 °C) in which certain BLC and η values are specified. The energy use for a particular bin of outdoor air temperature can then be calculated as:

$$Q_{bin,h} = N_{bin} \times BLC \times (T_{bal} - T_o)/\eta_h \tag{13.10}$$

where N_{bin} is the number of hours in the temperature interval (bin) during the year and T_o is the center value of the outdoor air temperature bin. BLC and η_h are specific for that bin.

13.2   Computer Simulation

13.2.1   Introduction

DD methods are the simplest methods for energy estimation and are appropriate if the building occupying and operating conditions are constant. If the conditions of the building and systems vary with outdoor temperature, the building energy consumption needs to be calculated for different values of the outdoor temperature and multiplied by the corresponding number of hours; this is the basic idea of various bin methods. More sophisticated models must be used when the situation becomes more complicated, such as varying indoor air temperature and interior heat gains. The manual methods provide a simple estimate of building annual loads, but they cannot, for example, be used for:

- evaluation of air-conditioning plant.
- evaluation of most control issues.
- medium or heavy weight buildings with significant diurnal fluctuations in internal temperature.

As more powerful computers have become available, computer modeling has been more and more important for the prediction of the energy and environmental performance of buildings and systems that serve them. Computer simulation is credited with speeding up the design process, enabling the comparison of a broader range of design variants, and leading to optimal designs. With reasonable physical assumptions and mathematical models, computer simulations provide more accurate and informative results than manual calculations. As a result, computer simulations provide a better understanding of the consequences of design decisions. The underlying mathematical models and numerical schemes of simulation tools distinguish them from each other, satisfying the different requirements of complexity and accuracy.

The development of computer energy simulation (ES) programs can be traced back to the 1960s and 1970s, when the groundwork of ES methods was laid (e.g. GATC 1967). After Mitalas and Stephenson (1967) published their milestone work on the response factor method to model the transient heat transfer through building envelopes and the heat transfer between internal surfaces and room air, ASHRAE published procedures for determining heating and cooling loads. The load calculation can then be used to size the system and compute the total energy cost.

Most ES programs adopt the Load, System and Plant (LSP) modeling strategy (Sowell and Hittle 1995), which subdivides the building ES into three sequential steps. The building's heating and cooling loads are first calculated for the entire analysis period (often a year) for an assumed set of indoor environmental conditions. These loads are then imposed as inputs to the second step of the simulation, which models the air handling and energy distribution systems (fans, heating coils, cooling coils, air diffusers, etc.). This second simulation step (also conducted for the entire analysis period) predicts the demands placed on the plant's energy conversion systems (boilers, chillers) and related equipment (cooling towers and circulation pumps). The third step is to calculate the source energy requirements in the central plant. Finally, one would estimate the costs of the source energy, sometimes introducing capital and other investment costs for a complete life cycle economic analysis.

The accuracy of the load calculation forms the base of the next two steps. The weighting factor method and heat balance method (ASHRAE 2021) are the two principal methods used for building load calculation in the past few decades. It is well known that heat gain is not the same as cooling load for a building. For example, the lighting energy in a room does not convert to 100% convective heat immediately. In fact, part of the heat is radiated and then will be absorbed by the building enclosures and furniture. This part of radiative heat may be released back to the room air at a later time because of the room's thermal capacity.

The weighting factor method estimates the ratio of convective heat to the total energy release in a time sequence. The weighting factors heavily depend on building material properties and may be precalculated and presented in tables for certain types of buildings. These tables can be directly used by an ES program, or even manual calculation, for the load estimate if the actual building is close to the one used to produce the weighting factors. The weighting factor method was popular in the 1970s because of the limited computing capacity at that time. Earlier building ES programs using weighting factors are the Post Office Program (GATC 1967), NESCAP (NASA 1975) and DOE-1 (Diamond et al. 1977).

The heat balance method was introduced in the 1970s (e.g. Kusuda 1976) to enable a more rigorous treatment of building loads. Rather than using pre-calculated weighting factors to characterize the thermal response of the room air to outdoor air temperature changes, solar radiation, and internal gains, this approach solves heat balances for the room air and enclosure surfaces to determine the loads. The enclosure surface temperatures calculated can be used to determine the radiant temperature. The heat balance method eliminates some significant linearity assumptions and allows building dynamic conditions to be modeled appropriately. For example, the convection coefficients that characterize heat transfer from interior surfaces to the room air could respond to thermal states within the room, rather than being treated as constant. NBSLD (Kusuda 1976) is probably the earliest program of this type. Other current programs that use the heat balance method include popular ones, such as BLAST (Hittle 1979), ESP-r (Clarke 1985) and EnergyPlus (Crawley et al. 2000).

Most weighting factor and heat balance programs use response factors (Mitalas and Stephenson 1967) and transfer functions (Stephenson and Mitalas 1971) to calculate transient conduction through walls, roofs, and floors with the assumption that the heat conduction is one dimensional. The response factors or transfer functions are based on control theory. The mathematical background is rather complicated. However, they determine heat conduction much faster than the finite-difference method. The finite-difference method does not have to assume one-dimensional heat conduction. It would yield much more accurate results for corner walls and would provide the temperature distribution in a wall that is useful for analyzing condensation (Chen et al. 1995). However, the computing time of the finite-difference method is still considerable. Hence, most current ES programs still use the response factor and transfer function methods with the fairly reasonable one-dimensional assumption.

Building ES has encountered incredible development since the 1970s. Recent years have especially witnessed the proliferation of building ES software for a broadening range of building performance assessment. Besides the continuous improvement on the well-noticed energy software such as BLAST, DOE-2, TRNSYS, ASHRAE Loads Toolkit, ESP-r, and CODYBA (Noel et al. 2001), many new energy programs are developed for research, education and design purposes, such as ColSim (Wittwer 1999), SIMEDIF(Larsen and Lesino 2001), and DOMUS (Mendes et al. 2001). To date, the ASHRAE bibliography of computer simulations of building has listed more than 200 programs.

Many of these building ES programs are reaching maturity, using simulation methods and even codes that originated in the 1960s. BLAST and DOE-2 are two of the most popular ES programs and are widely used in building design practices around the world. DOE-2 (Winkelmann et al. 1993), sponsored by the US Department of Energy (DOE), was developed from the Post Office program written in the late 1960s for the US Post Office. BLAST (Building System Laboratory 1999), sponsored by the US Department of Defense (DOD), has its origins in the NBSLD program developed at the US National Bureau of Standards (now NIST) in the early 1970s. The main difference between the two programs is the load calculation method – DOE-2 uses a room weighting factor approach while BLAST uses a heat balance approach. These two programs each have pros and cons and have wide utilities in various environments, but both have begun to show their ages in a variety of ways (Crawley et al. 2001). The simulation methodologies in both programs are often difficult to trace due to the decades of development (and multiple authors). To maintain, support, and enhance either program, a developer must have many years of experience working with the codes and knowledge of code unrelated to the task (due to a significant amount of "spaghetti" code). Without substantial redesign and recoding, expanding their capabilities has become difficult, time-consuming, and expensive (Crawley et al. 2001). As a result, DOE eventually decided to start developing a new ES program named EnergyPlus (E+) in 1996. EnergyPlus, initially developed by the Lawrence Berkeley National Laboratory (LBNL) and now mostly by the National Renewable Energy Laboratory (NREL), is an all-new heat-balance-based program with best efforts to combine the most popular features and capabilities of BLAST and DOE-2. Compared to the legacy programs, the highlights of EnergyPlus are:

1) simulation management structure that eliminates the interconnections between various program sections. As a result, it eliminates the need to understand all parts of the code in order to make an addition to a very limited part of the program.
2) modularity that allows other developers to quickly add other component simulation modules with only a limited knowledge of the entire program structure.
3) integration of loads, systems, and plants that overcome the most serious deficiency of DOE-2 and BLAST: an inaccurate space temperature prediction due to a lack of feedback from the HVAC module. The integration solution also allows users to evaluate a number of processes that neither DOE-2 nor BLAST can simulate well, such as realistic system control and interzone airflow.

Table 13.2 Comparison of major features and capabilities of three ES programs.

Features and capabilities	DOE-2	BLAST	EnergyPlus
Integrated loads/systems/plant solution	×	×	√
Heat balance calculation	×	√	√
Multiple time step for interaction between environment, zones, and systems	×	×	√
Moisture absorption/desorption in building envelope	×	×	√
Interior surface convection dependent on temperature and airflow	×	√	√
Anisotropic sky model	√	×	√
Advanced fenestration calculations	√	×	√
Daylighting illumination and controls	√	×	√
Thermal comfort model	×	√	√
User-configurable HVAC systems	×	×	√
Air and fluid loop in HVAC systems	×	×	√
Links to SPARK, TRNSYS	×	×	√

Source: Modified from Crawley et al. (2001).

Table 13.2 compares the major features and capabilities of EnergyPlus with those of BLAST and DOE-2. It is obvious that EnergyPlus is superior to its ancestors. This chapter, therefore, will introduce the fundamentals of the EnergyPlus program.

13.2.2 Fundamentals of EnergyPlus (E+)

13.2.2.1 General Descriptions of EnergyPlus

EnergyPlus (Crawley et al. 2000) is a new-generation building energy analysis and thermal load simulation program, with roots in both the BLAST (Hittle 1979) and DOE-2 (Winkelmann et al. 1993) programs. Based on a user's description of a building from the perspective of the building's physical makeup, associated mechanical systems, etc., EnergyPlus can calculate the heating and cooling loads necessary to maintain thermal control set points, the conditions throughout a secondary HVAC system and coil loads, and the energy consumption of primary plant equipment. The program has inherited many simulation characteristics from the legacy programs of BLAST and DOE-2, as well as created many features to overcome the shortcomings of its parent programs. In particular, the special simulation management philosophy and modular nature of the program eliminate the interconnections between various program sections and the need to understand all parts of the code in order to make an addition to a very limited part of the program. As a consequence, the program allows the easy expansion of functions and linkages to other programs, as illustrated in Figure 13.1.

Figure 13.1 shows the overall structure of the EnergyPlus program. It has three basic components – a surface heat balance manager, an air heat balance manager, and a building systems simulation manager – all under the control of the integrated solution manager. The surface and air heat balance managers calculate the outside and inside surface and room air heat balance and act as an interface between the heat

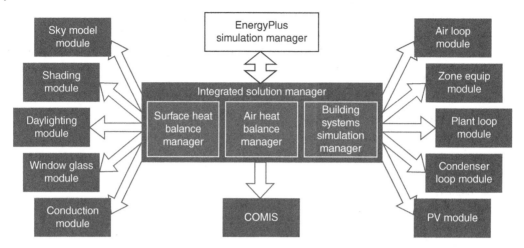

Figure 13.1 Overall EnergyPlus structure. *Source:* LBNL (2001).

balance and the building systems simulation manager. The building systems simulation manager handles communication between the heat balance managers and various HVAC modules and loops, such as coils, boilers, chillers, pumps, fans, and other equipment/components. With this program structure, more independent model modules, such as a daylighting module, can be easily plugged into the individual manager component without cross-influencing the others. Therefore, it can maximize the number of developers who can quickly integrate their work into EnergyPlus for the minimum investment of resources. This feature of EnergyPlus particularly highlights its feasibility for the present coupling study.

13.2.2.2 Heat Balance Method of EnergyPlus

As shown in Figure 13.1, EnergyPlus is a heat-balance-method-based building ES program. Chapter 11 introduces the fundamentals of the heat balance method, which provides more flexibility and generality with less physical and numerical assumptions compared to the conventional weighting factor method.

The heat balance equations for room air and surface heat transfer are two essential equations solved by EnergyPlus to determine the building heating/cooling load.

The heat balance equation for room air is

$$\sum_{i=1}^{N} q_{i,c} A_i + Q_{lights} + Q_{people} + Q_{appliances} + Q_{infiltration} - Q_{heat\_extraction} = \frac{\rho V_{room} C_p \Delta T}{\Delta t} \tag{13.11}$$

where

$\sum_{i=1}^{N} q_{i,c} A_i$ = convective heat transfer from enclosure surfaces to room air

$q_{i,c}$ = convective flux from surface i

i, N = index and number of enclosure surfaces

A_i = area of surface i

Q_{lights} = heat gains from lights

Q_{people} = heat gains from people
$Q_{appliances}$ = heat gains from appliances
$Q_{infiltration}$ = heat gains from infiltration
$Q_{heat\_extraction}$ = heat extraction rate of the room
$\dfrac{\rho V_{room} C_p \Delta T}{\Delta t}$ = internal energy change rate of room air
ρ = air density
V_{room} = room volume
C_p = specific heat of air
ΔT = temperature change of room air
Δt = sampling time interval, normally one hour

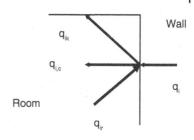

Figure 13.2 Energy balance on the interior surface of a wall, ceiling, floor, roof, or slab.

The heat extraction rate is the same as the cooling/heating load when the room air temperature is maintained as constant ($\Delta T = 0$). The convective heat flux from a wall is determined from the heat balance equation for the wall surface, as shown in Figure 13.2. A similar heat balance can be obtained for each window. The heat balance equation for a surface (wall/window) can be written as:

$$q_i + q_{ir} = \sum_{k=1}^{N} q_{ik} + q_{i,c} \tag{13.12}$$

where

q_i = conductive heat flux on surface i
q_{ir} = radiative heat flux from internal heat sources and solar radiation
q_{ik} = radiative heat flux from surface i to surface k

EnergyPlus uses the conduction transfer functions (CTF) method (Meyers 1980; Seem 1987) to compute surface conductive heat fluxes q_i through the building enclosures. The radiative heat flux is determined by

$$q_{ik} = h_{ik,r}(T_i - T_k) \tag{13.13}$$

where

$h_{ik,r}$ = linearized radiative heat transfer coefficient between surfaces i and k
T_i = temperature of interior surface i
T_k = temperature of interior surface k

And the convective heat flux is

$$q_{i,c} = h_c(T_i - T_{room}) \tag{13.14}$$

where

h_c = convective heat transfer coefficient
T_{room} = room air temperature

The convective heat transfer coefficient, h_c, is unknown, usually determined by empirical correlations in ES programs. EnergyPlus allows one to choose one of the following models (LBNL 2001) to calculate the convective heat transfer coefficients (unit: $W/m^2 \cdot K$):

13.2.2.2.1 Detailed Natural Convection Model

Based on flat plate experiments, the detailed natural convection model correlates the convective heat transfer coefficient to the surface orientation and the temperature difference between the surface and zone air (where ΔT = Surface temperature - Air temperature).

- If $\Delta T = 0.0$ or a vertical surface, then

$$h = 1.31|\Delta T|^{1/3} \tag{13.15}$$

- If $\Delta T < 0.0$ with an upward facing surface or $\Delta T > 0.0$ with a downward facing surface (enhanced convection), then

$$h = \frac{9.482|\Delta T|^{1/3}}{7.283 - |\cos \sum|} \tag{13.16}$$

where Σ is the surface tilt angle.

- If $\Delta T > 0.0$ with an upward facing surface or $\Delta T < 0.0$ with a downward facing surface (reduced convection), then

$$h = \frac{1.810|\Delta T|^{1/3}}{1.382 + |\cos \sum|} \tag{13.17}$$

where Σ is the surface tilt angle.

13.2.2.2.2 Simple Natural Convection Model

The simple convection model uses constant coefficients (all in SI units) for each of three heat transfer configurations as follows.

- For a horizontal surface with reduced convection,

$$h = 0.948 \tag{13.18}$$

- For a horizontal surface with enhanced convection,

$$h = 4.040 \tag{13.19}$$

- For a vertical surface,

$$h = 3.076 \tag{13.20}$$

- For a tilted surface with reduced convection,

$$h = 2.281 \tag{13.21}$$

- For a tilted surface with enhanced convection,

$$h = 3.870 \tag{13.22}$$

13.2.2.2.3 Ceiling Diffuser Model

The ceiling diffuser model correlates the convective heat transfer coefficient to the supply mass flow rate (ACH).

- For floors,

$$h = 3.873 + 0.082 \times ACH^{0.98} \tag{13.23}$$

- For ceilings,

$$h = 2.234 + 4.099 \times ACH^{0.503} \tag{13.24}$$

- For walls,

$$h = 1.208 + 1.012 \times ACH^{0.604} \tag{13.25}$$

After obtaining the expressions of the convective, radiative, and conductive heat fluxes on each envelope surface, the interior surface temperatures, T_i, can be determined by simultaneously solving Eq. (13.12), if the room air temperature, T_{room}, is assumed to be uniform and known. Space cooling/heating loads can then be determined from Eq. (13.11) with the calculated convection heat from the enclosures. Thereafter, the coil load is determined from the heat extraction rate and the corresponding air handling processes and HVAC equipment selected. With a plant model and hour-by-hour calculation of the coil load, the energy consumption of the whole HVAC system for a building can be determined.

On the other hand, if there is no HVAC system running in the space, iteratively solving Eqs. (13.11) and (13.12) then predicts the change of uniform indoor air temperature and enclosure surface temperatures during a day, season, or year of concern. It is obvious, in both air-conditioned and non-air-conditioned scenarios, that the interior convective heat transfer from the building enclosures to the indoor air is the explicit linkage between the room air and surface heat balance equations. Its accuracy will directly affect the energy calculated.

Figure 13.3 demonstrates the executive streamline of the EnergyPlus program. Given the building materials and geometry information, the ES program first calculates the conduction transfer factors. Under the weather condition at each time step, the ES program then simulates the heat balance of building envelope surfaces and room air to obtain the building heating/cooling load and thermal behaviors of the building. Based on this information, the second system and plant models can be operated for the system energy consumption and the total building operating cost.

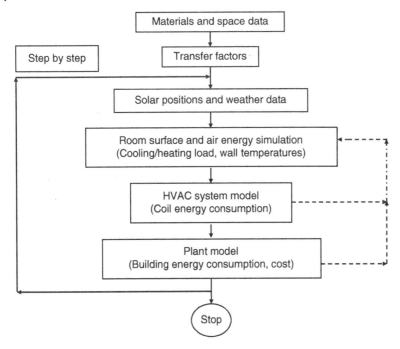

Figure 13.3 Executive streamline of the EnergyPlus program.

However, unlike the conventional ES programs, EnergyPlus allows the system and plant output to directly impact the building thermal response at the current time step, rather than calculating all loads first, then simulating systems and plants. This integrated simulation approach can help to more accurately investigate the effect of under-sizing fans and equipment on the thermal comfort of occupants within the build. By using the integrated solution technique, EnergyPlus can solve the most serious deficiency of the BLAST and DOE-2 sequential simulations: an inaccurate space temperature prediction due to a lack of feedback from the HVAC module to the load calculations. An accurate prediction of space temperatures is crucial to energy efficient system engineering solutions, such as for estimating heating/cooling loads, sizing systems and plants, and achieving occupant comfort and health. Integrated simulation also allows users to evaluate a number of processes that neither BLAST nor DOE-2 can simulate well, including realistic system controls, moisture adsorption and desorption in building elements, radiant heating and cooling systems, and interzone airflow (Crawley et al. 2001).

13.2.3 A Case Study of EnergyPlus (E+)

The Bang & Olufsen (B&O) Headquarters building is located in Struer, Denmark, and is shown in Figure 13.4. The building is a raised three-story structure. Each floor consists primarily of open-plan office space. Stairwells connect each floor to central roof-mounted exhaust fans. The building's northern façade contains the primary natural ventilation inlets. Other inlets on the southern façade are used for night ventilation only. The building's northern façade is entirely glazed and contains the building's

Figure 13.4 Bang & Olufsen headquarters (north & south facades). *Source:* Jimmy Baikovicius/Flickr.

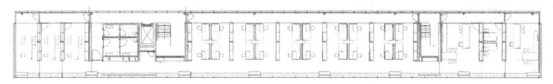

Figure 13.5 Bang & Olufsen headquarters floorplan.

primary natural ventilation inlets. Other inlets on the southern façade are used for night ventilation only. A floorplan of the building is shown in Figure 13.5.

Figure 13.6 illustrates the natural ventilation principle incorporated into the building. During the daytime, if the interior set point (25 °C) is reached, air enters through the building through low-level automatically controlled vents in the northern façade. At night, additional vents open on the southern façade, and the set point is 23 °C. Radiative heaters temper the incoming air if the outdoor air temperature is below 18 °C. The ventilation air exits through roof-level vents located in the top of two centrally located stairwells. Frequency-controlled exhaust fans at the roof-level vents ensure a design ventilation rate of 1.5 ACH during the daytime and 3.0 ACH during night ventilation mode. Other pertinent building information is shown in Table 13.3.

A very comprehensive set of measurements from the first floor of this building was obtained from the AIVC (Air Infiltration and Ventilation Center 2008), including hourly measurements of fan operation, vent operation, air temperatures, internal loads, and locally measured weather data. These measurements were recorded for periods of approximately one week during the winter, fall, and summer seasons.

The following modeling simplifications are required due to EnergyPlus limitations:

- Simulated exhaust fan control scheme
- Manual resize of windows
- Inlet heaters

Figure 13.6 Natural ventilation in the B&O headquarters.

Table 13.3 Bang & Olufsen headquarters information.

Parameter	Value	Units
Floor area	1680	m^2
Floor height	4	m
Loads		
Lighting	10	W/m^2
Equipment	14	W/m^2
Occupancy	27	People/floor
Glazing		
Type (N façade)	Double-pane, low-e, krypton-filled	
Type (S façade)	Double-pane, low-e, argon-filled	
U-value	1.0	W/m^2-K
Area	N = 100%; E,W = 10%; S = 30%	
Thermal mass	Exposed concrete ceiling	
Inlet vent area (gross)	6	Percent
Outlet vent area (gross)	0.75	Percent

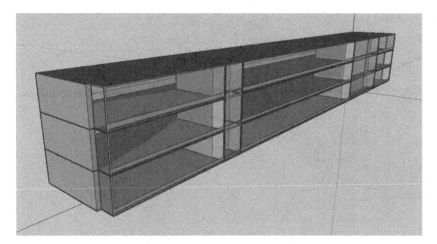

Figure 13.7 Sketchup rendering of B&O headquarters as modeled in EnergyPlus.

The vent control capabilities in EnergyPlus are much simpler than those in use in the building. In order to simulate temperature-controlled exhaust fans, separate fans for daytime and nighttime use were modeled. Custom schedules were developed to ensure that the fans switch on whenever a zone temperature rose above the set point. For this building, the openings must be manually resized to account for effective area. A rendering of the model is shown in Figure 13.7.

Since little information is known about the inlet heaters, adding them to the building model required great simplification. In order to approximate their effect, a very small zone was created at the interface between the inlet vents and the outdoors, and the temperature in this zone was manually scheduled to match the measured inlet temperature. Figure 13.8 shows a rendering of inlet heaters as modeled.

13.2.3.1 EnergyPlus Model Input Uncertainty

Because of the comprehensiveness of the available dataset, input uncertainty for this building is somewhat less than for general buildings. In addition to the inherent uncertainty in each building parameter described by publications on the building, the following areas of significant uncertainty are recognized at this time.

Net effective vent area: Only the gross vent area is available for the building. Due to the use of louvers, bug screens, convectors for heating, etc., the effective vent area is likely much lower. A value of 30% of gross was initially used, though varying this parameter greatly affects results, as will be shown below.

13.2.3.2 EnergyPlus Model Calibration

The available dataset for this building includes measurements taken during several seasons throughout the year. During the fall period, from 1 November 2000 to 12 November 2000, maximum daily outdoor temperatures were in the range of 7–10 °C. During this period, the natural ventilation system was not needed for cooling or for maintaining indoor air quality. Radiative heaters were used sparingly, and the heater power consumption was measured hourly. Modeling the building during this period allows

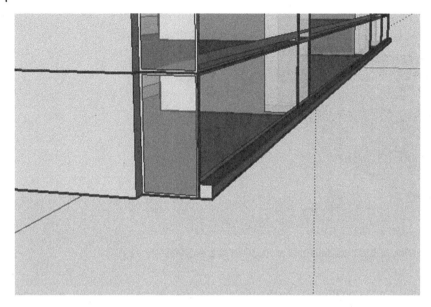

Figure 13.8 Modified EnergyPlus model with inlet heater approximation.

for insight into the accuracy of the EnergyPlus model for all components (e.g. infiltration, solar gains) of the building without the influence of the natural ventilation airflow network.

Figure 13.9 shows the predicted internal temperatures as compared to the measured internal temperatures. As shown, the percent error throughout the calibration period is generally less than 10%. Figure 13.10 shows the two temperature lines plotted against each other to highlight the accuracy of the model.

13.2.3.3 EnergyPlus Model Results

Initial temperature predictions from the EnergyPlus model for the summer period (19 June–26 June) are shown in Figure 13.11. The percent error is higher than 10% in several hours. The predicted air change rate varies with wind conditions but is in the range of 8 ACH when the vents are open. This ventilation rate may be too high, as the fans are designed to enforce 3 ACH exhaust airflow. While additional airflow may be exiting through the inlet vents on the 3rd floor, it is unlikely that the volume of air is so much higher. However, without measured ventilation rate data from the building, no definitive statement can be made about the airflow network model's capability to predict ventilation rate.

Figure 13.12 shows the same period of results in a thermal comfort format, sorted by hours in excess of 24 to 27 °C. This format highlights the significant overprediction of internal temperatures. For hours in excess of 27 °C, only three hours were measured but EnergyPlus predicts 26 hours.

To test the effect of the extra zone added to simulate the inlet heaters, the building was simulated during a period of extremely warm outdoor conditions when the heaters were not used during the day

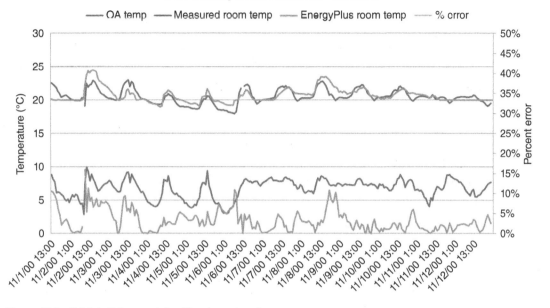

Figure 13.9 *B&O building model calibration – hourly.*

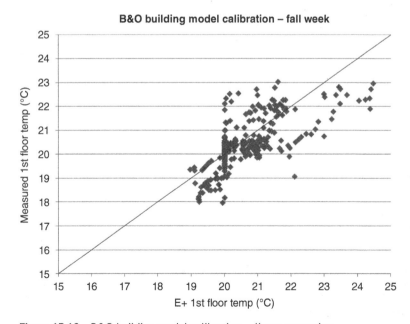

Figure 13.10 *B&O building model calibration – linear regression.*

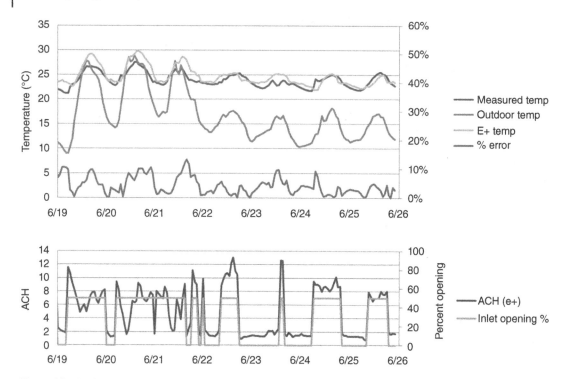

Figure 13.11 B&O building – temperature results.

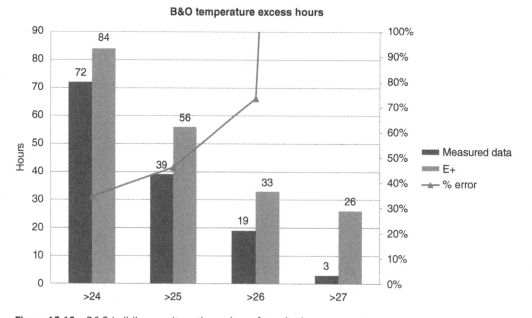

Figure 13.12 B&O building results – thermal comfort criteria.

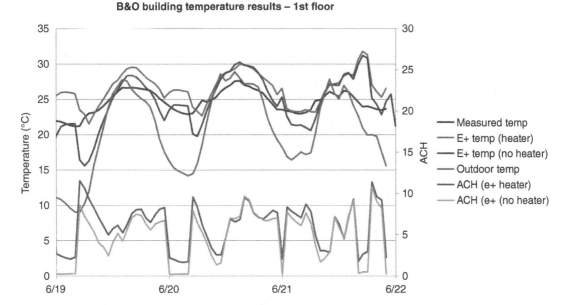

Figure 13.13 B&O building – effect of heater zone on predicted air change rate.

(19 June–21 June). The heaters were used during night cooling. Figure 13.13 shows the results predicted by EnergyPlus both with and without the heater zone. Some differences in predicted air change rates are noted but they are generally insignificant. Without the use of heaters at night, the predicted nighttime temperatures are lower but the daytime maximum temperatures are the same, both with and without the heater zone. Thus, the simulated heater zone's effect on the predicted air change rate is not significant.

The predicted ventilation rate when the vents are open, around 8 ACH, cannot be directly evaluated because of the lack of field data. However, to show the impact of varying ACH on the predicted thermal performance of the building, a sensitivity study was conducted by varying the effective vent area. Results of this study are shown in Figure 13.14. Note that varying the effective vent area does not linearly affect the ACH, in large part because of the temperature-based control scheme in place in the EnergyPlus model. During the afternoon periods of the first three days, when overheating is predicted, increasing the effective vent area has little effect, but decreasing it causes even more overheating.

The impact of level of internal thermal mass was investigated by increasing the amount and density of the thermal mass in the zones. Figure 13.15 shows that increasing the thermal mass from the originally used value of twice the floor area in a wood construction to four times the floor area in concrete has a major effect. The additional thermal mass serves to decrease the peak overheating predictions during the first three days. However, during the later cooler days, the nighttime temperature predictions become inaccurate in the model with the highest mass.

Because the locally recorded weather data included wind speed, wind direction, and solar radiation, in addition to outdoor temperature, a very accurate weather input file was constructed. To test the effect of

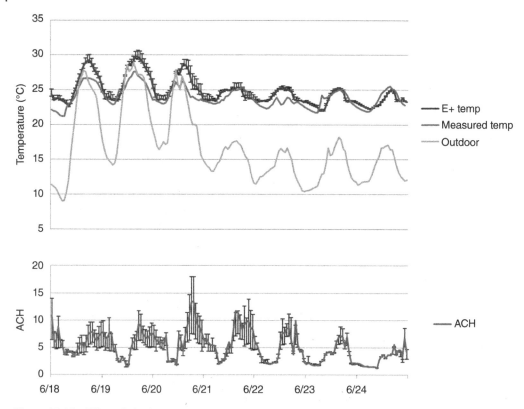

Figure 13.14 Effect of air change rate on temperature predictions – B&O.

B&O building temperature results – 1st floor

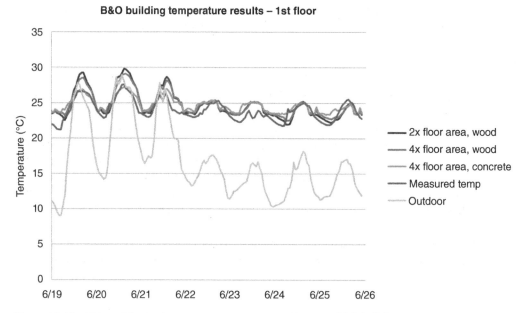

Figure 13.15 Effect of thermal mass on temperature predictions – B&O building.

using weather station data rather than locally measured data, simulations were run using four different weather files:

1) Locally measured wind & solar data
2) Locally measured solar, weather station wind data
3) Locally measured wind, weather station solar data
4) Both solar and wind data from weather station

Figure 13.16 shows the temperature excess hours predicted by EnergyPlus for the four different weather files as compared to the measured data. Having locally measured solar data is more important for this building than locally measured wind data, particularly in the higher temperature categories. This is due to the relatively high contribution of solar gains to the building's total internal loads.

The effect of using the simple heat transfer coefficient model versus the detailed model options in EnergyPlus was also investigated. Figure 13.17 shows the predicted heat transfer coefficient for internal surfaces using both models. Because the simple model predicts higher values for heat transfer coefficient under some conditions, the accompanying predictions for room temperature are lower. Thus, using the detailed heat transfer coefficient model for this building results in even worse overpredictions of internal temperature.

13.2.3.4 Summary

Modeling results for the B&O building show that even with a nearly complete dataset, uncertainties in model input parameters result in hourly error in the range of 10% even without the use of the airflow network. During ventilation periods, error increases significantly, up to 30% during some hours. The error

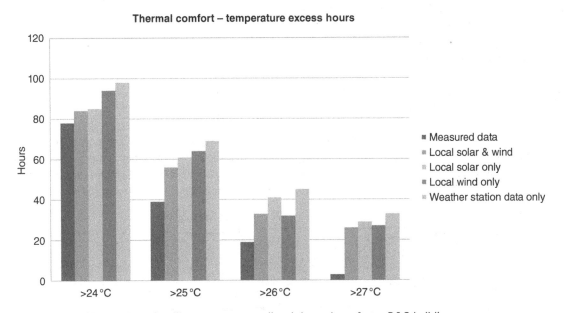

Figure 13.16 Effect of weather file accuracy on predicted thermal comfort – B&O building.

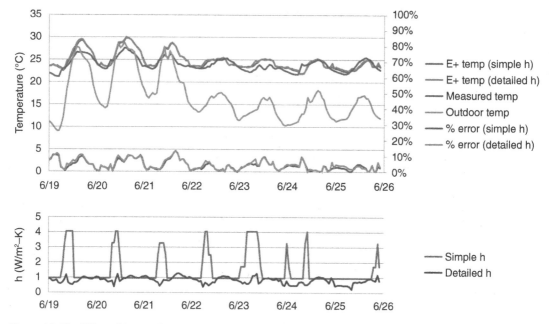

Figure 13.17 Effect of h model on temperature response – B&O building.

is in overprediction of internal temperatures. Without measured airflow data for the building, it is not possible to further evaluate the accuracy of the airflow network model. Studies on the impact of the available heat transfer coefficient models in EnergyPlus show that these parameters are insignificant to the overall EnergyPlus model for this building. Investigation into the effect of weather file accuracy showed that having locally measured solar radiation data is more important than locally measured wind data to improve hourly simulation accuracy.

Homework Problems

1   A classroom is a thermal system and its envelope is the boundary. Consider some energy transfer in a classroom (V = 2000 ft³, T_{room} = 77 °F, $T_{outdoor}$ = 67 °F) as shown below. What is the balance point temperature of the room? If the heating furnace efficiency is 92%, please size the system capacity using Denver conditions. Estimate the annual energy consumption using Denver Heating Degree Day.

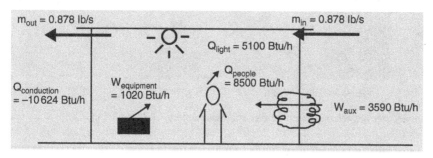

2 **Staged Project Assignment (Teamwork)**

Estimate the annual energy consumption of the design, with justified assumptions, by using both hand/manual calculation and computer simulation tools (e.g., eQUEST).

References

Air Infiltration and Ventilation Center (2008). AIVC Publication CD.

ASHRAE (2021). *ASHRAE Handbook—Fundamentals*. ISBN: ISBN(s): 9781947192898.

Building Systems Laboratory (1999). BLAST 3.0 users manual. Urbana-Champaign, Illinois: Building Systems Laboratory, Department of Mechanical and Industrial Engineering, University of Illinois.

Chen, Q., Peng, X., and van Paassen, A.H.C. (1995). Prediction of room thermal response by CFD technique with conjugate heat transfer and radiation models. *ASHRAE Transactions* 3884: 50–60.

Clarke, J.A. (1985). *Energy Simulation in Building Design*. Bristol: Adam Hilger.

Crawley, D.B., Lawrie, L.K., Pedersen, C.O., and Winkelmann, F.C. (2000). EnergyPlus: energy simulation program. *ASHRAE Journal* 42 (4): 49–56.

Crawley D.B. Lawrie, L., Winkelmann, F.C., Pedersen, C.O. (2001). EenergPlus: new capabilities in a whole-building energy simulation program. *Seventh International IBPSA Conference (BS2001)*, Rio de Janeiro, Brazil (13–15 August 2001). 1, 51–58.

Diamond, S.C., Horak, H.L., Hunn, B.D., et al. (1977). DOE-1 program manual. Report ANL/ENG-77-04, Los Alamos Scientific Laboratory.

GATC (1967). *Computer Program for Analysis of Energy Utilization in Postal Facilities: Volume 1 User's Manual*. Niles, USA: General American Transportation Corporation.

Hittle, D.C. (1979). Building loads analysis and system thermodynamics (BLAST) users manual (Version 2.0). Technical Report E-153, Vol. 1 and 2, U.S. Army Construction Engineering Research Laboratory (USA-CERL), Champaign, IL.

Kusuda, T. (1976). *NBSLD, the Computer Program for Heating and Cooling Loads in Buildings*, NBS Building Science Series No. 69-R. Washington, DC, USA: https://nvlpubs.nist.gov/nistpubs/Legacy/IR/nbsir74-574.pdf.

Larsen, S.F. and Lesino, G. (2001). A new code for the hour-by-hour thermal behavior simulation of buildings. *Seventh International IBPSA Conference (BS2001)*, 1, 75–82, Rio de Janeiro, Brazil (13–15 August 2001).

LBNL (Lawrence Berkeley National Lab) (2001). EnergyPlus users manual.

Mendes, N. de Oliveira, R.C.L.F. and dos Santos, G.H. (2001). DOMUS 1.0: a Brazilian PC program for building simulation. *Seventh International IBPSA Conference (BS2001)*, 1, 83–90, Rio de Janeiro, Brazil (13–15 August 2001).

Meyers, G.E. (1980). Long-time solutions to heat conduction transients with time dependent inputs. *Journal of Heat Transfer* 102: 115–120.

Mitalas, G.P. and Stephenson, D.G. (1967). Room thermal response factors. *ASHRAE Transactions* 73: 2.

NASA (1975). NECAP: NASA energy/cost analysis program. Vol. I-User's manual and Vol. II-Engineering manual. NASA Report CR-2590, Houston.

Noel, J., Roux, J., and Schneider, P.S. (2001). CODYBA, a design tool for buildings performance simulation. *Seventh International IBPSA Conference (BS2001)*, 1, 67–74, Rio de Janeiro, Brazil (13–15 August 2001).

Seem, J.E. (1987). Modeling of heat transfer in buildings. Ph.D. Thesis, University of Wisconsin-Madison, USA.

Sowell, E.F. and Hittle, D.C. (1995). Evolution of building energy simulation methodology. *ASHRAE Transactions* 101: 850–855.

Stephenson, D.G. and Mitalas, G.P. (1971). Calculation of heat conduction transfer functions for multi-layer slabs. *ASHRAE Transactions* 77 (2): 117–126.

Winkelmann, F.C., Birdsall, B.E., Buhl, W.F., et al. (1993). DOE-2 Supplement, Version 2.1E, LBL-34947, November 1993. Lawrence Berkeley Laboratory, Springfield, Virginia: National Technical Information Service.

Wittwer, C. (1999). ColSim - simulation von regelungssystemen in aktiven thermischen anlagen. Ph.D. thesis, http://www.ubka.uni-karlsruhe.de.

14

Building Energy Analysis and Optimization

14.1 Overview

The need to keep humans comfortable and healthy in buildings dates back centuries. One of the earliest mechanical ventilation systems designed and installed was in the House of Commons (Addis 2015) in 1734. The ventilation machine for this particular building system was powered by a man, the ventilator, whose job was to "turn the fan" when the House of Commons chamber was in use. Modern buildings use mechanical systems powered by electricity or natural gas to provide heating, ventilation, cooling, and artificial lighting. To reduce energy use in both modern and historic buildings, there should be focus on studying how to reduce the energy consumed in providing ventilation and thermal comfort.

Building systems use energy to counter the unintended heat gain and loss through the building envelope and internal energy gains. The two main methods of energy transfer are through envelope heat transmission and infiltration, as explained in Chapter 11. Improving the building envelope reduces the transmission and infiltration loss. A window with a clear glazing provides a view and natural light as well as allowing heat to pass into the building. Depending on the climate zone, increased heat gain from glazing and daylight can reduce heating and lighting energy use. On the other hand, cooling systems use more energy to remove the added solar heat gain. Lighting and equipment power usage account for a large percentage of a building's energy use and also produce heat. Energy use reduction in buildings requires the adaption of energy efficiency measures (EEMs). EEMs can be separated into five broad categories: building envelope, HVAC systems, lighting, plug loads, and operation controls. Over the years, national laboratories and professional societies have produced detailed energy efficiency design and retrofit recommendations.

The first critical factor in determining proper EEMs for a specific project is to consider both the technical and economic performance of each EEM by using the life cycle cost analysis (LCCA) approach. The cost increase from adopting advanced technologies and/or products for energy efficiency purposes must be offset and justified by the lifetime cost reduction. To achieve a higher level of energy savings, it is essential to understand not only the impacts of each potential EEM on the total building energy consumption/cost but also the inherent interactions among individual EEMs. For instance, a high-performance window may appear as a non-cost-effective solution due to its high initial cost when evaluated independently. However, the energy savings due to this high-performance window may lead to a significant reduction in heating and cooling loads, which results in the need for a much smaller HVAC systems than in

Energy Efficient Buildings: Fundamentals of Building Science and Thermal Systems, First Edition. Zhiqiang (John) Zhai.
© 2023 John Wiley & Sons, Inc. Published 2023 by John Wiley & Sons, Inc.

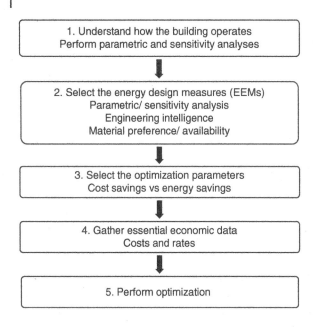

conventional design. The cost and energy savings from the HVAC systems offset the cost increase in the window. Only through considering these solutions in conjunction with each other can an optimal solution be ultimately achieved.

The procedure for identifying potential energy-saving strategies using a building simulation tool such as EnergyPlus is proposed, as illustrated in Figure 14.1. Several optimization tools were developed such as Opt-E-Plus (for commercial buildings) or BEOpt (for residential buildings). Both Opt-E-Plus and BEOpt are software interfaces to EnergyPlus and were developed by the US National Renewable Energy Laboratory (NREL). This chapter will use a case study to demonstrate the procedure of optimizing building energy performance by considering both LCCA and the systematical interactions of EEMs. The case study is the office building benchmark model for Harbin, China, representing a cold climate region. The simulation results can be used to improve existing building energy efficiency codes.

China's growth rate is astonishing, and future predictions of the country's energy consumption will have horrendous consequences for China and the rest of the world if business continues as usual. Total energy rose 3.8% during a 22-year period, from 603 million tonnes of coal equivalent (MTCE) in 1980 to 1320 in 2001. Unfortunately, coal, the leading source of CO_2 emissions, accounted for nearly 72 and 67% of the primary energy consumption in 1980 and 2001, respectively. Buildings make up a large fraction of this energy consumption; it was estimated that the total building stock accounted for about 24% of the nation's total energy use in Mainland China in 1996 and rose to 27.5% in 2001. This figure was expected to rise to 35% by 2020 (Yang et al. 2008). Currently, public buildings alone account for about 18% of China's total CO_2 emissions (Hong 2009). There is no doubt that the building sector will continue to be a key energy end-user in the future and that building energy efficiency will be key in sustaining China's growth.

Improving energy efficiency is one of the most cost-effective strategies for reducing CO_2 emissions, but, as the World Business Council for Sustainable Development (WBCSD) pointed out, economic incentives are not enough to curb the rapid growth rate of energy consumption. The WBCSD is urging policy makers and governments worldwide to impose stricter energy efficiency requirements and commit to enforcing tight building code standards (https://www.wbcsd.org/). Using code-compliant building models along with energy modeling software, building designers can identify the measures needed to achieve rigorous building code updates.

14.2 Simulation Tools

The building energy simulation tool EnergyPlus was developed by the US Department of Energy and is selected as the simulation engine for this research for two reasons (https://www.energy.gov/eere/buildings/downloads/energyplus-0). First, EnergyPlus computes building energy use based on interactions between building components, climate, location, and renewable energy systems such as photovoltaics (PV). Secondly, the optimization tool Opt-E-Plus can be used as an interface to find alternative building designs that lead to energy savings. Opt-E-Plus allows the user to select a wide range of design options to test on a baseline building model. These design options are referred to as energy efficiency measures (EEMs), as their applications are intended to impact the building's energy use (Hale et al. 2009). Opt-E-Plus also allows the user to compare the performance of these optional designs to the baseline model with respect to different performance metrics including energy savings, carbon savings, and a variety of economic functions. Further discussion on the Opt-E-Plus optimization process will be presented in later sections.

14.3 Benchmark Model Development

14.3.1 Developing the Benchmark Model

Before any energy analysis can be done, a baseline model must be developed. The baseline is a theoretical building with the same size, equipment, and operating schemes as the proposed building. It is used as a fixed reference point for comparing the annual energy consumption of different designs options which are created during the optimization (Deru et al. 2005). It is very important to develop an accurate baseline model so that the optimization results are useful and meaningful. Accurate results are especially important when dealing with the associated economics impacts as these are often key drivers in the decision for or against the implementation of EEMs.

14.3.2 Chinese Office Benchmark Description for the Cold Climate Region

The office benchmark model of Harbin represents a typical office building in the cold climate region of China and complies with the country's regional and national code standards and defines the baseline in the optimization (Zhai and Chen 2009). Building components not described in either code, including window-to-wall ratios, number of floors, and aspect ratio, are taken from common design practice. This model is a 6000 m^2 building. The envelope includes concrete exterior walls and an insulated concrete roof. The five-zone layout includes four perimeter zones and one core zone with heating supplied by hydronic

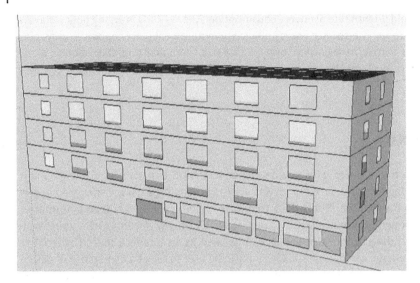

Figure 14.2 Harbin office benchmark model.

radiant baseboard heaters. No mechanical cooling system exists in the model; it is assumed that cooling and ventilation requirements are met by operable windows. Figure 14.2 illustrates the computer model of the building. Tables 14.1–14.3 summarize the input parameters that describe the benchmark model.

There are a few important points that should be addressed regarding some of the input parameters. First, the occupancy density is quite high; this value may be acceptable for the working space, but it is not certain whether it takes into account open spaces such as stair cases, corridors, or other common areas. Secondly, it is uncommon for all electrical loads to be powered down at night, as there may be small loads associated with equipment on standby and security lighting. Nevertheless, these assumptions will

Table 14.1 Office envelope and window construction standards.

Envelope construction standards	Units	Value
Exterior wall R-value	m^2K/W	1.11
Roof R-value	m^2K/W	1.39
Ground floor R-value	m^2K/W	0.56
Window U-value	W/m^2K	5.33
Window solar heat gain coefficient	Fraction	0.58
Window visual transmittance	Fraction	0.29
Window-to-wall ratios	Fraction	North/south: 25% East: 10%; west: 21%

Table 14.2 Office building HVAC components and standards.

HVAC components	Benchmark inputs
Zone equipment type	Hydronic radiant baseboard heaters
Baseboard heaters: radiant energy fraction	0.3
Hot water supply equipment	District heating
Variable-flow pump head (Pa)	600 000
Infiltration (ACH)	0.3

Table 14.3 Office building loads and set points.

Building loads and set points	Units	Occupied hours	Unoccupied hours
Occupant density	Occupants/100 m^2	25	0
Lighting density	W/m^2	11	0
Equipment density	W/m^2	20	0
Cooling supply air temperature	°C	7	7
Heating supply air temperature	°C	60	60
Cooling set point temperature	°C	26	37
Heating set point temperature	°C	20	12

hold until direct measurements are made. Precise measurements would improve the accuracy of the model as they have a significant impact on energy consumption.

14.3.3 Chinese Office Benchmark Performance

The benchmark building annual energy consumption is listed in Table 14.4. The three primary energy consumption categories in the order of significance are electrical equipment, heating energy, and lighting. The radiant heating system pumps accounts for less than 1% of overall energy consumption.

Table 14.4 Annual end-use energy intensity.

End-use category	Total annual energy consumption (kWh/m^2)	Percent of total
Heating (district heating)	44.97	38.1
Lighting	27.47	23.3
Electrical equipment	45.51	38.5
Pumps	0.19	0.16
Total energy per unit floor area (kWh/m^2)	118.13	–
Total energy per occupant-hour (kWh/occ.-hour)	0.18	–

14.4 Parametric Analysis

A building is an integrated system; its components impact one another. Therefore, a parametric analysis is an imperative step to carry out early in the energy-analysis route to understand how each building component interacts within the system and to determine which variables have the greatest impact on overall building energy consumption. This also helps identify which energy design measures should be included in the optimization. A parametric study can be performed by removing each source of heat gain and each thermal component from the baseline model separately.

The results of the parametric analysis are shown in Figure 14.3. Adiabatic walls reduce heating energy by about 80%, suggesting that the existing thermal properties of the walls contribute significantly to heating requirements. Infiltration also has a notable effect on energy consumption. When infiltration/outdoor air is eliminated from the model, heating energy is reduced by 66%. The results show that part of the heating load is met by internal and solar gains and that the windows and slab are not major sources of heat loss for this case.

14.5 Energy Efficiency Measures

14.5.1 Selecting Energy Efficiency Measures for the Initial Optimization

It is important to formulate a case-specific list of EEMs, as the number of EEMs selected increases the number of simulations in the optimization exponentially. Simulation time and memory requirements can be significantly reduced if a small number of EEMs are selected in each applicable category that span

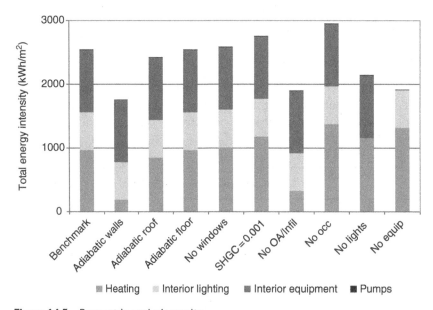

Figure 14.3 Parametric analysis results.

a large range of values. The results of initial optimization can identify which EEMs have the biggest impact on energy savings. It should also be noted that the EEMs are limited to the available technologies that can be modeled in EnergyPlus and Opt-E-Plus. There are two other filters that should be applied to filter out inapplicable EEMs:

- **Engineering intelligence:** Some EEMs that have the potential to save considerable amounts of energy may not be appropriate. For example, reducing the window-to-wall ratio of an office to 10% could significantly eliminate solar heat gains and reduce cooling loads, but the user must consider the indoor working environment as well as visual aesthetics. Sensible limits should be considered when selecting EEMs.
- **Material/technology preferences:** There may be certain building materials or technologies that the user may prefer not to use due to availability, cost, or skill limitations. Selecting only approved building materials eliminates extraneous options that waste time and computing power in the simulation.

14.5.2 Energy Efficiency Measures for the Initial Optimization

Twenty-two EEMs are selected across twelve categories from a US database of existing materials and costs. A US database is used in this situation due to the lack of specific material cost data for China. It should be mentioned that only material properties are specified in the database, not the specific products themselves. The EEMs selected for the initial optimization include:

- Reduced equipment power density (10% reduction) and lighting power density (20% reduction)
- Daylighting controls (400 lux set point)
- Skylights (4, 8, and 12% of roof area)
- Window-to-wall ratio options (70% of baseline model on the south façade)
- Shading (0.5 south façade overhang projection factor)
- Additional exterior wall insulation (R-2.6, 6.6, and 8.8 m^2K/W)
- Additional roof insulation (R-3.4 and 7.1 m^2K/W)
- Selective windows with varying solar heat gain coefficients (SHGC) and visual transmittances (VT)
- Infiltration/ventilation rate options (double outdoor air per person, increase outdoor air per person by half)
- PV application (50% of roof area)

14.6 Initial Optimization

14.6.1 Optimization Fundamentals

Opt-E-Plus uses an iterative solving method to create a variety of building models by selecting different combinations of EEMs and simulating the building's energy performance in EnergyPlus. The results are shown on a plot of two user-selected objective functions (e.g. as seen in Figure 14.4). Each building model is represented by a point on the plot. A solid black line connects the baseline model to the building designs with the lowest cost. This line extends upwards to the model with the greatest percent energy savings.

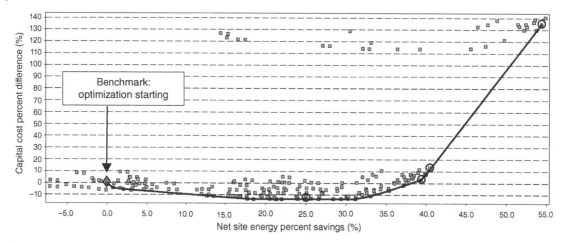

Figure 14.4 Initial optimization results.

This line is referred to as the *optimization curve*. The points along the curve that begin at the lowest cost building and move toward the building with the greatest energy savings define the *Pareto front* and are defined as the *Pareto points* [4].

14.6.2 Chinese Office Benchmark Initial Optimization

The results of the initial optimization are plotted on a graph of net site energy savings vs capital cost percent difference shown in Figure 14.4. A negative capita cost percent difference represents an economic savings. This economic parameter was selected due to the fact that accurate material costs, utility data, and other important economic figures are not available at this point in the research. It should be emphasized that the Opt-E-Plus is primarily used here to find potential energy savings and the economic results should not be regarded as accurate for this particular case. However, the results do provide insight into the economic trends associated with each design. As mentioned earlier, the results of the initial optimization show which EEMs have the biggest impact on energy savings; Opt-E-Plus provides a list of the EEMs selected for each design option. The results suggest that reducing the electrical loads, implementing day-lighting controls, using windows with lower U-values and lower SHGC, and increasing insulation levels are important measures for achieving high energy savings. The impact of these measures will be analyzed in more detail in the sensitivity test that follows.

There are four points of particular interest identified on Figure 14.4, and they will be described from left to right. The first point is defined as the optimal point as it corresponds to the building design with the greatest energy savings at the least capital cost. This point is of great interest if the user is concerned with saving energy and initial cost. The second point defines the design that is closest to being cost neutral in terms of initial cost. It is important for users who are most concerned with saving energy. The third point represents the building design that leads to the greatest potential energy savings before the application of PV. It is interesting to note that the cost increase associated with this design is small relative to the potential energy savings. The fourth case shows that applying PV to 50% of the roof area in the building design offsets energy by an additional 13% for a total energy savings of 54% over the benchmark. However, the cost of PV drastically increases the capital cost of the building. Determining the annual energy cost savings and simple payback period could be easily calculated for any design option if accurate utility cost data was determined. This would be particularly interesting for case four, as a 40% savings in annual energy will undoubtedly lead to notable energy cost savings.

14.7 Sensitivity Analysis

A sensitivity analysis is performed by applying each significant EEM individually to the benchmark model over a large range of values. The results show how each EEM is related to energy savings and identifies the point of diminishing return. This insight helps the user narrow the range of EEMs to analyze in a second optimization.

There are six categories of EEMs identified in the initial optimization that show the greatest impact on energy savings; they are exterior wall and roof insulation, lighting and equipment power densities (LPD and EPD), daylighting controls, and window types. Additional wall insulation has the highest potential for achieving energy savings.

Energy savings increase considerably up to an R-value of 4.0 m^2K/W. The energy savings beyond this point are small. This is important to note because additional insulation will increase cost. Additional roof insulation results in very little energy savings as the existing roof insulation level is considerably higher than the wall insulation level. The results of these two tests are shown in Figure 14.5 as an example. Reducing lighting and equipment power densities and implementing daylighting controls have linear effects on energy savings: 1.6 and 2.5% energy savings are achieved with every 10 W/m^2 reduction in lighting and equipment power density, respectively. Implementing daylighting controls with set points between 200 and 600 lux saves 7–8% of end-use energy. This shows that incrementing set point values does not have much impact on energy savings. Careful attention should be given to lighting power density and daylighting set point EEMs to assure proper lighting levels are met. It was shown in the initial optimization that windows with lower U-values lead to greater energy savings, so SHGC are varied in the window-type

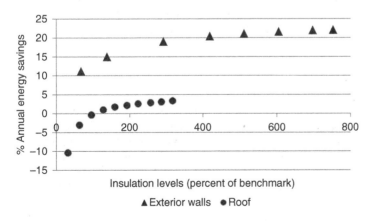

Figure 14.5 Sensitivity of insulation on energy savings.

sensitivity test using two similar U-values. Larger SHGCs show a reducing in heating energy as more solar energy is provided to help meet the heating load. However, a more in-depth study is recommended to weigh the benefits of this measure as higher SHGC could push zone temperatures out of the comfort zone during the cooling season. This value can be adjusted if necessary when passive cooling strategies are analyzed as this parameter shows little sensitivity to energy savings.

14.8 Second Optimization and Recommendations

The optimization is performed again by varying some of the most significant EEMs identified in the sensitivity analysis over a small range of values. This test serves two purposes: first, it confirms the optimal building design and second, it provides the user with different building design options which have similar potential energy savings. In this case, some EEMs are held constant based on best engineering practice and the information gathered from the initial optimization and sensitivity analysis. The EEMs held constant include the lighting power density (80% of benchmark), window U-value (U-2.671 W/m^{2}K), and daylighting control set point (400 lux). The variable EEMs include wall and roof insulation levels (R-6.6, 7.7, and 8.8 m^{2}K/W and R-7 and 8.8 m^{2}K/W, respectively), window SHGC (0.46, 0.61, and 0.72), and equipment power density (90 and 80% of benchmark). The results of the second optimization led to the recommended building design described in Table 14.5. Figure 14.6 shows the potential energy savings compared to the benchmark case. The projected end-use energy reduction potential by category is summarized in Table 14.6.

Window shading fins and alternative HVAC systems are applied to the recommended model to identify any additional savings. These measures were applied after the optimization due to the limitations of Opt-E-Plus. Both measures led to increased energy consumption. The alternative HVAC system applied uses a multi-zone variable air volume (VAV) fan coil system with electric air-conditioners, a natural gas boiler, and energy recovery ventilators (ERVs). Although this case shows an overall energy savings compared to the benchmark, heating energy increases by 51% and overall energy increase by 47% compared to the recommended case. Therefore, passive cooling should be implemented to eliminate the need for additional mechanical cooling energy. It is recommended that more research be done to ensure comfort levels and proper indoor air quality standards are met with passive cooling methods.

Table 14.5 Recommended energy efficiency measures.

	Wall insulation: R-value (m^{2}K/W)	Roof insulation: R-value (m^{2}K/W)	LPD (W/m^2)	EPD (W/m^2)	Window: U-value (W/m^{2}K)	Window SHGC	Daylight set point (lux)
Benchmark	1.33	6.37	11	20	3.06	0.700	None
Recommended	6.37	8.77	8.8	16	2.64	0.734	400
% +/−	+379	+138	−20.0	−20.0	−13.7	+5	+100

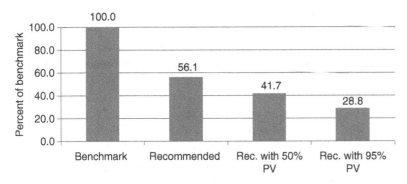

Figure 14.6 Potential energy savings.

Table 14.6 End-use energy reduction potential by category.

	Heating	Interior lighting	Interior equipment	Pumps	Total end-use
% Energy savings	60.0	56.6	20.0	59.0	43.8

14.9 Conclusions

A procedure has been developed for using the building energy optimization tool Opt-E-Plus to identify potential energy savings. This procedure has been applied, as a demonstration, to a Chinese public building benchmark model: an office benchmark model for Harbin, which is designed to meet the building codes for the cold climate region of China. The results are presented with recommendations for improved building code standards.

The optimization methodology includes the following steps:

1) Develop a benchmark model that represents the proposed building,
2) Performing a parametric analysis to understand the interactions between building components,
3) Formulate a list of appropriate EEMs and select the optimization parameters,
4) Run an initial optimization to identify the most important EEMs,
5) Perform a sensitivity analysis to determine points of diminishing returns,
6) Run a second optimization to confirm the optimal building design, and
7) Perform additional tests, if necessary, which could include alternative HVAC systems.

The results of this optimization process show that an annual energy savings of 43.9% can be achieved. The energy design measures which lead to these savings include increase wall and roof insulation, lighting and equipment power density reductions, higher efficiency windows with slightly larger SHGC, and daylighting controls. An alternative HVAC system, which uses a VAV fan coil system with mechanical cooling and ERVs, shows a 47% increase in annual end-use energy over the recommended design due to additional heating, cooling, and fan energy. However, this design does use 17% less energy than the benchmark building, even with mechanical cooling. It is recommended that passive cooling strategies be implemented. Applying PV to 95% of the roof area can offset nearly half of the electrical energy consumption in the recommended building design, leading to a 71% savings over the benchmark. This also shows that a net-zero energy building is not achievable due to the amount of available roof area compared to total floor area.

There are three recommendations for further research. First, an economic analysis should be carried out using accurate material costs, utility rates, and other important economic data to calculate a valid life cycle cost and net present value for the recommended building design. These economic results could play key roles in the decision making process. Second, more consideration should be given to building integrated PV to further offset energy consumption since a net-zero energy building cannot be achieved with the available roof space. Lastly, more attention should be given to passive cooling strategies, as this is a favorable method for eliminating the need for extraneous energy consumption.

Homework Problems

Staged Project Assignment (Teamwork)

Final Design Process

1 Refine and justify your **energy goal** (including overall energy demand, HVAC energy demand, etc.), and check with the final energy calculation results. Justify why you choose that specific percentage reduction goal (such as to obtain LEED energy points)

2 Finalize **building massing, orientation and layouts**
   (a) Indoor layout and elevation (computer drawing)
   (b) Outdoor landscaping if any
   (c) Connection to surroundings/terrain
   (d) 3D rendering

3 Finalize **envelope features**
   (a) Optimize key parameters such as insulation, glazing considering both cost and performance, using your own excel tool or energy computer model
   (b) Final ComCheck for compliance (https://www.energycodes.gov/comcheck)

4 Demonstrate **shading features and performance** for typical days and seasons

5 Calculate **ventilation requirements** for each zone. Demonstrate compliance with ASHRAE Standard 62.1-2019 (for both natural and mechanical ventilation)

6 Use eQUEST (or similar) to calculate **heating and cooling loads** and building **energy usage** (monthly and annual electricity and gas)
   (a) Use the final design features
   (b) Finetune/optimize key parameters and systems (e.g., WWR on each façade, shading, insulation, glazing, infiltration, different mechanical systems, etc.) (show comparison results)

7 Size **major equipment** (e.g., boiler, chiller, air handler, and heat pump)

    (a) Utilize load calculations and ventilation requirement calculations to size major equipment. Note that you may need to perform psychrometric calculations to ensure correct humidifier and coil sizing.

    (b) Identify locations for major equipment and main building outdoor air intakes and building exhaust if any.

8 Determine the **zone level terminals** (or indoor units) and their locations, and layout main ducts/pipes from major equipment to the zone terminals (in drawing)

9 Coordinate and demonstrate **heating, cooling, and ventilation schemes,** for typical days and seasons (integrating both passive and active systems)

10 Details on **other features** (e.g., green roof, PV, solar hot water, etc.) if any (with extra credits)

Deliverable (Electronic Version in PDF)

1 Energy Design Goal and Benchmarks – 5%

2 Building Massing and Layout Analysis and Design – 15%

3 Building Envelope and Passive Designs – 15%

4 Building Load and Energy Calculation and Optimization – 20%

5 Heating and Cooling System Design and Operation Schemes – 20%

6 Computer Drawing and Rendering Quality – 15%

7 Completeness/Clarity/General Organization – 10%

8 Other Features – 5% [*extra credits*]

Design Report Format

1 Cover

2 Executive Summary

3 Table of Contents

4 List of Tables and Figures

5 Project Overview

6 Design Goals and Code Compliance Requirements

7 Building Massing and Layout

8 Passive Building Designs

9 Building Envelope Design and Compliances

10 Comfort and Ventilation Requirements

11 Building Load and Energy Calculation

12 Heating, Cooling, and Ventilation System Design

13 Building Operating Schemes

14 Summary

15 References

16 Appendices (e.g., Calculation details, Major reference tables/figures used, Additional renderings, Product cut sheets, Large drawings, etc.)

References

Addis, B. (2015). *Building: 3,000 Years of Design, Engineering, and Construction*. Phaidon Press; Reprint edition, ISBN-13: 978-0714869391.

Deru, M., Torcelline, P., and Pless, S. (2005). Energy Design and Performance Analysis of the BigHorn Home Improvement Center. NREL/TP-550-3493.

Hale, E., Leach, M., Hirch, A., Torcelline, P. (2009). General Merchandise 50% Energy Savings. NREL/TP-550-46100.

Hong, T. (2009). A close look at the China design standard for energy efficiency of public buildings. *Energy and Buildings* 41 (4): 426–435.

Yang, L., Lam, J., and Tsang, C.L. (2008). Energy performance of building envelopes in different climate zones in China. *Applied Energy* 85: 800–817.

Zhai, J. and Chen, Z. (2009). Development of Simulation Data Sets and Benchmark Models for the Chinese Commercial Building Sector. US DOE-Asia-Pacific Partnership Report.

Index